Women Spacefarers

Sixty Different Paths to Space

Umberto Cavallaro

Women Spacefarers

Sixty Different Paths to Space

Published in association with
Praxis Publishing
Chichester, UK

Umberto Cavallaro
Italian Astrophilately Society, AS.IT.AF
Villarbasse, Torino
Italy

SPRINGER-PRAXIS BOOKS IN SPACE EXPLORATION

Original Edition: Donne nello spazio. © Ugo Mursia Editore S.r.l. - Milan 2015
Springer Praxis Books
ISBN 978-3-319-34047-0 ISBN 978-3-319-34048-7 (eBook)
DOI 10.1007/978-3-319-34048-7

Library of Congress Control Number: 2016955846

Cover design: Jim Wilkie

Printed on acid-free paper

This Springer imprint is published by Springer Nature
The registered company is Springer International Publishing AG
The registered company address is: Gewerbestrasse 11, 6330 Cham, Switzerland

To Lorenzo and Martina
Aim high and chase your dreams!

Acronyms

ASI	Agenzia Spaziale Italiana (Italian Space Agency)
ASP	Astronaut Support Personnel (see *Cape Crusader*)
CAIB	Columbia Accident Investigation Board
CCCP	Советская космическая программа (Soviet space program)
CNES	Centre National d'Études Spatiales (French Space Agency)
CNSA	China National Space Administration
CSA	Canadian Space Agency
DLR	Deutsches Zentrum für Luft- und Raumfahrt (German Aerospace Center)
EMU	Extravehicular Mobility Unit, i.e., a spacesuit
ESA	European Space Agency
EVA	Extravehicular Activity
ISS	International Space Station
JAXA	Japan Aerospace Exploration Agency
JSC	NASA Johnson Space Center, Houston, Texas
KARI	Korean Aerospace Research Institute
KSC	NASA Kennedy Space Center, Florida
MCC	Mission Control Center
NASA	National Aeronautics and Space Administration
ROSCOSMOS	Федеральное космическое агентство (Russian Federal Space Agency)
SAS	Space Adaptation Syndrome or Space Sickness
STS	Shuttle Transport System or Space Shuttle
TRDS	Tracking and Data Relay Satellite
UNOOSA	United Nations Office for Outer Space Affairs
USAF	United States Air Force

Contents

Foreword

Writing an introduction to a book is always challenging, mainly if the goal is to say something new and special every time. Writing the introduction to a book on space is even more challenging because, as space is my day-to-day life, extremely high attention must be paid on what I write and how I do it. But writing the introduction to *Women Spacefarers—Sixty Different Paths to Space* by Umberto Cavallaro has been difficult and inspiring at the same time. Why? Let's start with some statistics. At the time at which this book is being printed, we can count 60 human beings of the female gender who have left Earth to fly into space—60 women out of about 550 astronauts, cosmonauts, and taikonauts in 55 years (from Yuri Gagarin in 1961 until today). Still, 60 out of 550 means about 11 %. Well, let's make a comparison between the total number of flights instead. We can count about 1100 flights made by men and about 130 made by women, for a total of about 43,700 days in orbit, against 3700. The only "first" for a woman comes with Susan Helms, who, together with James Voss, still holds the record for the longest extravehicular activity (i.e., 8 h and 56 min, during STS-102 on March 11, 2001). And, just to underline another unique event, let's go back to 2007 when Peggy Witson, ISS commander, welcomed Pamela Melroy, Shuttle commander, aboard the International Space Station (ISS). What sounds strange here is that these events are so rare. And they merit mention exactly because they are so rare.

These figures represent well the overall situation in the scientific domain, and in particular in the aerospace field, where we can count an average of about 15 % of experts of the female gender in the workforce.

Let's look at this situation from a different viewpoint. Indeed, the ultimate goal of sending humans into space is to perform scientific experiments in microgravity conditions, and it is nowadays crystal clear that men and women have different reactions to weightlessness. NASA and the National Space Biomedical Research Institute (NSBRI) conducted a study, published in 2014, on the basis of a request stemming out of a 2011 National Academy of Science decadal survey report. The main result of this study is that there are key differences between men and women in cardiovascular, immunologic, sensorimotor, musculoskeletal, and behavioral adaptations to human spaceflight. Among various recommendations, the study suggested selecting more female astronauts for future spaceflights and designing scientific experiments considering the gender aspect. As a natural consequence, the last NASA astronaut class, selected in 2015, was made up of eight human beings: 50 % male and 50 % female. All in all, this balanced selection is nothing other than a scientific result.

Apart from these female heroes, every mission into space has been possible because of hundreds of men and women, working as a team, all over the world. All of them are passionate about their jobs and about space, even if they never flew in space, and probably never will. During my career, I moved to be an astrophysicist to become a manager, working always in the development of programs with international cooperation. I never flew in space, but I allowed astronauts to fly, missions to be launched successfully, and experiments to be performed in space to improve our quality of life on Earth. In this respect, one of the questions I get asked the most is: What brought you into the space field and in your current position as Director of the Office for Outer Space Affairs at the United Nations? I always say that actively participating in the aerospace field gave me a feeling of fulfillment and pride in human achievements. It changed my perspective on humanity, detaching myself from the daily routine and thinking about the world as a whole, understanding our hyperconnectivity and interdependence. In addition, I have been an advocate for the use of space research and technology as a tool for sustainable development on Earth for a long time. A stronger and narrower cooperation in the space sector among all involved stakeholders is necessary to achieve the most cutting-edge results. At the same time, I firmly believe that the rise of new actors is transforming the current playground and the need for shared international rules is crucial. What have I been looking for in my career then? Well, those factors, coupled with the fact that my position at the UN perfectly satisfies my need for continuous intellectual impetus, my willingness to learn something new every day, to meet new people, lead me to this challenge. It is really important to have those feelings when choosing your career. My inspiration every day comes from the clear and tangible outputs of the use of space research, technology, and applications for the improvement of the quality of life on Earth. An example? Space-based technology and applications are used to gather information that can help us to understand global water cycles, map water courses, and monitor and mitigate the effects of floods and droughts.

And, as women suffer disproportionably from poverty, they are more likely to be affected by natural disasters as well as climate change and its impacts. We should think of women in these environments not only as victims, but rather as powerful agents of change, who possess specific knowledge and skills to effectively contribute to climate change adaptation and mitigation, and to the prevention of and education on natural disasters.

At the UN, we do capacity building and we are now in the process of changing the approach to it after 50 years since the start of the space era. We are working therefore to define new innovative and effective approaches to overall capacity building and development needs as a fundamental pillar of global space governance, strengthening comprehensive outreach activities. We are also working to promote efforts to encourage science, technology, engineering, and mathematics education, especially for women in developing countries. And, in doing so, we really need role models. The 60 women in this *Women Spacefarers* book are heroes—role models to be followed. But also, all the space experts of the female gender who are not leaving Earth, but for 50 years have allowed space missions to become a reality together with their male colleagues, of different nationalities, in different places in the world, all working for the same goal, merit recognition as role models—evidence that yes, we can!

After more than 30 years of my rising professional career, I can say, with no doubt, that true passion and professionalism have no gender, and focusing on our goals and dreams helps us to overcome gender and cultural boundaries. *Women Spacefarers* is therefore a book for everyone, for each and every one of us who wants to understand and believe that, yes, we can! We can follow our dreams, we can help others to do their jobs, we can believe in a better future, which will be better due also to our contribution, and this is not linked to our gender, but only to our willingness to contribute to the development of the human race. We have to keep in mind a famous sentence if we want to succeed and never give up in life, which you can find in StarCity in the Shepherd's house, when going down into the basement, which has become my preferred one: *the last easy day was yesterday!*

Simonetta Di Pippo
Director of the United Nations
Office for Outer Space Affairs
(UNOOSA)

Simonetta Di Pippo is the Director of the United Nations Office for Outer Space Affairs (UNOOSA). Her experience includes serving as Director of the Observation of the Universe at Agenzia Spaziale Italiana (ASI) (2002–2008) and Director of Human Spaceflight of the European Space Agency (ESA) (2008–2011). Prior to her appointment at UNOOSA in March 2014, she was Head of the European Space Policy Observatory at ASI Brussels.

Ms. Di Pippo holds a master's degree in Astrophysics and Space Physics from University "La Sapienza" in Rome, and an Honoris Causa Degree in Environmental Studies from St. John International University in Vinovo. Ms. Di Pippo was knighted by the President of the Italian Republic in 2006. In 2008, the International Astronomical Union (IAU) named asteroid 21887 "Dipippo" in honor of her contribution to this field.

Preface

Women in Space: A Soviet Record

"Hey sky, take off your hat, I'm coming!" So shouted the euphoric Valentina Tereshkova, the first Soviet cosmonaut, while blasting off from Baikonur to begin her ride to the stars, at 12:29 on June 16, 1963, when, at the height of the Cold War, the requirements for the missions were dictated by the state propaganda: what was important was to be the first in everything and at any cost. Her 49 orbits around Earth had huge media coverage and she became an instant celebrity throughout the world and the new symbol of the Soviet space progress.

It took 19 years to see another woman flying in space, and once again it was a Soviet female cosmonaut: Svetlana Savitskaya.

Once more—in the late 1970s' progressive deterioration of East–West relations—political scenarios played an important role, together with the will to excel of the Soviet Union, needing to reaffirm their superiority in space at a time when the USA had launched the new Shuttle program and rumors began to circulate that NASA had opened to women and was selecting the first female astronauts.

As Sally Ride, in April 1982, was assigned to her first Shuttle mission scheduled for the following year, the name "Svetlana Savitskaya" almost magically appeared in the crew of the Soyuz T-7. The launch of the Soyuz was planned in August, and so Svetlana would fly before Sally Ride.

As in the days of the Cold War, the competition engaged in between the two superpowers was reflected even in space. Again, the Soviets arrived first and overshadowed the pre-announced flights of the six female "Shuttlenauts."

Propaganda played a role even in the second assignment of Savitskaya to Soyuz T-12, as David Shayler and Rex Hall outlined.[1] Her appointment was in fact announced in December 1983, 3 weeks after NASA had officially announced the crew of the STS-41G mission, with Kathy Sullivan performing the first "spacewalk" by a woman. In the same

[1] Shayler, D.J.; Moule, I. *Women in Space—Following Valentina*, p. 219. Springer/Praxis Publishing, Chichester, UK (2005).

mission, even Sally Ride would fly—the first American astronaut on her second mission in space. Valentin Glushko—the former designer of rocket engines, who at the time headed the Soviet space agency—immediately decided to assign Savitskaya again and to schedule her extravehicular activity (EVA). Thus, he managed to beat once again the Americans in sending a female Soviet cosmonaut who, at the same time, was on her second venture into space and performed an EVA. To carry out his "women in space" program, Glushko had to fiercely fight against Georgi Beregovoij, the director of the Cosmonaut Training Center in Star City, who suddenly stepped in twice, using his authority to end the training program of the female cosmonauts. Both times, the program was restored again after an arduous struggle between Glushko and Beregovoi.[2]

The emancipation of women was a cornerstone of Communist propaganda: in all countries of the Soviet bloc, most of the women could study, work, have abortions, and divorce; women were also admitted into the armed forces (especially in aviation). And this is perhaps the key point. This inclusion of women in the armed forces (who ran the Soviet space program) was the bridge that allowed, in the pioneering era, the admittance of a woman into the Soviet space program well in advance of their American competitors: women in aviation were in fact well established in Russia following World War II.

Despite this, however, the Soviet Union failed to build on that promising start and, in more than 50 years, of the 20 women who have trained as Russian or Soviet spaceflyers, only four followed Valentina and went into space.

During the debate following a crowded conference that I attended in Berlin in 2013—for the celebration for the 50th anniversary of the mission of Tereshkova—in response to the question of why, after sending the first woman to orbit Earth 50 years ago, Russia has sent so few women into space, Russian cosmonaut Vladimir Kovalyonok jokingly replied that Russian spacecrafts hadn't as many places as the American Shuttle and, after all, space is a man's affair. The joke—which is still recurring—says a lot about the views of two worlds that continue to be distant from each other.

The Soviets had always accepted, even depended on, women in the workplace. But acceptance did not necessarily mean respect, much less equality. The Soviet reluctance to fly women had probably also to do with competition for slots. There have always been fewer cosmonauts than astronauts and fewer seats to fill. With only Soyuz, the cosmonauts understandably did not want to relinquish their few places to women.[3]

"Sexism has played an important role in limiting the number of Russian female cosmonauts," said Elena Dobrokvashina[4] in an interview with the Russian agency, RIA Novosti, on the occasion of the celebration of the 50th anniversary of the flight of Tereshkova. In the 1980s, Dobrokvashina was selected to train as a cosmonaut along with Svetlana Savitskaya and had followed the same training program all the way but never had the chance to fly in space. She says:

---

[2] Gibson, K.B. *Women in Space: 23 Stories of First Flights, Scientific Missions and Gravity-Breaking Adventures*, pp. 45, 59–60. Chicago Review Press, Inc., Chicago (2014).

[3] Kevles, T.H. *Almost Heaven: The Story of Women in Space*, p. 137. The MIT Press, Cambridge, MA, and London, UK (2006).

[4] Interview by Makarov, A. "Sexism Limited Female Space Flights," *en.ria.ru* (RIA Novosti, June 14, 2013).

> The Russian cosmonauts are scared that if women were to go into space their aura of heroism would be lost. It's part of our mentality. Although they always say that everyone—men and women—is equal, it's no secret that we live in a man's world, where high-profile professions are reserved to them.

The chronicles of the time were filled with the resistance that the highly decorated cosmonaut Valeri V. Ryumin—veteran of three Soviet flights, in space for 362 days—opposed to the space missions of his young wife, Elena Kondakova.[5]

Right Stuff[6] but Wrong Sex

The story of the Mercury-13 shows that, although the cultural context was very different in the 1960s, the time wasn't ripe for women in space in the USA either, and enabling gender-independent access to the final frontier was a long process that would take a couple of decades.

During World War II, it seemed that *force majeure* had opened new spaces for women when they were called to replace in their daily tasks the men who had gone to the front. But these spaces shut again after the war when the situation returned to normal.

Despite that, at the turn of the 1960s, 13 talented American women pilots successfully underwent—more or less in secret—the same tests as the original Mercury-7 went through (and in some cases also outshined and outperformed the male astronauts—68 % of the women passed with "no medical reservations" compared with 56 % of the men) and, despite their excellent credentials (Jerrie Cobb, who had started flying at 12, had logged 10,000 h piloting a huge variety of aircraft, twice that accumulated by John Glenn, the most experienced Mercury astronaut), for NASA, the social role of the woman was the one firmly assigned by a consolidated stereotype masterfully represented by *LIFE* magazine that had the exclusive rights on the astronauts and their families: the woman was to take care of the house and look after the children, quietly awaiting the return of her astronaut husband.[7]

The fact that the 13 women—known today as "Mercury-13"—never had (or agreed to have) a common denomination contributed for a long time to their media (and historical) invisibility. The name itself, "Mercury-13," was suggested only in 1995 by the Hollywood producer, James Cross.[8]

Although some of them knew each other, they never met all together. Most had never even met before 1986 when one of them, Beatrice Steadman, convened them to celebrate their 25th anniversary. Not all of them came. In 1994, another of the "Mercury-13," Gene Nora Jessen, a former president of "Ninety-Nine," the international organization of women pilots, tried again. This time, it was to celebrate an important event: the fact that a woman, Eileen Collins, was appointed by NASA as first pilot astronaut—actually the first

[5] Kevles, *Almost Heaven*, pp. 148–149.

[6] This expression became common saying after the publication of the successful book *The Right Stuff* in 1979, in which Tom Wolfe celebrated the famed Mercury astronauts.

[7] Kevles, *Almost Heaven*, p. 47.

[8] Shayler and Moule, *Women in Space*, p. 92.

woman to pilot the Space Shuttle. Nine of the 13 arrived, and 7 of them attended the Kennedy Space Center for her first launch the following year. Eight came 4 years later, in July 1999, to witness her departure when Eileen became the first commander of a spaceflight.[9]

In the early 1960s, America was gripped by space fever. The news that women were undergoing astronaut tests leaked soon. In a conference held in August 1960 at the Space and Naval Medicine Congress in Stockholm, Dr. Lovelace presented a paper on the performance of the award-winning pilot Geraldyn "Jerrie" Cobb, arguing that "females require less oxygen than the average male, and women's reproductive organs, being internal, are less vulnerable to radiation." This last point resounded promisingly to the assembled physicians as, in 1960, many people were sensitive to the issue of radioactive fallout from nuclear testing.[10]

The news immediately spread and, in the USA, the *Washington Post* ran a story entitled "Women qualifies for space training" on August 19, 1960. Jerry Cobb became an instant celebrity. In their stories, reporters called her everything: astro-nette, feminaut, astronautrix, space-girl,....[11] The term "astronaut" apparently carried such a masculine connotation that even a potential female candidate for space travel required the coining of a new label.[12]

In mid-1961, two Americans had flown in space, in ballistic flights of only a few minutes each. It was too little to seize on what was really important from a physical point of view. So candidates had to endure every possible test, even the most uncomfortable. Donald Kilgore, who was conducting the tests at the Lovelace Clinic, reported that women performed very well, and that they complained far less than their male colleagues did: "Women are more tolerant of pain and discomfort than men."

But the excitement was suddenly toned down when, in September 1961, 2 days before most of them were scheduled to start their final tests in the military naval base of Pensacola, a telegram from Lovelace arrived—like a cold shower—suspending all tests with immediate effect.

Despite Janey Hart, one of the "Mercury-13," being the wife of the powerful senator from Michigan and Jerrie Cobb, the first and only female astronaut candidate to successfully complete all the tests, knowing many of the big names in government, they didn't succeed in making contact with President Kennedy. Cobb managed, however, to meet with Vice President Johnson, who, in the late 1950s, had pushed the space race. As she later reported in an interview, Johnson, with some embarrassment in the end, shared with Jerrie:

> If we let you or other women into the space programme, we have to let black in, we'd had to let Mexican American in, we have to let every minority in, and we just can't do it.[13]

---

[9] Bush Gibson, *Women in Space*, pp. 10–11.

[10] Kevles, *Almost Heaven*, p. 11.

[11] Bush Gibson, *Women in Space*, p. 21.

[12] Weitekamp, M.A. *Right Stuff, Wrong Sex: America's First Women in Space Program*, p. 78, Johns Hopkins University Press, Baltimore, MD (2006).

[13] "Women Astronaut Predicted," *New York Times* (June 26, 1962), quoted in Bush Gibson, *Women in Space*, p. 33.

Mercury-13

It was a program almost ambiguous and covert, started on the initiative of some researchers in the Lovelace Foundation who had carried out the medical tests for the selection of the Mercury candidates. Twenty-five female pilots in 1961 went through many of the same three-phase tests (physical, psychological, and space simulation) as the Mercury astronauts.

Scientists regarded women with interest, since they are smaller and lighter than men, weigh less, and breath less oxygen, and they speculated that they might make good occupants for cramped space vehicles. Thirteen out of them passed the exams and were then named "Mercury-13," as they were selected immediately after the original Mercury-7 astronauts.

The project was far from being sponsored by NASA or having any official status, and all was based on a "bootleg" effort: it was the private project of a few of the medical experts—who had tested the Mercury astronaut candidates—hinting that, if women did especially well, then NASA might consider some of them as candidate astronauts.[14]

In fact, this wouldn't even cross the mind of NASA managers. First of all, social attitudes of the time weren't favorable to the introduction of women in the space program and, mostly, NASA had already made some basic decisions that—for other reasons entirely—required astronauts to have quite specific qualifications. After preparing a draft public tender to recruit the first astronauts, NASA had second thoughts and, with the approval of President Eisenhower, decided to select candidates of the Mercury program from among the military test pilots.

Spaceflight, especially in the Mercury spacecraft, clearly wasn't going to be much like flying an airplane. But *test pilots* in particular were already doing vaguely similar work: risky testing of new high-tech vehicles. They were physically fit; they already had NOS security clearances; many of them had strong engineering backgrounds. And, most importantly, there were only a few hundred active-duty military test pilots and the first pass of selection could be done by just going through their military records. This looked a *lot* easier than sorting through thousands of applications from the public. At the end of 1958, President Eisenhower approved this pragmatic change of plans. At that time, there were no female military test pilots, so the question of whether to accept women as astronauts never even came up. The good test results obtained in the private and secret project of the Lovelace Foundation meant nothing; the exclusion of women from the early space program was mostly an accidental side effect of NASA's selection criteria: the female pilots didn't qualify and the issue simply could not arise, and never ever came up. The "Mercury-13 Program" never involved NASA and, in NASA's History Program, is referred to as "Lovelace's Woman in Space Program."[15]

---

[14] Spencer, H. "Why NASA Barred Women Astronauts," *newscientist.com* (October 8, 2009).

[15] *http://history.nasa.gov/flats.html.*

And, when the medical testing started to attract too much media attention, pragmatically, the project was abruptly ended. Realistically, it was going nowhere anyway. In that context, it was unlikely that NASA could accept a female astronaut as equal partners to the male astronauts. And also the political token of launching the first woman in space was far from NASA's vision. That idea never got very far and died completely when the Soviets beat NASA by launching Valentina Tereshkova in 1963.

As later groups of astronauts were recruited for Gemini and Apollo, the criteria were loosened up, but the choice of military test pilots as the first astronauts had set the pattern. After all, nobody thought that launching the *second* woman in space was a worthwhile political gesture.

In the subsequent selections of "scientists–astronauts" of 1965 and 1967, no gender requirement was explicitly indicated. Applicants were, however, informed that they had to attend a US Air Force jet pilots training course (USAF opened to women only in 1976). A few women applied, but were rejected in the preselection phase and, again, the problem did not arise.

FLATs (Fellow Lady Astronaut Trainees)

Until the late 1970s, NASA had always refused to consider the idea of including women in the space program. The original decision of Eisenhower to choose the astronauts from the ranks of the best military test pilots had definitely excluded women and included only a small minority of white men, without worrying whether this would threaten some basic principles of democracy. Many complained that, by setting aside the "Mercury-13" operation, America had missed an incredible opportunity. As a matter of fact, although some of the Edwards best test pilots—proud soldiers with crew cuts—in their arrogance, were reluctant to become "spam in a can" in a Mercury capsule, the first 73 astronauts selected between 1959 and 1967 were military or former military, typical WASPs (White Anglo-Saxon Protestants) and the space program remained a men's club "with the right stuff."

At that time, women could not fill the roles that they cover today in the US armed forces, and this is why they had not even been considered. No one, for decades, had ever questioned the role of men in space or men on the Moon, long before Armstrong set foot there. The statement of President Kennedy in 1961 solemnly pledged America to put *a man* on the Moon before the end of the decade and to return *him* safely to Earth. At the time, the issue of gender neutrality did not arise. No one objected that he had not spoken about *a person* on the Moon and had not talked about *his/her* healthy return. When people thought about astronauts, they simply thought of men, and perhaps WASPs.

In 2008, a letter written on February 26, 1962 (a few days after the historic mission of John Glenn), by O.B. Lloyd—who was NASA Director of Public Information from 1961 to 1979—popped up on the *Reddit.com* Web site in response to a letter from an unidentified "Miss Kelly" of the University of Connecticut. In a few words, he bluntly says that NASA had no place for women: "Your offer to go on a space mission is commendable and we are very grateful. This is to advise that we have no existing program concerning women astronauts nor do we contemplate any such plan."

NATIONAL AERONAUTICS AND SPACE ADMINISTRATION
WASHINGTON 25, D.C.

February 26, 1962

IN REPLY REFER TO: AFP

Spencer A
The University of Connecticut
Storrs, Connecticut

Dear Miss Kelly:

This is in response to your letter of February 20, 1962.

Your offer to go on a space mission is commendable, and we are very grateful.

This is to advise that we have no existing program concerning women astronauts nor do we contemplate any such plan.

We appreciate your interest and support of the nation's space program.

Sincerely,

O. B. Lloyd, Jr.
Director
Public Information

"Letter to Miss Kelly." This rejection letter sent by NASA in 1962 was first published on *www.reddit.com* in July 2013

Only 16 months after this rejection, the USSR gave the USA a more resounding slap and grasped another record: after sending the first man into space, they sent also the first Soviet woman, Valentina Tereshkova, into space.

Hillary Clinton, during her presidential campaign of 2009, claimed that she had received from NASA a similar rejection letter (although it has been highlighted that there is no record of such letter, and the political context in which this claim was disclosed would legitimize some doubt).[16]

[16] Oberg, J. "'We Don't Take Girls': Hillary Clinton and Her NASA Letter," *thespacereview.com* (June 10, 2013).

Traditionally, in the USA, women could at most work as teachers, nurses, or secretaries, without aspiring to important positions, and were almost totally excluded from the "STEM careers," dealing with science, technology, engineering, and mathematics.

The attitude of NASA had, however, to change, albeit very slowly. Already, in 1970, Deke Slayton, the legendary "indecipherable" boss of the Astronaut Office, wrote to Marsha Ivins, who a long time after would become an astronaut herself (see Chap. 14 on Marsha Ivins): "I do not envision needing additional astronauts for a number of years." And, in this, he was right: the next selection was made in 1978. He added: "The exact time when we would seriously consider women is indefinite, but I am sure it is inevitable."

CA/File

Astronaut Selection - Misc.

CA MAY 4 1970

Miss Marsha Ivins
Farrand Hall Box 77
Boulder, Colorado 80302

Dear Miss Ivins:

Regarding your interest in becoming an astronaut, I believe your pursuit of an education in aerospace engineering is as good as any at this time. I do not envision needing additional astronauts for a number of years. The exact time when we would seriously consider women is indefinite, but I am sure it is inevitable. Therefore, you should continue your studies and do the best you can.

We will publicly announce a program for the selection of additional astronauts, and if you meet the criteria you should apply at that time.

Best of luck.

Sincerely,

ORIGINAL SIGNED BY:
D. K. SLAYTON

Donald K. Slayton
Director of Flight
Crew Operations

CA:DKSlayton:sms 4/30/70

Marsha Ivin's letter from Deke Slayton. This letter was found in the JSC History Collection, UHCL Archives (University of Houston-Clear Lake), and was first published on *neumannlib.blogspot.it* in October 2011. The Letter was in reply to the letter by Marsha Ivins reproduced at page 102

And he concluded: "We will publicly announce a program for the selection of additional astronauts, and if you meet the criteria you should apply at that time."

Every Country Has Its Own Astronauts

The term "astronaut" derives from the Greek words *ástron* (ἄστρον), meaning "star," and *nautes* (ναύτης), meaning "sailor." The first known use of the term "astronaut" was in *Voyage dans la Lune* (*Journey to the Moon*, 1657) by French poet, Cyrano de Bergerac (1619–1655).[17] The term is referred to as a spacecraft in the science fiction novel *Across the Zodiac* by Percy Greg (UK, 1880). In the modern sense, it is found in *Les Navigateurs de l'Infini* by Josephi Henri Honoré Boex (Belgium, 1925). The word may have likely been inspired from the novel *Auf Zwei Planeten* (*On Two Planets*) by the German philosopher and writer, Kurd Laßwitz, who, in his 1897 science fiction novel, refers to the first encounter with the highly advanced Martian civilization. The term itself may have originated from "aeronaut," an older term for an air traveler, which was first applied to balloonists in 1784. An early use in a nonfiction publication is Eric Frank Russell's poem "The Astronaut" published in November 1934 on the *Bulletin of the British Interplanetary Society*. The term was used liberally during the infancy of the rocket plane, as in a *New York Times* article that opens with this sentence: "Evidently the astronauts who dreamed of kicking themselves from the Earth to Mars were not mad."[18] The term entered then into the official name of the International Astronautical Federation, which, since 1950, has held its yearly International Astronautical Congress.

The term made its fortune when, at the beginning of the 1960s, it was adopted by NASA and eventually by the European Space Agency (ESA), though at the beginning it wasn't welcomed very much by the "Mercury-7" astronauts who—themselves test pilots—had preferred the term "spacecraft pilot." According to journalist James Scheftern,[19] the name was picked by Bob Gilruth (the first director of NASA's Manned Spacecraft Center, later renamed the Lyndon B. Johnson Space Center).

The Russians diversified also in this respect from American competitors and, when designing their space explorer, they coined their own term—"cosmonaut" (*Космонавт*), where the word космос (from Greek κόσμος) has a wider meaning encompassing the concepts of "universe," "order," and "organization." While "stars" are loaded up by classical Western mythology with references to "divine" and "celestial," the Soviet космос, laical, and Communist refers to the order of the universe.

To refer to a Chinese space explorer, in the West, we use the phrase "taikonaut": a hybrid expression indeed putting together the Greek word *nautes* and the Chinese

[17] Wells, H.T.; Whiteley, S.H.; Karegeannes, C.E. "Origin of NASA Names," The NASA History Series, SP-4402, Washington (1976), p. 200.

[18] "The Rocket Plane Is Here," *New York Times* (January 8, 1944), 12—quoted in Dickson, P. *A Dictionary of the Space Age*, pp. 26–27. Johns Hopkins University Press (2009).

[19] Quoted in *ibid.*, at p. 27

tàikōng (meaning "space"). It was first used in 1998 by a Malaysian newsgroup. In Hong Kong and Taiwan, 太空人 (*tài kōng rén* meaning "spaceman": a concept very similar to "astronaut") is often used. This word isn't, however, used in China, where they prefer instead the term 航天亮 (*háng tiān yuán* meaning "space navigating personnel") when referring to Chinese space travelers and the term 宇航亮 (yǔ háng *yuán*, with very similar meaning) when they refer to American astronauts or Russian/Soviet cosmonauts.

While no nation other than the Russian Federation (and previous Soviet Union), the USA, and China has launched a manned spacecraft, several other nations have sent people into space in cooperation with one of these countries. Other synonyms for "astronaut" have entered occasional usage in the different countries. In France, for example, the term *Spationaut* or "space sailor," encapsulating the Latin term *Spatium*, is often used.

It is easy to predict that, in the future, yet more terms with similar meanings will appear, such as the term *Vyomanaut*—coined from the Sanskrit word for space—that has already begun circulating in India.

The media have occasionally used terms like "Austronaut" during the flight of Franz Viehböck, the first Austrian astronaut, or "Afronaut" during the flight of Mark Shuttleworth, the South African billionaire who was the first astronaut from the Black Continent.

With the rise in space tourism, NASA and the Russian Federal Space Agency agreed to use the term "spaceflight participant"—which was applied for the first time for the "Teacher in Space" mission—to distinguish the profile of those space travelers from professional astronauts. The same name was eventually adopted even by Russians, who translated it as Участник космического полёта (*učastnik kosmičeskogo polyota*).

In the USA, a candidate becomes an "astronaut" after completing 20 months of basic training, while, in Russia, he/she becomes "cosmonaut" after his/her first successful spaceflight. It is the same in China.

NASA Opens to Women

Although NASA was formally established as a civilian agency, it suffered an internal conflict with its military roots from the beginning. Not only were the astronauts military, but many of the administrative and technical staff and their bosses were also military or former military; the language was military (astronaut goes "in mission") and the command chain was military. Also, the organization and habits were borrowed from the military world.

The US Navy began to accept women for their first pilot training courses in 1974. The US Air Force opened up to women 2 years later, in 1976, although they didn't accept women for the test pilot career until 1988.

After the opening of aviation to women, even NASA threw in the towel. On the other hand, in the new Shuttle scenario, the needs were changing and not only fighter pilots were required. A new type of astronaut was introduced: the mission specialist, who was a researcher with a deep technical and scientific background. At this point, there wasn't a reason to keep women away, also because, in the second half of the 1970s, the percentage of women with tech/sci university education had significantly grown. In its 1977 call for astronauts, 10 years after the previous one, NASA—overcoming quite a lot of internal resistance—finally opened the Astronaut Corps to women.

One of the hardest opponents to the new policy was Chris Kraft, the Director of Human Flight at the NASA Marshall Space Center in Houston. The authoritarian Chris Kraft, for better and for worse, had been, together with the impenetrable Deke Slayton, one of the stars of the NASA golden age in the days of Mercury, Gemini, and Apollo, and was the architect of the Mission Control Center.[20] And, ironically, it just happened to fall to him to manage the new deal and to pave the way for the entrance of women into NASA.

In response to the much-anticipated announcement for new astronauts, at NASA headquarters in Washington, they expected to get a flood of applications from female candidates but, after 6 months, only 93 had arrived. Until then, at every astronaut public appearance, people wondered why the space program didn't include also women and minorities. Now that NASA wanted to recruit them, the women and minorities seemed to have vanished. It was embarrassing: the public, and especially the women, had lost interest in the space program. The headquarters asked Kraft to develop a plan to recruit women and minorities. He had to establish an Astronaut Selection Board. The first suggestion he got from this committee was "not to ask female candidates questions that were not asked to men," especially questions about their marital status or family plans: the tactics of recruitment were targeted by feminists.

The committee sent thousands of letters to various public agencies, to university faculties, and to technical and scientific associations, especially where female scientists and engineers worked. He also asked for the help of Nichelle Nichols, the successful Afro-American actress who had played Lieutenant Uhura in *Star Trek* and, to publicize the new campaign, NASA awarded US$49,900 to her production agency, Women in Motion. After 6 months, 8000 additional applications had arrived, including 1000 from women.

And, from there, NASA selected, for the first time, six women to be integrated into its Astronaut Corps.

When the women arrived at NASA, the Mercury astronauts were gone, but still there were several Gemini and Apollo astronauts and crowds of girls running after them as if they were movie stars. The women found a different welcome. There weren't crowds of boys behind them. The fatal attraction that astronauts practiced on girls did not work with the female astronauts with respect to the boys, who seemed rather to suffer from what Mary Cleave called "PWS" (Professional Women Syndrome): men who were not astronauts saw them as unapproachable and were frightened.[21]

The women found it easier to socialize with their fellow astronauts. But not all. Some astronauts of the old guard greeted them well, while others saw them with annoyance. Rhea Seddon says:

> They thought we would not be able to do things well and thought we occupied the place that could occupy some other guy. On the other hand, some did not even know well how to deal with women who came to be at their same level.

---

[20] During the NASA golden age, as head of the Mission Control Centre, Chris Kraft played an important role in deciding who would not fly anymore. The exclusion of Carpenter from future missions and of the entire crew of Apollo 7 was attributed to him, as reported by Walter Cunningham in his book *The All-American Boys*, p. 191. iBooks, Inc., New York (2004).

[21] Kevles, *Almost Heaven*, p. 71.

The first six women astronauts selected by NASA. From left to right: Shannon Lucid, Rhea Seddon, Kathy Sullivan, Judy Resnik, Anna Fisher Fisher, and Sally Ride. Credit: © NASA

At first, the newcomers also had to face the hostility of the old astronauts' wives. They were military wives who had accepted—as part of their status—to stay in the background and play a supporting role. When Patricia Collins, wife of Michael Collins, was offered by a local newspaper to write a regular column, NASA recommended—or rather, ordered—her not to accept. These wives were threatened by the presence of the women who would train with, work with, and fly with their husbands, at the same level (considering also the reputation of being unruly that always had accompanied some of them).[22]

Also, NASA management found it hard to deal with the new issues. Rhea Seddon, the first astronaut to become pregnant, recalls when she went with her husband, fellow astronaut Hoot Gibson, to announce the news to her bosses:

> I didn't want to be held back on jobs or flight assignments. We went to tell the chief of the Astronaut Office. John Young, and he didn't seem to know what to say except congratulations. We talked to Mr. Abbey, the head on Flight Operations, and got his usual taciturn response. We decided there might be outside questions about it, so we

[22] One of the most realistic and truthful testimonies of the atmosphere at the NASA Astronaut Corps during the Gemini and Apollo era is found in the aforementioned book by Walter Cunningham, *The All-American Boys*, which was judged by *Los Angeles Times* as "the best of all astronaut books."

also talked with our friend and Center Director Dr. Chris Kraft. He seemed pleased and comfortable with my continuing my current career path. We left their offices feeling like: "No sweat, they aren't worried." Almost before I could get back to my office, my phone started ringing. The Flight Medicine Clinic called to tell me no more T-38 flying if I was pregnant.[23]

The Long Road to Integration in the East

Leaving for Star City in the early 1990s to train for the Shuttle–Mir program, Americans noticed a cultural division between Russian men and Russian women that reminded them of what had happened in the USA a generation before.

On board Mir in the 1990s, there wasn't great gender discrimination between Russian men and foreign women, but there was, however, a marked division between Russians: males—rightly or wrongly—never had a great concept of their female cosmonauts.

Cosmonauts at Star City preferred not to comment on the performance of Tereshkova, whom they felt was an embarrassing case. On the other hand, Glushko himself had done everything to keep her away from the active scene. When, in 1974, Glushko—already an honorary member of the powerful Central Committee of the Communist Party—had taken the reins of NPO Energia, he wanted to overthrow the Soviet policy on women in space. In search of new records, he wanted to have an only-women crew and started a complex selection of ten candidate female cosmonauts. But he did not want to have to deal either with Valentina or with any of her group again. To avoid any misunderstanding, he lowered the maximum age requirement to 33 years. "Tereshkova," he diplomatically asserted, "is a national asset; flying into space is risky. We must protect her. It's better that the risk is taken by someone else."

On her part, Tereshkova, who wanted to go into space herself, never showed any sympathy for the new class of female cosmonauts; she never paid them a visit during their training or attended the launch of one of them, Svetlana Savitskaya, "due to illness"—the official reason for her absence.

At Star City, they even remembered conflicts between Svetlana Savitskaya and his Salyut 7 colleagues and disagreements between Elena Kondakova and the Mir crew.

Before Elena Serova's mission, out of 19 female cosmonauts who had completed all of the hard training, only three could actually fly in over 50 years. The Soviet program did not have a great concept of female cosmonauts.[24]

Nor did the Soviet female cosmonauts appreciate each other that much. We have already mentioned the relationship between Savitskaya and Tereshkova. Also, Elena Kondakova is very judgmental with regard to Svetlana Savitskaya, whom she condemns for her "typical American attitude" and for her determination to do everything a man could do on the International Space Station (ISS)—an "attitude that caused much misunderstanding on MIR." Kondakova says:

---

[23] Seddon, R. *Go for Orbit*, p. 133. Your Space Press, Murfreesboro, TN (2015).

[24] Pultarova, T. "Much Ado about Liu Yang," *spacesafetymagazine.com* (June 21, 2012).

> In summary—strong personality but lack of diplomacy. Savitskaya is a good pilot and a good engineer. Nothing to complain about her technical expertise. But when working with men was a typical case of 'diamond cut diamond'. She wanted to have more responsibilities than men were willing to grant her. One of the most important things in space is to be able to find a compromise.[25]

The Soviet cosmonauts used to treat their women in one way and foreign women—whose countries had paid for the trip—as anyone would treat valued customers. Or at least that's how it happened in space. On the ground—in Star City—more than one experience was not exactly comforting. Dunbar, suddenly thrown into a culture light years away, experienced at first hand the aversion of the Russian space environment against women. She hung in there for over 1 year. At the final party held at Star City, she remembers:

> Jurij Kargapolov, the head of training there, who had been in charge of training and had been hardest on me in the previous twelve months, got up and, looking at me, devoted me a toast for my perseverance. Then General Genibechov came up to me afterward and said 'You know, you American women are tough! We would like to have you fly a long-duration mission with us anytime!' … I was glad because, apart from anything else, I had done what I had set to do, which was to finish my job, graduate with… and who knows maybe they would think a little bit differently even about women in their program![26]

Full Integration in the West

In a little more than 50 years, a total of 60 women have ventured into space, and they represent about 10 % of the total of 552 space travelers (according to the United States Army Ground Forces (USAGF) definition, as of April 2016).

Today, in the Western world, a woman in space has become almost routine—inasmuch as confrontation with a hostile and risky environment like the space may be considered "routine." They are now fully integrated as ordinary members of this out-of-the-ordinary club that is the Astronaut Corps.

Pilots, doctors, scientists, engineers, single, married, divorced, mothers—all try to reconcile the demands of work with those of families, sometimes succeeding well, sometimes having a hard time, more or less as happens to everybody. Payload Specialist, Mission Specialist, Flight Engineer, Pilot Commander, Mission Commander, Head of Astronaut Office: there is no position that NASA female astronauts haven't occupied in the recent years, both in the Space Shuttle and on the ISS, and also on Earth in the NASA organization.

And, even now that, after the end of the Shuttle era, the number of astronauts has again decreased, when a woman leaves for a mission, her name is just mentioned along with the other colleagues, as an "astronaut," not as a "woman astronauts." The departure of a

---

[25] Kevles, *Almost Heaven*, pp. 148–149.

[26] Interview with Bonnie Dunbar, "Oral History," *nasa.gov* (June 16, 1998).

woman for space doesn't make headline news anymore. And this means that now their integration in the space program has completely taken place.

Men and women work together on the ISS with interchangeable roles: no difference in training, or in the operation, or in responsibility.[27]

Six American female astronauts have flown in space five times: Shannon Lucid, Bonnie Dunbar, Marsha Ivins, Tamara Jernigan, Susan Helms, and Janice Voss, who recently passed away. It is actually a record, especially if we consider that the American female astronauts began to fly in space just 30 years ago and that the overall record of flights in space, held by the two American astronauts Jerry L. Ross and Franklin Chang-Diaz, is seven space missions.

Susan Helms holds the record for the longest spacewalk in history (8 h and 56 min), while Peggy Whitson—who was the first female Commander of the ISS and is considered by NASA the most experienced woman astronaut—holds the record for a female astronaut in zero gravity, having spent, over two long-duration missions on the ISS, more than 1 year of her life off the planet: in fact, over 376 days (a record that will be soon surpassed as she has been assigned to a new long-duration mission on the ISS). She is closely followed by Sunita Williams, who has spent a total of over 321 days in space and also holds the record for the number of spacewalks performed by a woman, with seven EVAs and 50 h spent in open space.

Women have also paid heavy costs, with four female astronauts as victims in the two tragedies of *Challenger* in 1986 and *Columbia* in 2003—four exceptional life stories broken. Christa McAuliffe, the legendary "Teacher in Space,"[28] and the "veteran" Judith Resnik, who was the second American in space and the first American Jewish astronaut, lost their lives in the disaster of the Shuttle *Challenger* STS-51L, which exploded 73 s after launch. Two more died tragically in the Shuttle *Columbia* STS-107, which disintegrated over the skies of Texas during re-entry at the end of a fruitful scientific mission of 16 days in space. They were Kalpana Chawla, the first Indian astronaut, in her second spaceflight, and Laurel Clark, Medical Officer of the US Navy, who had been initially assigned to a mission on the ISS that would lead her to becoming the first woman on a long-term mission and was then diverted to this fatal flight.

---

[27] How men and women adapt differently to spaceflight has been investigated by NASA and NSBRI (National Space Biomedical Research Institute). A comprehensive report on sex and gender differences related to human physiology and psychology in spaceflight has been published on the *Journal of Women's Health*, November 2014, and is available online in PDF format at the Web site of the Mary Ann Liebert, Inc. publisher: *http://online.liebertpub.com/toc/jwh/23/11*. Although there is an imbalance of data available for men and women, primarily due to fewer women who have flown in space, a long list of differences is reported by the six Sex & Gender Work Groups that participated in the research. A summary of the reports can be found online at www.nasa.gov/content/men-women-spaceflight-adaptation.

[28] Strictly speaking, Christa McAuliffe should not be regarded as an astronaut because the *Challenger* that was bringing her for the first time into space exploded 73 s after launch, at a height of just over 14,500 m—long before it crossed the Kármán line. But we cannot do without mentioning her here, since, more than many of her colleagues, during her intense preparation, the "Teacher in Space" captured the imagination of the USA and of the entire world and gave a great contribution to the revival of interest in the space program.

The women who fly into space have an average age of 40 years and have a degree, typically in engineering, but many are also doctors, biologists, biochemists, or physicists. On the other hand, today, astronauts fly into space to learn, to experience, to investigate; and the most significant research contributions are expected from doctors, biologists, and physicists. Often, women astronauts have more than just one degree, although it is difficult to surpass the Canadian Roberta Bondar who, after graduating in—among others—zoology, pathology, neurology, and neurobiology, was awarded with 24 honorary degrees by US and Canadian universities. Even Claudie Haigneré—who, at 20, had already completed her university studies in Medicine—in terms of academic qualifications is not joking around, and she earned among her friends the nickname "BAC + 19" (referring to her 19 university degrees).

Many are married to an astronaut, almost perpetuating the "space marriage" of Tereshkova and Nikolayev, personally sponsored by Khrushchev for propaganda purposes.

Many are mothers. Anna Lee Fisher made history when she was the first mother to fly in space, leaving at home a daughter just a few months old: the emblem of her mission features six stars—five representing the crew and the sixth for the newborn.

Often, they are civilians, but a few are also military pilots, with impressive career records.

On the other hand, safely going to space and returning requires special skills: the return of the Shuttle, which landed like a glider, began from a speed of 17,000 miles per hour—something like four and a half miles per second!

I recently met the astronaut Kathryn Hire, who was the first American woman to be assigned to a combat aircraft and, between 2001 and 2003, she participated in the Enduring Freedom and Iraqi Freedom Operations. But this is not the only case. US Navy pilots are Susan Kilrain, with 3000 h of flight on 30 different fighter jets, and Sunita Williams, who scored 3000 flight hours and, in 1992, operated as a combat helicopter pilot in the Desert Shield Operation; not to mention Colonel Eileen Collins, the US military pilot who became the first instructor of T-38 and other high-performance fighter jets, and test pilot: she was then the first pilot of the Shuttle, on which she flew four times, and finally the first Shuttle Commander. Or Pamela Melroy, who, after participating in the Gulf War with more than 200 combat hours, also became a military jet test pilot, with a backlog of over 5000 h on 50 different types of aircraft, and joined NASA, where she became the second woman to command the Space Shuttle. Even the Italian astronaut Samantha Cristoforetti is a military pilot certified for combat.

"Female Quotas" Are Rising

NASA active female astronauts in April 2016 number 12 and account for over 26 % of the entire Astronaut Corps, totaling 46 active astronauts:[29] indeed, they constitute one in every four astronauts. The percentage is not bad if we consider that, for example, in the US Police Corps, there is 1 woman in 7, and 1 in 20 in the Air Force. With the end of the Shuttle era, the NASA Astronaut Corps sharply decreased from 149 astronauts in 2000[30] to just one-third today, but the number of women is growing. Fifty percent of the

[29] See "*astrobio*" at *www.jsc.nasa.gov*.

[30] Rhian, J. "How Many Astronauts Does NASA Need?," *universetoday.com* (December 7, 2010).

astronauts selected by NASA in the last selection in June 2013 were women: four out of the eight—the highest percentage of women ever selected so far in a group.

In recent years, China has launched its first two female "taikonauts," as Chinese "space travelers" are called in the West. Currently, out of 10 taikonauts who have flown so far, two are women.

Only one female astronaut, the Italian Samantha Cristoforetti, is currently in the Astronaut Corps of the ESA, which consists of 14 active members, from Belgium (1), Denmark (1), France (3), Germany (2), Italy (4), Netherlands (1), the UK (1), and Sweden (1). Only two other women have preceded her in the ESA Astronaut Corps: Claudie Haigneré (France), already mentioned above, who left the ESA in 2002, and Marianne Merchez (Belgium), who never flew in space. They are joined by the British Helen Sharman, who never entered the ESA Astronaut Corps.[31]

Women, Space, and Philately

Philately has always been used as a showcase for the excellence of a country and, especially in the Soviet Union during the Cold War, stamps were exploited as a powerful means of propaganda. Even female astronauts have been portrayed in the stamps of many countries. Some of them, especially the first Soviet female cosmonauts, have dozens of stamps featuring them. A small sample is provided.

Villarbasse, Italy Umberto Cavallaro

---

[31] Cavallaro, U. "La Donna e lo Spazio" [Women and Space]. *Astronomia*, **39**(1), 32–34 (2014); Cavallaro, U. "Women in Space". *The Bridge (Quarterly Journal of International Women's Association of Prague)*, 25 (Summer 2016), pp 14–16.

1

Valentina Tereshkova: The Icon of Soviet Female Emancipation

Credit: RioNovosti

U. Cavallaro, *Women Spacefarers*, Springer Praxis Books, DOI 10.1007/978-3-319-34048-7_1

Mission	*Launch*	*Return*
Vostok-6	June 16, 1963	June 19, 1963

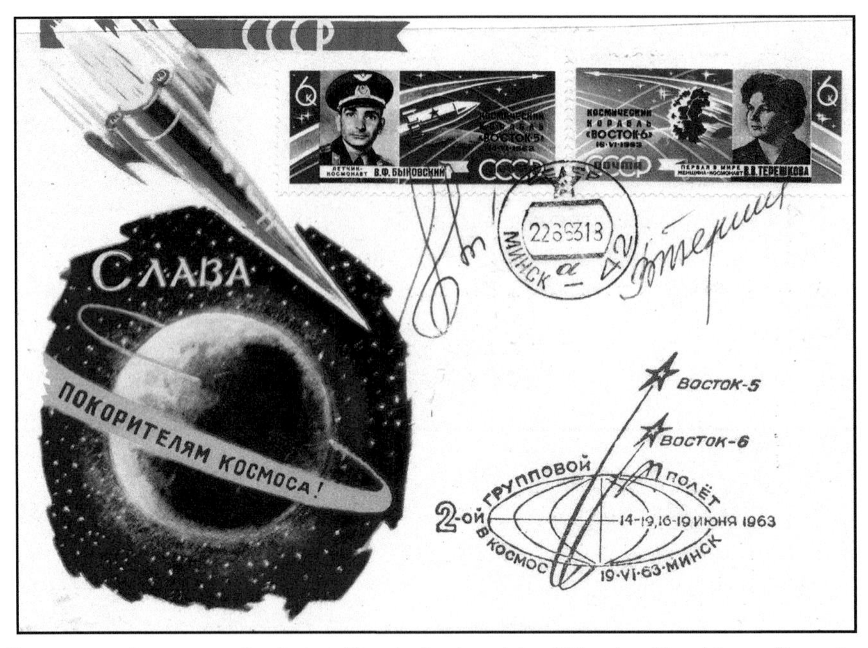

Commemorative cover of mission Vostok 6, signed by Valentina Tereshkova (from the collection of Umberto Cavallaro)

The 26-year-old Valentina Vladimirovna Tereshkova "orbited over the sex barrier" on June 16, 1963. In this way, the Soviet Union scored another point in the race to space after putting Yuri Gagarin into orbit—the first man ever to fly in space. This time, the leading actor was a woman, who spent 3 full days in a Vostok 6 spacecraft and orbited Earth 49 times, thus becoming a new symbol worldwide of Soviet superiority in space and in history as the Female Achiever of the twentieth century (Fig. 1.1).

Valentina Vladimirovna Tereshkova (Валенти́на Влади́мировна Терешко́ва) was born in Russia on March 6, 1937, in Maslennikovo, a small village in Tutayevsky District, Yaroslavl Oblast, in central Russia. Her childhood was affected by two consecutive wars. In 1939, her father, Vladimir Tereshkov, a tractor driver, was killed in action during the Russian–Finnish war, a year-long armed conflict in the north-west of the country. Two years later, World War II came to the USSR, and the Nazi forces were stopped only 50 km from the Yaroslavl region. Her mother, Elena Fyodorovna Tereshkova, a worker at the Krasny Perekop cotton mill, single-handedly raised Valentina, her brother Vladimir,

Fig. 1.1 Valentina Tereshkova holds many honors including the Hero of the Soviet Union (1963), Order of Lenin (1963, 1981), Order of the October Revolution (1971), Order of the Red Banner of Labour (1987), Order of the Friendship of Peoples, Order of Merit for the Fatherland, and Order of Honour (2003). Credit: © Alexander Mokletsov and RIA Novosti

and her sister Ludmilla under economically trying conditions. Valentina helped her mother at home and was not able to begin school until she was 10. After finishing the seventh grade, she moved to her grandmother's home in nearby Yaroslavl, where she worked as an apprentice at a tire factory in 1954. In 1955, she joined her mother and sister as a loom operator at the cotton mill, but continued her education by correspondence and graduated from the Light Industry Technical School in 1960. An ardent Communist (believer that there should be no private property), she joined the mill's Komsomol (Young Communist League) and soon advanced to the Communist Party.

In 1959, Valentina unknowingly made a decision that completely changed her life: her friend talked her into joining the local Yaroslavl Aviation and Skydiving Club, associated with the All-Union Voluntary Society for Assistance to the Army (in Russian, DOSAAF). She made her first jump at age 22 on May 21, 1959, although her mother did not approve of the hobby, saying it seemed too boyish.

When, in 1961, inspired by the flight of Yuri Gagarin, she volunteered for the Soviet space program, she already had a backlog of 126 parachute jumps.

The requirements to be a cosmonaut were very simple: parachuting experience, under 170 cm tall, under 70 kg in weight, under 30 years of age, being fit and being "ideologically pure," preferably workers without a high education. Khrushchev wanted to show the world that, in the USSR, anyone could pilot a spacecraft.

The list of 400 female candidates was agreed on January 15, 1962. In reviewing this list, Kamanin, responsible for the training of the cosmonauts, had asked DOSAAF to help him to find suitable candidates and wrote in his memoirs:

> Yesterday I considered the files for the fifty-eight female candidates. Generally disappointed and dissatisfied. The majority are not suitable for our requirement and have been rejected. Only twenty free will be brought to Moscow for medical tests because DOSAAF did not examine their credentials correctly. I told them that I needed girls who were young, brave, physically strong, and with experience of aviation, who we can prepare for space flight in no more than six months. The central objective of this accelerated preparation is to ensure that the Americans do not beat us to place the first woman in space.

Many of these candidates remained unknown.

In the cut-throat USSR–US unproclaimed competition, NASA had just established a new American record for time spent in space with the mission of Gordon Cooper, who launched on May 15, 1963, aboard the Mercury MA-9 on a flight that lasted for 34 h and 19 min and completed 22 orbits of Earth.

Although the Soviets were already firmly ahead with the solo-spaceflight records established the previous year by Andrian Nikolayev and Pavel Popovich, Premier Krushchev wanted to establish once more the Soviets' leadership in space with a new set of exploits and ordered the launch of the first woman into space and a new "rendezvous" of two spacecraft which would route the same orbit. The enterprise was presented to the press as a "long-duration joint spaceflight."

Not only was any ordinary factory worker of the socialist nation able to fly in space, but he/she also could stay there in one single flight for more than twice as long as all of the degree-educated Mercury Seven counterparts combined.

Khrushchev demanded that three different spacecraft should be sent simultaneously (and sending Komarov also was immediately considered), but there were only two Vostok spacecraft available and there was not enough time to build a new one.

The launch of Vostok 5 with the Soviet Air Force Lieutenant Valery Fyodorovich Bykovsky on board was initially planned for June 12. Because of excessive solar flare activity, the mission had to be delayed by 2 days. During the countdown on June 14, suddenly the Attitude Control Handle stopped working—a cable became unplugged and lodged under the seat of the cosmonaut. At the explicit request of Bykovsky, both problems were fixed without interrupting the countdown.

And the mission planned for a record 8 days began.

Soon, the Control Center realized that, due to errors in calculations, the flight path was significantly lower than expected. Troubles arose also with the survival system: the internal temperature decreased from the initial 30 to 10 °C. It was decided to return the spacecraft after only 5 days in space.

Bykovsky pulverized Cooper's record, however, and the Vostok 5 mission still holds the world duration record for a single-crew spaceship.

Two days later, from the secret base of Baikonur, Vostok 6 was launched into space with the first woman on board: Valentina Tereshkova. Her call sign during the flight was Chaika (Ча́йка in Russian means "Seagull") (Fig. 1.2). On its first orbit, Vostok 6 came within about 5 km of Vostok 5, the closest distance achieved during the flight, and Valentina established radio contact with Valery. In the following orbits, their distance progressively increased without any possibility for the cosmonauts to influence the trajectory. From the second day, communication between the two spacecraft was only possible through the Control Center.

Fig. 1.2 In honor of the 25th anniversary of Valentina Tereshkova's flight, artist cosmonaut Alexei Leonov (the first man to walk in space in 1965) painted "Seagull" which was the call-sign of Valentina during her historical mission. The 11 seagulls represent the women flown in space at that time. The painting is signed by Alexei Leonov and Valentina Tereshkova. Credit: © A. Leonov

In an interview published in *Komsomolskaya Pravda* in 2007, Tereshkova reported that Vostok 6 had launched faultlessly and the flight was going as planned until entering Earth orbit. After she had orbited Earth a few times, Ground Control realized that there was a dramatic, potentially fatal miscalculation in the orbit and provided a new automatic flying algorithm. The information about the mistake was classified. Only 40 years later was this story disclosed to the media.

But the troubles for Valentina were not over. The spacecraft was very tiny and she was ordered to remain strapped to her seat for all of the 70 h of the flight, wearing spacesuit and helmet. After several orbits with weightlessness, Valentina started to experience space adaptation syndrome (also known as SAS), with nausea, physical discomfort, and vomiting. On Day 2, she developed a cramp in her right leg; on Day 3, the pain had become unbearable. Her helmet put pressure on her shoulder and a sensor on her head gave her an itchy sensation. Due to the consequences of SAS, she was physically unwell and in a poor situation, without any possibility of cleaning herself. During the landing procedure, she was ejected at 7000 m, as planned, to land under her personal parachute. "To my horror," she reported, "I saw that I was heading for a splashdown in a large lake instead of on solid ground. We were trained for such circumstance, but I doubted I would have had the strength to survive." However, a high wind blew her over the shore. This resulted in a heavy landing. She hit her nose on her helmet, making a dark-blue bruise. She was in pain, dirty, and almost unconscious. She was immediately hospitalized. The honor of the Soviet Union required that the re-entry of the first spacewoman was a triumph. After recovering, she was brought back to the landing site, cleaned up, and given a pristine spacesuit to be filmed for the official news releases.

The idea of training female cosmonauts was launched by General Nikolai Kamanin—head of cosmonaut training—in 1961, after the historical mission of Yuri Gagarin, when he thought that the US was considering sending women from the Mercury Program into space. This spurred the Soviets on to select a number of women for their own space program, with the aim of getting them into space before the US.

Initially, the project was opposed by both military and bureaucratic powers.

The idea was, however, supported by Nikita Khrushchev, who recognized the propaganda value that would have shown both the reliability of the Soviet spacecraft and the gender equality under the Communist regime.

As few female pilots existed, the search for possible candidates was extended to parachutists: on the other hand the Vostok spacecraft was completely automatic and nothing during the flight would depend on the ability of the pilot.

Since it was a political mission, the only requirements were knowing how to use a parachute and having a strong Communist spirit.

Among the 58 candidates, five female cosmonauts—namely engineer Irina Solovyova, mathematician and programmer Valentina Ponomaryova, weaver Valentina Tereshkova, teacher Zhanna Yerkina, and secretary and stenographer Tatiana Kuznetsova—passed the assessment test and were selected as the second USSR cosmonaut group, which was kept secret until the late 1980s. None of the other four women in Tereshkova's group flew and the pioneering female cosmonaut group was dissolved due to "lack of utilization" in October 1969.

As happened for Gagarin, the final candidate was chosen by Khrushchev in person. Valentina was endowed with all the typical traits of the New Soviet Woman: a committed Communist, ideologically pure, a textile factory worker, daughter of a war hero who had been declared MIA (presumed dead) during the war, true proletarian. And a "pretty girl."

As she was a civilian, she was inducted into the Soviet Air Force as a Lieutenant so that she could become a member of the cosmonaut corps.

The psychosis of the secrecy surrounding the Soviet space program at the time was such that Tereshkova was ordered to keep the secret even from her mother. When saying her goodbyes and leaving for the training phase, she told her that she had been selected for an acrobatic parachuting course—her mother discovered the truth only by radio on the day of lift-off.

After 5 days of the mission, Bykovsky was ejected from his spacecraft 7000 m (23,000 ft) above ground and landed separately from his spacecraft at 14.06 (Moscow time) on June 19, 1963, 540 km from Karaganda—today in Kazakstan. A few dozen people living there rushed over to help him and, using a car, brought him to his spacecraft, which had landed 2 km farther away, in order to take the ritual pictures.

Tereshkova had landed only 2 h before.

With the first woman in space, the Soviet Union gained immense esteem worldwide that the Politburo was then able to exploit in a masterly manner for propaganda purposes to confirm the Soviets' supremacy: the achievement was heralded as a triumph and Tereshkova, a few days later, was awarded Pilot-Cosmonaut of the Soviet Union, Hero of the Soviet Union, and Gold Star of the Order of Lenin. The Telegraph Agency of the Soviet Union (TASS) emphatically declared that:

> Valentina Tereshkova was set forth in the same glorious path of progress already walked through by the most reknown Russian woman of the past. The triumph of the first spacewoman is obliged to the extraordinary scientific success of Sophie Kovaleskova, the first female professor and to Sophie Peroskaja, the Russian revolutionary who was executed by hanging after attempting assassination of the czar. Without the social emancipation the path to science would be barred to women.

Five months later, the General Party Secretary Khrushchev could announce worldwide another unexpected event: the first marriage of two cosmonauts—the "space family." Valentina Tereshkova and Andriyan Grigoryevich Nikolayev, the third man in space and the only bachelor cosmonaut to have flown, married on November 3, 1963.

The marriage ceremony took place at the Moscow Wedding Palace and had huge propaganda resonance. It was rumored that this marriage was another expedient of Khrushchev. The couple were assigned a luxury apartment on the *Kutuzovskij Prospekt*. On June 8, 1964, Tereshkova gave birth to their daughter, Aljenka Andrianovna Nikolaeva-Tereshkova. But this marriage was not as idyllic as the press had claimed: Nikolayev, an ethnic Chuvash, was quite a gruff man and the "space family" fell apart within a few years. However, as was the case with the American astronauts of that era, a divorce would mean the end of their careers, so the couple remained together. The marriage was only officially broken up in 1982.

References

Briggs, C.S. *Women Space Pioneers*, pp. 30–35. Lerner Publications, Minneapolis (2005).
Cavallaro, U. *Propaganda e Pragmatismo, in gara per la conquista della Luna* [*Propaganda and Pragmatism, in the Race to the Moon*], pp. 73–76. Impremix, Turin, Italy (2011).
Dragosei, F. "Un bluff le immagini del rientro dallo spazio", *Corriere della Sera* (March 7, 2007), 24.
Gueldenpfenning, S. *Women in Space Who Changed the World*, pp. 30–38. The Rosen Publishing Group, New York (2012).
Kevles, T.H. *Almost Heaven: The Story of Women in Space*, pp. 29–37. The MIT Press, Cambridge, MA (2006).
Petro, J.: The real rocket women: all-female astronaut panel represents international cooperation. rocket-women.com (2014). Accessed 23 July 2014
Quine T. "Tereshkova's Secret Sisters", *Spaceflight*, 54(6) 216–217 (2012)
Shayler, D.J.; Moule, I. *Women in Space – Following Valentina*, pp. 44–67. Springer/Praxis Publishing, Chichester, UK (2005).
Vis B., "Soviet Women Cosmonaut Flight Assignments 1963-1989", *Spaceflight*, 41(11), 474–480 (1999).
Woodmansee, L.S. *Women Astronauts*, pp. 36–37. Apogee Books, Burlington, Ontario, Canada (2002).

2

Svetlana Savitskaya: Twice in Space—The Second Soviet First

Credit: RIA Novosti

U. Cavallaro, *Women Spacefarers*, Springer Praxis Books, DOI 10.1007/978-3-319-34048-7_2

Launch		*Return*	
Soyuz T-7	August 19, 1982	Soyuz T-5	August 27, 1982
Soyuz T-12	July 17, 1984	Soyuz T-12	July 29, 1984

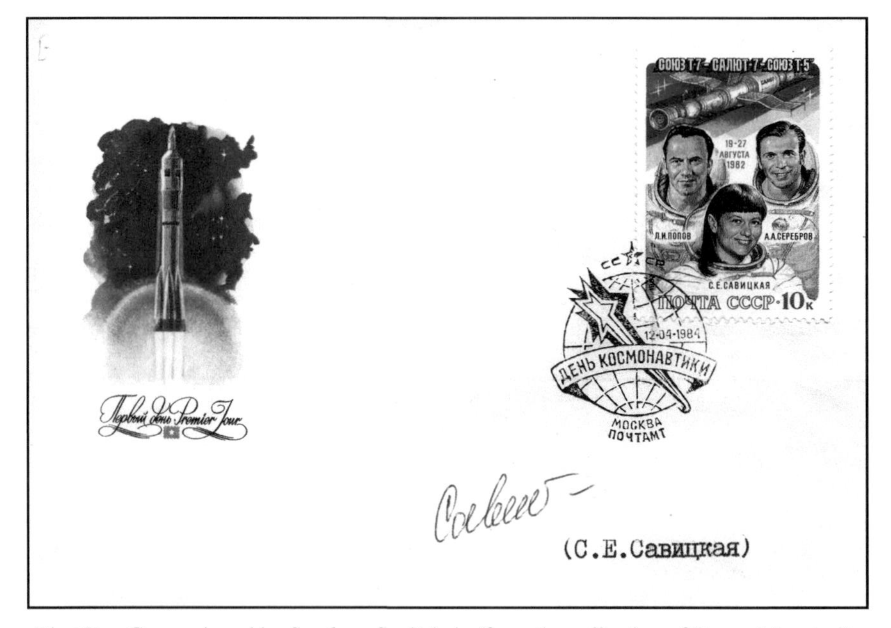

First Day Cover, signed by Svetlana Savitskaia (from the collection of Renzo Monateri)

"When in orbit, one thinks of the whole of the Earth, rather than of one's country, as one's home. One may even land on water, somewhere in the world Ocean, still the planet is his home," says the second woman in space.

Svetlana Yevgenyevna Savitskaya (Светлáна Евгéньевна Савúцкая) flew in space almost 20 years after Valentina Tereshkova. She flew twice to the Salyut 7 space station: in 1982 and 1984—and became the first woman to fly in space twice and also the first woman to perform an extravehicular activity (EVA), spending more than 3½ h in outer space.

Svetlana was born in Moscow on August 8, 1948. Daughter of a fervent activist of the Communist Party of Moscow and of a General—Yevgenii Savitsky was a World War II veteran flying ace, twice decorated Hero of the Soviet Union, and Deputy Commander of the Soviet Unions' Air Defense Forces—her lifelong dream was to follow her father into flight. After rejection from a flying school due to her young age, she secretly began to take parachute training instead, pretending to be 16 (which was the minimum age for admission). Her father discovered her parachute knife stashed in her school bag. Far from being angry, he actively encouraged her interest and, by her 17th birthday in 1965, the fearless parachutist

had already logged 450 jumps and set a world record of stratospheric parachuting by skydiving from 14,252 m and opening the chute a mere 500 m from the ground after a free fall of almost 14,000 m.

Supported by her father, at age 18, she enrolled at the prestigious Moscow Aviation Institute (MAI). In 1970, by age 20, she soloed in a YaK-18 training aircraft and joined the Soviet National Aerobatics Team. She emerged that year as world champion at the Sixth FAI World Aerobatic Championship, the world aerobatics competition held at Hullavingstone, UK, and drew the attention of the British press, who hailed her as "Miss Sensation."

She graduated in 1972 and became a flying trainer at DOSAAF (Central Technical Flying School of the USSR Voluntary Society for the Promotion of the Army, Air Force and Navy) and was accepted in the test pilots school. She qualified in 20 different types of aircraft.

After completing her test-pilot training and membership in the Communist Part of the Soviet Union, she establish many world records in turbo-prop and supersonic aircraft, including the women's world speed record set on June 22, 1975, by reaching the speed of 2683 km per hour in a MiG-21 supersonic aircraft. The following year, she became a test pilot for the Yakovlev design bureau.

She was selected as a cosmonaut and entered the Soviet Cosmonauts Corps in 1980. She knew well the powerful Valentin Glushko—the veteran rocket engine designer and now head of NPO Energia and responsible for assignments—as he had been her instructor in Moscow, and she was assigned to the Soyuz T-7 mission, to be launched to the Salyut 7 in August 1982. On the other hand, Glushko was not insensitive to the powerful position held by her father, then at the top of the Soviet Air Force.

She was the only female cosmonaut of her class to fly into space. The assignment was emphasized as the triumph of the emancipation of women under Soviet rule. The official Soviet press seized the opportunity for proudly highlighting that, for the second time, a woman was flying into space and once again she was a Soviet woman—further proof of the liberation acquired by women in socialist society. And, it pointed out, she would fly 8 months before Sally Ride, chosen to be the first US female astronaut, to be launched as a mission specialist on the Space Shuttle. To mark the event, the Soviet Post also dedicated a stamp to her.

As already mentioned, in a climate of steady decline in East–West relations in the early 1980s, a sort of new cold war, Savitskaya was undeservedly exploited for propaganda purposes, to restore the sense that the Soviets were ahead in space. The assignment to her first spaceflight in April 1982 was the response by the powerful and ambitious Glushko to the NASA announcement, released a few days beforehand, that Sally Ride was assigned to a Shuttle flight and would fly on STS-7 early in 1983. Savitskaya appeared, as if by magic, in the crew for Soyuz T-7, a mission scheduled for August of that same year. She would therefore launch before the American female astronaut.

Despite her being an accomplished pilot, her role in the spaceflight was as a researcher. The perfunctory acceleration imposed by Glushko did not allow the normal rules that psychologists of the Moscow Institute of Biological Problems would request to ensure the harmony of the crew during the mission to be followed. Aleksandr Serebrov, the rookie who was flying with her, did not miss out on showing his hostility against women cosmonauts who "do the job that should be done by a man." An episode connected to her arrival on the Salyut 7 space station, on the one hand, underlines the ambiguities in Soviet attitudes toward women, which combine a vaunting pride in the advancement of women with a degree of sentimentality and deference that would jar many Western feminists and, on the other hand, confirms the spirit that pervaded the deeply sexist Cosmonauts Corps.

When the Soyuz T-7 space vehicle carrying Savitskaya and the Soviet crew into orbit docked with the Salyut 7, the orbital space station had been Berezovoi and Lebedev's home for a month—long enough for them to grow a bouquet of Arabidopsis (flowers from a scientific experiment) which Commander Berezovoi presented to the female guest.

As referred to in the article issued in the *New York Times* on August 29, 1982—which in turn quoted a report published by the Russian newspaper *Sotsialisticheskaya Industriya*—the Salyut flight engineer Valentin Lebedev welcomed Miss Savitskaya on board by handing her an apron: "Feel like you're at your house!", said Lebedev, pointing to the "kitchen." Savitskaya bit the bullet. After that, somewhat jokingly, aprons showing her image began to appear on the market (Fig. 2.1).

In his diary—published following the fall of Communism—Lebedev reports that, after the T-7 crew had safely docked, Popov and Serebrov entered the station as soon as the hatch was opened, but Savitskaya remained in the Soyuz. When the others had looked into the docking module to see where she was, they saw her combing her hair with typical feminine vanity and only when she was comfortable with her hair did she float into the station.

Anatoli Berezovoi, the Salyut Commander admitted that "the presence of a woman on the Station severely limited their freedom and complicated daily life." He was also very frank about the lack of teamwork between Serebrov and Savitskaya, who were "like a cat and a dog" at each other, with Serebrov voicing his concerns about his female colleague and Savitskaya doing little more than was expected of her from the flight program.

During their 211 days on the station, Berezovoi and Lebedev lived moments of tension, spending whole days without talking to each other and working at opposite ends of the station. With the arrival of the Soyuz T-7 with the guests, and especially with a woman, the hassle for Berezovoi increased. He unleashed his resentment against her in a letter to his wife, sold at Sotheby's auction in New York in 1993:

> Isn't there a superstition among sailors that a woman aboard a ship is a bad omen? At the moment, I don't know for sure, but I can't imagine all the difficulties that we will have with this mixed crew. I'm afraid this will not be limited to just the problem

Fig. 2.1 The “Savitskaya apron.” Credit: zazzle.com

> of shared facilities … I will say nothing of Svetlana. She is a woman and that says it all. It will not be easy.

On August 27, Savitskaya and her two crewmembers said goodbye and boarded the Soyuz T-5 for return to Earth. Neither Berezovoi nor Lebedev was particularly sad when they left. Lebedev would later note that hosting the visiting crew was extremely tiring and the Salyut crews would need 2–3 days to get life back to normal after the visit.

Publicly, of course, the crew reported that everything went ideally and colleagues eventually said that she had been “as good as a man!”, expressing themselves with the best compliment they could do, while the daily newspaper *Izvestia* depicted her as “charming and gentle, a hospitable hostess who likes to make patterns and sew her own clothes when she has spare time.”

As happened to Valentina Tereshkova, Svetlana Savitskaya, on her return to Earth, also became a favorite interview subject for the media but, unlike Valentina, she did not make much of an effort to appear charming. Savitskaya tried to promote the results of the flight with the medical and scientific data obtained: during her mission, she had performed experiments on the cardiovascular system, space sickness, and the movement of the eyes. She also performed the first experiment of electrophoresis on human cells. But she was extremely difficult to interview. She presented herself as just one of the crew and did not like questions that were not to the point or were to do with the fact that she was a woman. One Western reporter who had arranged to talk to her in the first place almost missed out on the interview when none of the interpreters was willing to sit down with Savitskaya, who was described as a tough lady, as Lebedev had mentioned in his diary. One of her colleagues, Georgy Grechko, described her as "an Iron Lady, like Margaret Thatcher!"

Hers was not only the second mission of a woman in space, but also the first space mission of a mixed crew. The usual secrecy that had surrounded the mission, with carefully controlled news reports, had a spicy side effect. In April 1983, the West German press agency Deutsche Presse Agentur (DPA) reported that a German physician who had connections with the T-7 medical team had revealed that, during her mission, it had been attempted to get Savitskaya pregnant as part of a medical experiment. Although the news was denied, and there has never been any evidence backing this story, it was hot news in several countries.

After the successful completion of Savitskaya's mission, it was decided that another female cosmonaut should be sent to the Salyut on a long-duration mission to be in orbit during the time that Sally Ride was flying, so scoring another coup over the West. Irina Pronina, who had served as Savitskaya's backup, was assigned to fly the Soyuz T-8 mission that was planned to launch in April 1983, to last for more than 3 months, landing in July or early August, so completely overlapping and hopefully overshadowing the flight of Sally Ride in June. Svetlana was appointed as her backup, but she was soon replaced by Serebrov. For 6 months, the crew prepared for this mission until, shortly before the launch, the plan was changed and—according to the space historians Dave Shayler and the late Rex Hall—"internal politics of the Soviet programme" led to "heavy pressure" to remove Pronina from the crew.

With the second mission on the Soyuz T-12, the 50th mission of a Soyuz, on July 25, 1984, Svetlana became—as already mentioned—the first woman to complete a second space mission and the first to make an EVA. Again, also, this second assignment was the Soviet response to the American mission STS-41G in which Sally Ride would fly for the second time and another female astronaut (Kathy Sullivan) would perform an EVA for the first time; in one fell swoop, Savitskaya appropriated these two firsts and pre-empted Ride's chance to be the first woman to return to space and Sullivan's chance to become the first woman to leave a spacecraft in outer space, becoming herself both the first woman to fly twice and the first woman to make a "spacewalk."

The cosmonaut Valeri Ryumin, who was directing the operation from the Mission Control Center—seeing how Svetlana was excited to exit for the EVA—ordered the Commander Dzhanibekov to let her go first. He "gallantly" floated aside to give her his place. That gesture to her womanhood seemed to be one of the few she was willing to accept. Incidentally, Valeri Ryumin many years later married the third Soviet cosmonaut, Elena Kondakova.

For more than 3 h, Svetlana worked hard in outer space and tested the Universalny Rabochy Instrument (or "URI"), the new general-purpose hand tool, a portable electron beam gun that had the capacity to cut metal, weld, and solder. Before returning to the station, she collected some samples to bring back to Earth to study the impact of micrometeorites and the effect of aging on structures exposed in open space.

In the Telegraph Agency of the Soviet Union (TASS) reports following the end of the EVA, Savitskaya's achievements were emphasized as a major contribution to the future participation of women in space missions: "Her successful performance of unique experiments in space conditions has clearly shown the possibility to make efficient use of women in complex research not only aboard orbital manned complexes but also in the open space."

Svetlana had to fight against the prevailing machismo in the Soviet Cosmonauts Corps, trying not to be intimidated or discouraged by those who wondered—during her training—why a woman should have been employed to perform welding and other maintenance tasks in orbit, with the risk of burning each other's suits and setting fire to the space station itself. "After my space flight, everyone had to shut up!" was Svetlana's comment.

During the press conference after the mission, she was unable to mask her irritation by references to the "pleasant atmosphere a woman brings to a Space Station." "We do not go into space to improve the mood of the crew," she snapped. "Women go into space because they measure up to the job. They can do it." She then added:

> At a minimum, women are equal to men in space; Women are actually better at some space tasks than men. They are better at dealing with precision tasks. They are more meticulous. They are more flexible at switching from one task to another. Men of course are better where heavy exertion is required.

She came back to the same issue often, whenever she had the opportunity. Yet, in an interview in 2011, she declared that "The ability to work in space depends on factors such as personal preparation, the physical and psychological, the ability to self-control, and personal goals .... When there is professionalism, gender does not matter."

After these missions, she was decorated twice as Hero of the Soviet Union. Her patron, Glushko, eventually appointed her as the Commander of the mission that was supposed to be an all-female Soyuz crew to Salyut 7. The departure would have to take place in November 1985, coinciding with the commemoration of the Bolshevik Revolution (probably aboard the Soyuz T-14). Svetlana would have been joined by engineer Yekaterina Ivanova and physician Elena Dobrokvashina. But the plan had to be changed when, as a result of the loss of contact with the Salyut 7 that had remained temporarily uninhabited, a hasty emergency flight had to be organized to bring the station back to life. The all-female flight was rescheduled for March 8, 1986—International Women's Day. Savitskaya's (and Ivanova's and Dobrokvashina's) final chance to fly in space went up in smoke when Vladimir Vasyutin fell ill aboard Salyut 7 and the crew was obliged to return to Earth in November 1985.

Officially, the mission was eventually cancelled as, due to the pregnancy of Savitskaya (her son Konstantin was born in October 1986), Star City was temporarily without a woman cosmonaut who had already previously flown and could, according to the rules, take the role of Commander. "She thought it was better making a child that doing a third flight!" sarcastically commented the third Soviet cosmonaut, Elena Kondakova. Besides Vasyutin's accident, kept under wraps for a long time, rumors suggested however that strong opposition against the idea of an all-female flight existed at the higher level of the Soviet space hierarchy. Ironically, the backup crew for the mission was all-male, further underlining the hypocritical reality that the all-female flight had little practical merit.

As a matter of fact, half a world away at NASA, flights of women on the Shuttle had become almost "routine," with more women at a time, and the mission of women only had lost its bite.

While remaining on active duty as a cosmonaut, she did not fly anymore and, in 1987, Glushko appointed her as Chief Designer at the NPO Energia. In 1993, she left the space program and retired from the Russian Air Force with the rank of Major. In 1996, she was elected a member of the State Duma representing the Communist Party of the Russian

Federation, and was re-elected four times after that. She currently serves as Deputy Chair of the Committee on Defense, and is also a member of the Coordination council presidium of the National Patriotic Union.

Svetlana Savitskaya, as the first and only female cosmonaut to walk in space, was one of the five cosmonauts who raised the Russian flag when the Sretensky Monastery choir sang the national anthem during the colorful opening ceremony of the 2014 Winter Olympics in Sochi. With her also was the last woman to leave for space few months later: Elena Serova.

Svetlana is married to Viktor Khatkovsky, an engineer and pilot at the Ilyushin aircraft design bureau, with the one son born in 1986.

References

Biography of Svetlana Savitskaya, *astronautix.com.*

Briggs, C.S. *Women Space Pioneers*, pp. 42–47. Lerner Publications, Minneapolis (2005).

Burns, J.F. "An Apron for Soviet Woman in Space", *New York Times* (August 29, 1982), 18.

Evans, B. "A Cog in a Political Machine: The Career of Svetlana Savitskaya", *americans-pace.com* (July 2013).

Gibson, K.B. *Women in Space: 23 Stories of First Flights, Scientific Missions and Gravity-Breaking Adventures*, pp. 57–61. Chicago Review Press, Inc., Chicago (2014).

Harnett, S. "Svetlana Savitskaya – Test Pilot and Cosmonaut", *australianscience.com.au* (March 12, 2013).

Hendrickx, B. Illness in Orbit. *Spaceflight*, **53**(March), 104–109 (2011).

Kevles, T.H. *Almost Heaven: The Story of Women in Space*, pp. 83–91. The MIT Press, Cambridge, MA, and London, UK (2006).

Mydans, S. "Female Soviet Astronaut Says that Women Have a Place in Space", *nytimes.com* (August 11, 1984).

Shayler, D.; Hall, R. *Soyuz: A Universal Spacecraft*, p. 294. Springer, London (2003).

Shayler, D.J.; Moule, I. *Women in Space – Following Valentina*, pp. 203–208. Springer/Praxis Publishing, Chichester, UK (2005).

Vis, B. Soviet Women Cosmonaut Flight Assignments 1963–1989. *Spaceflight*, **41**(November), 474–480 (1999).

Woodmansee, L.S. *Women Astronauts*, pp. 51–52. Apogee Books, Burlington, Ontario, Canada (2002).

3

Sally Ride: America's Pioneering Woman in Space

Credit: NASA

U. Cavallaro, *Women Spacefarers*, Springer Praxis Books, DOI 10.1007/978-3-319-34048-7_3

Mission	*Launch*	*Return*
STS-7	June 18, 1983	June 24, 1983
STS-41G	October 5, 1984	October 13, 1984

Commemorative cover of mission STS-7, signed by the Crew, including Sally Ride (from the collection of Umberto Cavallaro)

There was a palpable excitement in the air: "The spirit and the substance of America's manned space program had changed irrevocably," emphatically proclaimed a reporter while Sally Ride, connected with Houston as the Shuttle was shooting skyward, was asking the capsule communicator (CapCom) Roy D. Bridges "Have you ever been to Disneyland? This is definitely an E-ticket"—in other words, the most exciting ride of all. As Rick Hauck recalls in his *Memories*, several months later Mickey Mouse visited the Kennedy Space Center where he presented each of the STS-7's team with a Mickey Mouse watch.

Sally Kristen Ride was born on May 26, 1951, in Encino, California, near Los Angeles, into a family with Norwegian roots. When Sally and Kathy Sullivan, as two of the first six women in NASA's Astronaut Corps, got to know each other, it turned out that in 1958 they had been in the first grade together at the Hayvenhurst Elementary School, though neither really remembered the other clearly, and they had a good laugh as they pieced this together 20 years later.

From an early age she started practicing softball and football. She wanted to join the Los Angeles Rams as a pro but, at the time, football was a sport precluded from girls and this helped to make her sympathetic to the feminist movement. She therefore decided to dedicate herself to the tennis that she had discovered at the age of 10 years. Her first coach was Alice Marble, the famous four-time US Open tennis champion in the 1930s, who also had won twice at Wimbledon. Sally devoted herself passionately to tennis until she earned a national ranking when aged 16 and broke into the US top 20 at 18. She played eventually in the Westlake High School tennis team during her 3 years at the school, and served as captain for the last year, leading a team which that year managed to play in the local interscholastic tournament without losing a set.

Sally continued to compete on the tennis courts while at Stanford. The tennis world champion Billie Jean King saw Sally playing and suggested that she should quit school and pursue a career as a professional tennis player. At the first major crossroads in her life, Sally chose however to trade tennis for academics and graduated in 1973 with a Bachelor of Science degree in Physics and a Bachelor of Arts degree in English. Between 1973 and 1975, she was part of Stanford's Master of Science Program in Physics, specializing in Astrophysics and focusing on X-rays given off by stars; in 1975, she received her master's degree in Physics. In 1978, she received a doctorate in Physics at the same university, where she was also involved in research and laser physics.

One day, she casually read in the student newspaper *Stanford Daily* an advertisement that would change her life: NASA was selecting scientists to conduct experiments on Space Shuttle flights. These scientists, called "mission specialists," would go into space as part of the Shuttle crews: for the first time, NASA also called for applications from women and minority representatives to join its Astronaut Corps, which until then was an "all-male–all-white club." She applied, together with 8370 candidates, including 1000 women.

The selection was tough. "It was of big help," recalled Sally, "the experience gained with tennis matches, which had given me a certain amount of self-esteem" and self-control, which was one of her distinctive characteristics. "What Sally really thinks of anything," her sister once said, "is a mystery."

On January 16, 1978, while she was still working on her doctorate thesis, very early in the morning, Sally got the call she was waiting for. George Abbey, the legendary and mysterious NASA director of flight crew operations at the time—and chairman of the selection board in 1977—was on the phone: "We've got a job here for you, if you are still interested." So she entered into the TFNGs (Thirty-Five New Guys),[1] the new NASA astronauts group which for the first time also included six women. Sally Ride told the agency's oral historian:

> There was a lot of media attention surrounding the announcement because not only was it the first astronaut selection in nearly 10 years, it was the first time that women were part of an astronaut class. There was a lot of press attention surrounding all six of us. Stanford arranged a press conference for me, the day of the announcement – the first time, of course, that I had even thought of being part of a press conference. I mean, my gosh, I was a Ph.D. physics student. Press conferences were not a normal part of my day.
>
> A lot of newspaper and magazine articles were written – primarily about the women in the group – even before we arrived at Johnson Space Center (JSC). The media attention settled down quite a bit once we got to Houston.
>
> NASA protected me while we were in training, and even the day that we did all our pre-flight press conference about a month before the flight, we did them in pairs. ... They did a really good job shielding me from the media so that I could train with the rest of the crew and not be singled out. There were still the occasional stories, and we definitely found ourselves being sent on plenty of public appearances.

Sally Ride was one of the most successful achievements of the glamorous African American actress Nichelle Nichols, who at the time was well known for the role of Lieutenant Uhura that she had played in the popular *Star Trek* television series. She was enlisted by NASA to recruit minority and female personnel for the space agency. Nichelle recalls:

> When NASA asked me to help them find the first qualified women and minorities to join the then all-male-all-white Astronaut Corp I did so with great enthusiasm. One of the first that I was able to reach was a beautiful, young, brilliant woman named Sally Ride. She not only joined the Astronaut Corps – she revolutionized it by blazing the trail that so many female astronauts followed. She became my inspiration to continue to search to find the next Sally Ride.

Together with her colleague, John Fabian, Sally was assigned to the Shuttle's robot arm, the RMS (Remote Manipulator System), that was still in the testing and development stage. She became heavily involved in the simulator work to verify that the simulators accurately modeled the arm and to prepare procedures for using the arm in orbit and to develop the malfunction procedures so that astronauts would know what to do if something went wrong. She worked on the robot arm for a couple of years and also spent a lot of time at the

---

[1] The class name has rather a "marine" origin explained by Mike Mullane in his book, *Riding Rockets* (p. 63): "In polite company TNFG translated to '*Thirty-Five New Guys*'. Not very creative, it would seem. However it was actually a twist of an obscene military term. In every military unit a new person was a FNG, a 'fucking new guy'. You remained a FNG until someone newer showed up, then they became the FNG. While the public knew us as the 'Thirty-Five New Guys', we knew ourselves as the 'The Fucking New Guys'."

contractor facility in Toronto, Canada, with the engineers who performed the hardware development and testing. She then worked with the STS-2 crew on the detailed procedures for the tests of the robot arm, as this mission would in fact be the first to carry the RMS. She also worked to help prepare the STS-2 Earth observations plan and train the crew on how to recognize the things that scientists were interested in. During the second and third Space Shuttle flights (STS-2 and STS-3), she served as the ground-based CapCom—a function that traditionally was exclusively performed by another NASA astronaut.

In 1982, she married fellow NASA astronaut Steve Hawley, who had entered the Astronaut Corps in the same TFNGs class. They divorced 5 years later in 1987.

After intensive training, in 1983, she was assigned as a mission specialist to the Space Shuttle *Challenger* STS-7 mission. She recalls in an interview:

> It was around 7 or 7:30 in the morning because I tended to get into the office early. So I got a call to go over to Mr. Abbey's office. And you know, that was generally taken as either a very good or a very bad sign. And so I walked over to Building One and went to the eighth floor. His assistant had me go right in, and it was me and George Abbey, and he said in typical George Abbey fashion, 'So do you still want to fly? Well, um, we're going to assign you to STS-7'. So George talked to me for a few minutes, and then took me up to the ninth floor to meet with Chris Craft (Johnson Space Center Director). And he said: 'Are you sure you want to do this? We're thrilled to have you on the crew, but we just want to make sure that you know what you're getting into'. And at that point, of course, all I cared about was getting a chance to fly. So I said 'Yes–Yes–Yes'.

And so Sally soared into history as the first American woman—and, at 32, the youngest American—in space, and captured the imagination of hundreds of young women who, in the US and the Western world, were attracted by her example and decided that they wanted to be astronauts.

Later, her colleague and crewmate, Kathy Sullivan, commented:

> 'Why was Sally Ride tapped to fly first?' is a question I've been asked many times in the past few days. The truth is, none of us ever knew, not even Sally. Whatever the reasons may have been, history will record that the selection turned out well indeed. Sally performed superbly on STS-7 and stepped into the role of first American woman to fly in space with intelligence, dignity and grace.

The *Challenger* commander, Robert L. Crippen, had chosen Sally Ride for that mission because she was extremely well qualified on the RMS systems and operations—as the robotic arm had to be used intensively—and also because of her "very good public presence": NASA leaders believed she could handle the pressure of being a celebrity. It was indeed predictable that "the first American woman in space" would receive notable media attention, and Sally was known for keeping her cool under stress. For months, she had to endure silly questions from the press and crazy articles in the tabloids. "There was a lot more attention on us," Sally recalls, "than there was on previous crews, probably even more than the STS-1 crew." The four male members of this crew would fade into the shadows in the public eye. The media were interested only in Ride. "The clamor from the press for interviews with Sally was unabated," Rick Hauck confirms in his *Memories*. Sally said in one interview:

> None of the astronauts who applied did it for publicity. Everybody applied because this is what they wanted to do. So the males in the group didn't really want to be spending their time with reporters; they wanted to be spending their time training and learning things. They didn't seem to mind at all that more of the attention was paid to the women astronaut candidates. In fact, they wished us well. And, frankly the women probably would have preferred less attention.

The pressure of the spotlights was intense and some of the coverage was laughable. During one pre-flight press conference, Sally was ridiculously asked: "Do you think you will cry during the flight?" She always exhibited grace under pressure and responded humorously: "Why you don't ask Rick [the pilot] that question?" When asked how much she was getting paid, she very frankly answered: "I mean, who cares? I'd pay them!" She managed to handle with grace the insanity of the media activity, which was not exactly congenial to her, similarly to her colleague Judy Resnik, who indeed loved it even less.

This was the second flight of *Challenger* and it was the first mission with a five-person crew. During the mission, the STS-7 crew deployed telecommunication satellites for Canada (ANIK C-2) and per Indonesia (PALAPA B-1). Sally was involved in ten experiments and maily operated the Canadian-built RMS to perform the first deployment and retrieval exercise with the Shuttle Pallet Satellite (SPAS-01): "Operating with the robotic arm," Sally explained, "was a little like playing with a complicated videogame." She was aware that her level of performance would be very important for the other female colleagues who would follow her.

Everything went smoothly and the flight was fun. "The thing that I'll remember most about the flight is that it was fun," Sally annotated in her website, *sallyridescience.com*. "In fact, I'm sure it was the most fun I'll ever have in my life."

After the mission, she was asked to take part in a goodwill tour of Europe, together with the pilot, Rick Hauck. They visited eight countries, where they met kings and queens and prime ministers, and visited research facilities—and gave innumerable presentations and interviews. The itinerary also included an appearance at the International Astronautical Federation conference in Budapest, Hungary. Russian cosmonauts were there and American astronauts were urged not to mingle with or be photographed enjoying drinks with them. Sally was invited to a private meeting with cosmonaut Svetlana Savitskaya. Although it had been suggested that she and Dick keep their distance from the cosmonauts, Sally accepted. No report of the meeting was ever made. It was a difficult moment, as US/USSR relations were strained due to the recent shooting down of Korean Airlines Flight 007.

In 1984, she served as a mission specialist on STS 41-G and totaled altogether 344 h in space. With seven astronauts, it was the largest crew to fly to that date and her friend, Kathryn Sullivan, also flew on the same mission: it was the first time that two women astronauts had flown together. During that mission, Kathryn became the first American woman to walk in open space.

In June 1985, Dr. Ride was assigned to the crew of STS 61-M, which would fly in July 1986, but mission training was halted in January 1986 following the Space Shuttle *Challenger* accident, in which all seven crewmembers were killed.

Dr. Ride served as a member of the Presidential Commission "Rogers" investigating the tragedy. She was the only astronaut assigned to that Commission and was responsible for the sub-committee *Operations*. As a member of the panel, Sally gained a reputation for asking tough questions. The panel learned from testimony and other evidence that there had been signs of trouble on earlier *Challenger* flights, but that they had been dismissed as not critical. One witness was Roger Boisjoly, an engineer who had worked for the company that made the Shuttle's rocket boosters and who had been shunned by colleagues for revealing that he had warned his bosses and NASA that the boosters' seals, called O-rings, could fail in cold weather. *Challenger* had taken off on a very cold morning. After Boisjoly's testimony, Sally, who was known to be reserved and reticent, publicly hugged him. She was the only panelist to offer him support.

Upon completion of the investigation, she was assigned to NASA Headquarters in Washington as Special Assistant to the Administrator for long-range and strategic planning and directed NASA's first strategic planning effort. There she wrote an influential report entitled *NASA Leadership and America's Future in Space* and became the first Director of NASA's Office of Exploration.

Sally retired from NASA in 1987. She had been the first woman to fly in space and was also the first woman to leave NASA. NASA required a stay in the Astronauts Corps of at least 6 years and she remained there for 8. She wanted to go back to the academic world and became a science fellow at the Center for International Security and Arms Control at Stanford University. The following year, she joined the faculty at the University of California San Diego (UCSD) as a Professor of Physics and Director of the University of California's California Space Institute.

A symbol of the ability of women to break barriers, in 2001, she founded her own company, Sally Ride Science, to pursue her long-time passion of promoting STEM (science, technology, engineering, mathematics) literacy and motivating young women to pursue careers in STEM. The company aimed at creating entertaining science programs and publications for upper-elementary and middle-school students and their parents and teachers. A long-time advocate for improved science education, Sally co-wrote seven science books for children: *To Space and Back* (with Sue Okie); *Voyager*; *The Third Planet*; *The Mystery of Mars*; *Exploring Our Solar System*; *Mission Planet Earth*; and *Mission Save the Planet* (all with Tam O'Shaughnessy). She also initiated and directed education projects designed to fuel middle-school students' fascination with science: EarthKAM (Earth Knowledge Acquired by Middle-school students). Originally called KidSat, the program allowed middle-school students to capture images of Earth using a camera aboard the Space Shuttle. The program operated during five Shuttle flights (STS-76, STS-81, STS-86, STS-89, and STS-99). KidSat was renamed EarthKAM in 1998. In 2001, the camera moved to the International Space Station and the program was renamed ISS EarthKAM. After Ride's death in 2012, NASA eventually renamed the program Sally Ride EarthKAM.

In 2003, Sally was asked to serve on the Columbia Accident Investigation Board. She was the only person to sit on both panels investigating the catastrophic Shuttle accidents that killed all astronauts on board: the *Challenger* explosion in 1986 and the *Columbia* crash in 2003. Sally said in an interview with *The Times* that part of the problem at NASA was that people had forgotten some of the lessons learned from the *Challenger* accident.

Sally Ride died on July 23, 2012, at 61, after courageously battling for 17 months against pancreatic cancer. In her obituary that she herself co-wrote, she publicly revealed that she was survived by Professor Tam O'Shaughnessy, "her partner of 27 years." Sally had started a relationship with Tam in 1985, which was kept secret and led to the dissolution of her marriage with the astronaut Steven Hawley in 1987.

Tam and Sally first met at the age of 12 when both were aspiring tennis players and, over tennis matches and camps, they embarked on a lifelong friendship. Their paths then separated. Sally chose the academic career and graduated in Physics and became an astronaut. Tam O'Shaughnessy went on to play on the women's professional tennis circuit and, at the beginning of the 1970s, competed in the US National Championships (now known as the US Open) and at Wimbledon. After retiring from tennis, she started a career as a professor at San Diego State University until 2007, when she decided to devote her time and energy to Sally Ride Science, as co-founder and chief operating officer and then chief executive officer.

She was committed to breaking down the cultural and educational barriers that block so many, especially young girls, from science and the opportunities it offers in life. And this she did through her work as a physics professor, mentor, motivator, and businesswoman, as Kathy Sullivan states.

After Sally Ride's death, President Barack Obama posthumously awarded her the Presidential Medal of Freedom, as "Sally Ride was to generations of young women a powerful role model and showed us that there are no limits to what we can achieve."

References

Briggs, C.S. *Women Space Pioneers*, pp. 35–42. Lerner Publications, Minneapolis (2005).

Evans, B. "Spaceflight Pioneer Sally Ride Loses Battle with Pancreatic Cancer", *americaspace.com* (July 23, 2012).

Gibson, K.B. *Women in Space: 23 Stories of First Flights, Scientific Missions and Gravity-Breaking Adventures*, pp. 81–90. Chicago Review Press, Inc., Chicago (2014).

Grady, D. "American Woman Who Shattered Space Ceiling", *nytimes.com* (July 23, 2012).

Gueldenpfenning, S. *Women in Space Who Changed the World*, pp. 39–48. The Rosen Publishing Group, New York (2012).

Harnett, S. "Women in Space: Sally Ride", *australianscience.com.au* (May 20, 2013).

Hauck, F. "Memories of My Space Flight with Sally Ride", *lightyears.blogs.cnn.com* (July 24, 2012).

Interview with Sally Ride, "*Oral Histories*" program, *jsc.nasa.gov* (October 22, 2002).

Interview with Sally Ride, "*Oral Histories*" program, *jsc.nasa.gov* (December 6, 2002).

Kevles, T.H. *Almost Heaven: The Story of Women in Space*, pp. 62–63, 66–68. The MIT Press, Cambridge, MA, and London, UK (2006).

Mullane, M. *Riding Rockets: The Outrageous Tales of a Space Shuttle Astronaut*. Scribner, New York (2006).

O'Shaughnessy, T. *Sally Ride: a Photobiography of American Pioneering Woman in Space*, Roaring Brook Press, New York (2015).

Official biography of Sally Ride, *nasa.com* (July 2012).

Official biography of Sally Ride, *Sallyridescience.com* (June 2014).
Shayler, D.J.; Moule, I. *Women in Space – Following Valentina*, pp. 198, 208–215. Springer/Praxis Publishing, Chichester, UK (2005).
Sullivan, K. "The human, funny side of Sally Ride", *cnn.com* (July 26, 2012).
Woodmansee, L.S. *Women Astronauts*, pp. 52–54. Apogee Books, Burlington, Ontario, Canada (2002).

4

Judith Resnik: The Second "Shuttlenaut"

Credit: NASA

U. Cavallaro, *Women Spacefarers*, Springer Praxis Books, DOI 10.1007/978-3-319-34048-7_4

Mission	*Launch*	*Return*
STS-41D	August 30, 1984	September 5, 1984
STS-51L	January 28, 1986	–

Commemorative cover of mission STS-41D, signed by Judy Resnik (from the collection of Umberto Cavallaro)

Judy Resnik was one of the victims of the tragic explosion of *Challenger* STS-51L—the 25th flight of the American Space Shuttle program and the 10th mission of the Space Shuttle *Challenger*—which disintegrated 73 s after lift-off over Cape Canaveral on January 28, 1986.

It was her second spaceflight after participating as a mission specialist, in August 1984, in the STS-41D mission and becoming the second American woman to fly in space after Sally Ride and the fourth woman worldwide. "I think," she said 1 year before, "something is only dangerous if you are not prepared for it, or if you don't have control over it, or if you can't think through how to get yourself out of a problem." Carl Glassman, who interviewed Dr. Resnik for a book, *Dangerous Lives*, referred to her telling him in a 1979 interview that "It does not enter any of our minds that we're doing something dangerous."

Judith Arlene Resnik, born in Akron, Ohio, on April 5, 1949, was the progeny of an upper-middle-class Jewish-Ukrainian family devoted to their religion. She attended Hebrew school for many years at Beth El Synagogue in Akron, even if she did not practice Judaism and disliked any reference to her as "the first Jewish astronaut."

Fig. 4.1 Judy Resnik on the middeck of the Space Shuttle *Discovery* during her first mission, STS-41-D (1984). Credit: NASA

Her father, Marvin Resnik—to whom she was so close as to choose to live with him after her parents divorced when she was 17 years old—practiced optometry and was a part-time cantor. He affectionately called his daughter *K'tanah* (which is Hebrew for "little one") and proudly hung in the reception area of his optometry office a picture of his daughter on the Shuttle, greeting him with a floating sheet of paper saying in large letters "HI DAD" (Fig. 4.1).

Judy was noticed early on for her brilliant intelligence. Soon after entering kindergarten, she was able to read and solve simple mathematical problems, so she was admitted to the elementary school 1 year in advance. Her talents became more evident over the years when—very diligent and methodical—she began to excel in mathematics, chemistry, and French. She also was a gifted musician and played piano with more than technical mastery, appreciated by all for her vividness, but also for her perfectionism—qualities that she brought with her all her life.

After high school, she wanted to become a professional classical pianist but, after achieving the highest possible score on the mathematics component of her SAT test, she decided to pursue a Bachelor of Science degree in Mathematics at the Carnegie-Tech (today Carnegie-Mellon University) in Pittsburgh, but she changed her plans in her freshman year after she began to date Michael Oldak, a fellow Engineering student, and visited some of his classes and realized that "she liked more of the practical aspects of science," as Michael recalls.

She therefore entered Engineering and graduated in 1970. Shortly after graduation, she married Michael. Both were hired by the RCA (Radio Corporation of America) and went to work in Moorestown, New Jersey. Judy was employed in the missile and surface radar division, working on custom integrated circuitry for phased-array radar control systems and engineering support for NASA sounding rocket and telemetry systems programs. Her paper concerning design procedures for special-purpose integrated circuitry caught the attention of NASA. The following year, the couple moved to Washington, DC, where Judy received her master's degree in Engineering from the University of Maryland and began working as a biomedical engineer at the Laboratory of Neurophysiology of the National Institute of Health in Bethesda, Maryland, where she performed biological research experiments concerning the physiology of visual systems.

In 1975, she and Michael separated but the two remained friendly and she continued to share milestones of her life with her former husband. Not long after her divorce, while she was completing her doctorate in 1977, she spotted an advertisement for NASA's astronaut program that began recruiting women and minorities to the space program. Though this was not her lifelong desire (until then, it would have been unthinkable for a woman to fly in space) and although she did not think NASA would select her, she suddenly decided to apply because "She was looking for a purpose in her life," her father said later. After receiving *magna cum laude* academic honors for her doctoral work in electrical engineering, she accepted a job with Xerox as a senior systems engineer and relocated to Redondo Beach near Los Angeles, California. To her amazement, in January 1978, she was called for a NASA interview. Her background in electrical engineering had made her a primary candidate for the mission specialists needed to conduct scientific experiments in space.

At age 29, Judy was one of six women accepted into the program. After completing a 1-year training and evaluation period in August 1979, she worked on a number of projects in support of orbiter development, including experiment software, the Remote Manipulator System (RMS), and training techniques. "She had a great sense of humor," recalls Henry Hartsfield, the Commander of her first mission. "We used to joke around all the time and Judy was right in the middle of it." "Judy was an astronaut's astronaut," Hartsfield adds. "She was not satisfied with second best."

Both of her missions were marked by technical problems and delays. The launch of the first—the maiden flight of *Discovery* STS-41D—was postponed four times. During the June 26 launch attempt, there was a launch abort at the last moment, at T–4 s, due to an anomaly in one of the main engines followed by a pad fire. The crew had to evacuate the Shuttle. This marked the first time since Gemini 6A that a manned spacecraft had experienced a shutdown of its engines just prior to launch. *Discovery* rolled back to the Vehicle Assembly Building and the faulty engine was replaced. The launch was delayed by over 2 months. The mission was rescheduled and merged with mission STS-41F, which was actually canceled. Mission STS-41D was finally launched on August 30, 1984, after a 6-min 50-s delay when a private aircraft intruded into the warning area off the coast of Cape Canaveral.

The primary payloads of STS-41F were included on the STS-41D flight. The combined cargo weighed over 41,184 pounds (18,681 kg)—a record for a Space Shuttle payload up to that time. The cargo bay payloads included three commercial communications satellites, originally scheduled to fly on mission STS-41F, for NASA, the US Navy, and Canada. The first task of Judy as mission specialist was to operate the robotic arm that she had helped to design, to open the OAST-1 solar array (Office of Application and Space Technology)—a device 13 ft (4.0 m) wide and 102 ft (31 m) high that carried five different types of solar cells. It was the largest structure ever extended from a manned spacecraft and demonstrated the feasibility of large lightweight solar arrays for future application to large facilities in space, such as the International Space Station.

Located in the middeck was an IMAX high-fidelity motion picture camera, making the second of three scheduled trips into space aboard the Shuttle. Footage from the Shuttle flights was then assembled into the documentary called *The Dream Is Alive*. The crew earned the name "Icebusters" after successfully removing hazardous ice particles from the orbiter using the RMS. With this flight, Judy completed 96 orbits and logged 144 h and 57 min in space.

Judy was also mission specialist on STS 51-L. The lift-off of the second mission, the ill-fated *Challenger* STS-51L, had been delayed five times for inclement weather and mechanical problems. STS-51-L was the 10th mission of *Challenger* and the 25th Shuttle launch in NASA's history. It was notable for its seven-person crew's diversity: in addition to Resnik, a Jewish woman, it included an African American astronaut (Ronald McNair), a native Japanese (Ellison Onizuka), and the first teacher in space (Christa McAuliffe). Judy's task was to operate the Shuttle's robotic arm to release SPARTAN, a platform with scientific instruments that would float in space to study Halley's Comet and then be recovered.

After several technical and weather-related delays, finally it was decided to launch *Challenger* on January 28, 1986, an atypically cold day for Florida and the coldest weather conditions under which a Shuttle launch had ever been attempted, with long icicles hanging from the launch tower, as shown in many impressive photographs, and a temperature of 27 °F (−2.7 °C). The low temperature had prompted concerns from Thiokol engineers, who had recommended a launch postponement: it was 1° below the minimum temperature for take-off (which was 33 °F)—beyond the tolerances for which the rubber seals of the O-rings were approved (Fig. 4.2).

There was lot of pressure to start this mission, both because it was expected that the weather would get worse in the following days and because there was some impatience to watch the first "Teacher in Space" who had to deliver two lessons followed by schoolchildren across the nation, providing a publicity boost for the space agency.

Fig. 4.2 The launch day of the ill-fated *Challenger* STS-51L was an atypically cold day for Florida, with long icicles hanging from the launch tower. Credit: NASA

The copious amounts of ice on Pad 39B forced an additional 2-h delay to permit thawing and finally the green light was given at 11:38. The decision to go ahead with the launch on such a cold morning had proven fatal. Years later, McAuliffe's mother, Grace Corrigan, insisted that the general atmosphere in the weeks leading up to *Challenger*'s fateful launch was that the Shuttle was far safer than an airliner, simply due to the higher number of precautions taken by NASA.

Fourteen seconds after take-off, Judy Resnik was heard to scan "LVLH" (local-vertical/local-horizontal), reminding all crew of a cockpit switch configuration change. These were her last words heard from the Control Center, which seconds later gave the order "Go at throttle-up" ("Full speed ahead!"). Commander Dick Scobee confirmed: "Go at throttle-up!"

A flame appeared and then the explosion (Fig. 4.3).

Investigators would later conclude that cold had caused the failure of both primary and secondary O-ring seals at the base of the right-hand booster that, under the pressure of acceleration required to reach the escape velocity, were no longer able to contain the exhaust gases, causing the catastrophic explosion 73 s after launch, when the most dangerous phase of the ascent was already over.

Fig. 4.3 Seventy-three seconds after launch, with millions of people watching on live TV, the Space Shuttle *Challenger* broke up in a forking plume of smoke and fire. Within seconds, the spacecraft broke apart and plunged into the ocean, killing the entire crew of seven astronauts. The devastating tragedy shocked the world and threw NASA's Shuttle program into turmoil. Credit: NASA

As the investigation demonstrated, Judy Resnik and Ellison Onizuka were still alive when the crew cabin separated from the rest of the spacecraft, and they were fully conscious after the breakup and during the entire descent until the impact with the ocean at a speed of roughly 207 miles per hour (333 kilometer per hour). But, unfortunately, there was no crew escape: this was one of the costs that NASA had cut.

Senator John Glenn, who was the first American astronaut to fly into Earth orbit, concluded his eulogy held in Akron, the birthplace of Judy Resnik, a few days after the incident by saying:

> As we reflect on Judy's life, and Challenger's last voyage in the days and weeks ahead, let's never forget the last words that came from that spacecraft: 'Go at throttle-up'. Those are far more than a courageous epitaph. They are America's history. They are America's destiny. And they will turn tragedy into triumph once again.

Judy Resnik is also remembered as the first American Jew to reach space and the second Jewish spacefarer, after the Soviet Boris Volynov, although it must be said that she once stated that "Firsts are only the means to the end of full equality, not the end itself" and she wanted to be remembered not as a Jewish astronaut, but as "just another astronaut—period."

A star, an asteroid, a crater on Venus, and a crater on the Moon all bear her name for future generations.

References

Anon. "STS 41D/Flight 12 Mission Report", *nasaspaceflight.com*.

Anon. "Judith Resnik: The 'Little One' Who Went to Space", *jspace.com*.

Cohen, L. "Judith Resnik", Jewish Women's Archive, *jwa.org/encyclopedia*.

Evans, B. "'Major Malfunction': The Final Launch of Challenger, 28 Years Ago Today", *americaspace.com* (January 28, 2015).

Gibson, K.B. *Women in Space: 23 Stories of First Flights, Scientific Missions and Gravity-Breaking Adventures*, pp. 91–97. Chicago Review Press, Inc., Chicago (2014).

Glenn, J. "Memorial Service for Judith Resnik", *nasa.gov* (Akron, February 3, 1986).

Kolbert, E. "Two Paths to the Stars", *nytimes.com* (February 9, 1986).

Official NASA biography of Judy Resnik, *nasa.gov* (December 2003).

5

Christa McAuliffe: "NASA Teacher in Space"

Credit: NASA

U. Cavallaro, *Women Spacefarers*, Springer Praxis Books, DOI 10.1007/978-3-319-34048-7_5

Mission	*Launch*	*Return*
STS-51L	January 28, 1986	–

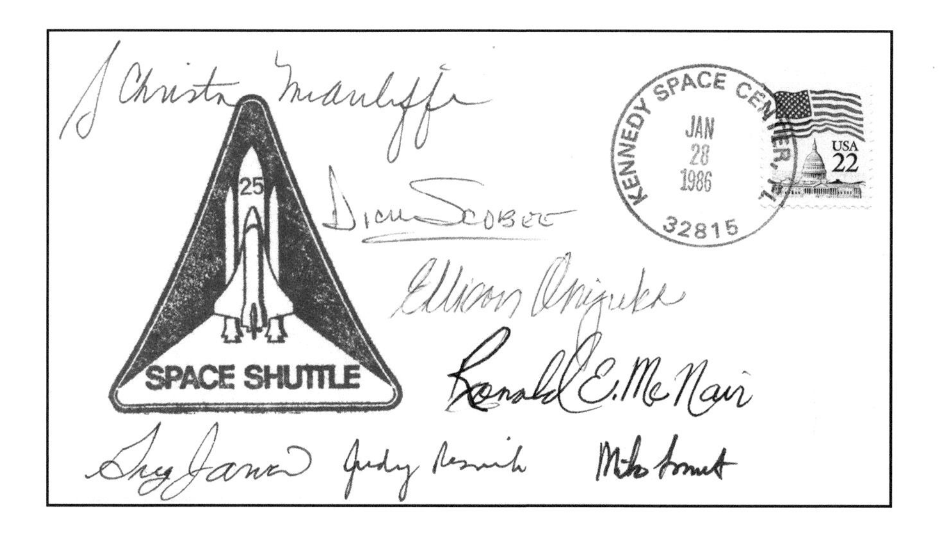

Commemorative cover of mission STS-51L, signed by The Crew, including Chista McAuliffe and Judy Resnik (from the collection of Bob Mc Leod)

Christa McAuliffe was chosen from nearly 10,500 candidates who had participated in "NASA Teacher in Space," the program that President Ronald Reagan wanted to use to attract students' interest in the sciences and space exploration, and to endorse teachers. On August 27, 1984, he proclaimed:

> I am directing NASA to begin a search in all of our elementary and secondary schools and to choose as the first citizen passenger in the history of our space programme one of America's finest: a teacher. When that shuttle lifts off, all of America will be reminded of the crucial role teachers and education play in the life of our nation. I can't think of a better lesson for our children, and our country.

This large "public relations" operation—at a critical time for NASA, whose budget was under pressure—had provided a much-needed publicity boost for the agency, aroused huge expectation, and attracted the attention of young people, and even of the general public and the media that, after many years, had returned to show interest in the space program. The smile of Christa filled the covers of magazines. Hers was an infectious enthusiasm: "If you're offered a seat on a rocket ship, don't ask what seat. Just get on!"

Eight hundred journalists were accrued on that morning of January 28, 1986, to watch the launch of the mission: twice the average of the previous launches. All of Christa's students, friends, parents, and her husband and two young children, aged 9 and 6 years

respectively, were at the Kennedy Space Center, anxiously watching and waiting for the *Challenger* Space Shuttle to take off.

Sharon Christa Corrigan McAuliffe was born in Boston, Massachusetts, on September 2, 1948. She was the eldest of the five children of Edward C. Corrigan of Irish descent and Grace M. Corrigan, whose father was Lebanese. As a youth, she was inspired by the race to the Moon and the Apollo Moon-landing program. The day after John Glenn orbited Earth in Friendship 7, she said to a friend: "Do you realize that someday people will be going to the Moon? Maybe even taking a bus, and I want to do that!"

She attended the private Catholic high school run by the Archdiocese of Boston. Christa graduated in 1970 with a Bachelor of Arts in Education and History at the Framingham State College in her hometown. A few weeks later, she married her longtime boyfriend Steve McAuliffe, whom she had known since high school, and they moved to Washington, DC, where she obtained her first teaching position as an American history teacher at Benjamin Foulois Junior High School in Morningside, Maryland, which she continued until the birth of her first baby, Scott.

In 1978, after completing a Master of Arts in Education Supervision and Administration from Bowie State University in Maryland, she moved to Concord, New Hampshire, when Steve accepted a job as an assistant to the New Hampshire Attorney General. After the birth of her second daughter, Caroline, she took a teaching post at Concord High School, where she developed innovative and engaging teaching techniques—and also delivered the course "The American Woman," which she had designed—and became very active in the local community, giving her time to the church, a tennis club, the local playhouse, the YMCA (Young Men's Christian Association), the Girl Scouts, and Concord Hospital.

In 1984, Christa learned about the "NASA Teacher in Space" program. NASA hoped that sending a teacher into space would increase public interest in the Space Shuttle program and also demonstrate the reliability of spaceflight at a time when the agency was under continuous pressure to find financial support. A bit hesitant, but encouraged by friends and acquaintances, at the last minute, she filled out the 11-page form and mailed it, so becoming one of the 10,463 applicants. She explained in her NASA application:

> As a woman, I have been envious of those men who could participate in the space programme and who were encouraged to excel in the areas of math and science. I felt that women had indeed been left outside of the one of the most exciting careers available. When Sally Ride and other women began to train as astronauts, I could

> look among my students and see ahead of them an ever-increasing list of opportunities. I cannot join the space programme and restart my life as an astronaut, but this opportunity to connect my abilities as an educator with my interests in history and space is a unique opportunity to fulfill my early fantasies. I watched the Space Age being born and I would like to participate.

During her interview at NASA, she added: "I want to demystify NASA and space flight. I want students to see and understand the special perspective of space and relate it to them" (Fig. 5.1).

There was fierce competition for the designation of "Teacher in Space." In every state, a committee was appointed that proposed two candidates each. Even when she knew that she was selected for the group of 114 finalists, she still did not believe that she would be the one picked because she felt she was an ordinary person and knew that other competitors could boast many publications and many more academic titles than she could. But, this time, NASA were not looking for a researcher to be sent into space—they already had those. They were looking for a teacher who most people could identify with as having had as young students: someone who had left an unforgettable mark in their memory. And Christa—unusually "ordinary"—was just what they were looking for.

Christa was shocked but ecstatic when she learned, in the summer of 1984, that she was going to make history as the first school teacher in space. Vice President George H.W. Bush

Fig. 5.1 Christa McAuliffe experiencing weightlessness during the KC-135 flight. Credit: NASA

delivered the good news at a special ceremony at the White House: he said that McAuliffe was going to be the "first private citizen passenger in the history of space flight."

After taking a year-long leave of absence, in September 1985, Christa arrived at the Johnson Space Center in Houston to begin her training. She feared the other astronauts would consider her an intruder and worked hard to prove herself. She discovered that the other crewmembers would treat her as part of the team and she quickly gained self-confidence and developed trust in the mission:

> A lot of people thought it was over when we reached the Moon. They put space on the back burner. But people have a connection with teachers. Now that a teacher has been selected, they are starting to watch the launches again.

Christa was assigned as a payload specialist to the *Challenger* STS-51L mission. During her mission—which she called her "ultimate field trip"—Christa planned to conduct two live lessons from the Shuttle, 15 min each, connected via satellite with students all over the world. The first lesson would detail daily life aboard the Shuttle and the second was aimed at helping students to understand the goals of space exploration.

Her infectious enthusiasm was condensed in her motto: "I touch the future, I teach." The motto expresses her vision of her profession as a teacher, aware of the great opportunity that was offered not only to her, but also to the whole category of teachers who—through her—could show the world the important job that, without the spotlight, they do in their daily life:

> I'm hoping that this is going to elevate the teaching profession in the eyes of the public and of those potential teachers out there, and hopefully, maybe one of the secondary objectives of this is students are going to be looking at me and perhaps thinking of going into teaching as professions.

Barbara Morgan, who lived with her for 2 years as her "backup" (and flew in space 20 years later), said:

> Christa was, is, and always will be our 'Teacher in Space,' our first teacher to fly. She truly knew what this was all about—not just bringing the world to her classroom—but also helping to show the world what all the good teachers do across our country day in and day out.

The launch of the craft was delayed three times. After over two decades of successful space missions, spaceflights were taken for granted and NASA's self-confidence had turned into an almost arrogant sense of infallibility. The press had worked to create an atmosphere in which the delays were seen more as components of the Shuttle-launching ritual than as necessary safety measures that would be less well tolerated. After three delays, the launch in the end was still postponed by 2 h, to allow time to remove the icicles hanging on the launch tower. The go-ahead was finally given at 11:38.

Seventy-three seconds later, *Challenger* exploded at a height of just over 14,500 m, well before it crossed the Kármán line that, according to the definition of the Fédération Aéronautique Internationale (FAI), conventionally represents the boundary between Earth's atmosphere and outer space, and lies at an altitude of 100 km (62 miles) above

Earth's sea level. Therefore, strictly speaking, Christa McAuliffe should not be regarded as an "astronaut." But we cannot do without mentioning her here since, for many of her colleagues, during her intense preparation, the "Teacher in Space" gave a great contribution to the revival of interest in the space program, and captured the imagination of the US and of the entire world, as President Ronald Reagan remarked in his eulogy:

> We remember Christa McAuliffe, who captured the imagination of the entire nation, inspiring us with her pluck, her restless spirit of discovery; a teacher, not just to her students but to an entire people, instilling us all with the excitement of this journey we ride into the future.

The investigating committee eventually established that the cause of the disaster was a seal ("O-ring") that, exposed to unusually cold temperatures, had failed. But perhaps the problem was more political than technical. McAuliffe's family respected the astronaut code of silence until after the death of her father, Ed Corrigan, in 1990. Eventually, Christa's mother, Grace, published a memorial that included some notes left by her husband, who had written:

> My daughter Christa McAuliffe was not an astronaut. She did not die for NASA and for the space programme. She died because of NASA and its egos, marginal decisions, ignorance, and irresponsibility … one of the Commissioners stated: 'It was no accident. It was a mistake that was allowed to happen'. NASA betrayed seven fine people who deserved to live. President Reagan said that the act was not deliberate, was not criminal. But I say that the sins of omission are no less sins that those of commission. I say 'they' deliberately neglected to make necessary corrections to the O-rings and are, therefore, as guilty as if they planned a deliberate criminal act.

A specialist in education at NASA reproduced Christa's lessons and exercises and made them available to allow teachers to replicate in the classroom what Christa did not have time to do from space. Besides her lessons, Christa was planning to keep a journal of her adventure: "That's our new frontier out there," she said, "and it's everybody's business to know about space."

Eventually, NASA managers were removed and promoted. The asteroid No. 3352, a crater on the Moon, and one on Venus, as well as 40 schools worldwide, bear Christa McAuliffe's name.

References

Barbree, J. "The Challenger Saga: An American Space Tragedy", *nbcnews.com*.

Cunningham, W. *The All-American Boys*, pp. 369–376. iBooks, Inc., New York (2004).

Evans, R. *Tragedy and Triumph in Orbit: The Eighties and Early Nineties*, pp. 440–443, 446. Springer-Praxis, New York (2012).

Gibson, K.B. *Women in Space: 23 Stories of First Flights, Scientific Missions and Gravity-Breaking Adventures*, pp. 158–164. Chicago Review Press, Inc., Chicago (2014).

Greene, N. "I Touch the Future: Christa McAuliffe, the Space Shuttle Challenger Astronaut Teacher", *space.about.com* (December 9, 2014).

Kevles, T.H. *Almost Heaven: The Story of Women in Space*, pp. 100–105. The MIT Press, Cambridge, MA, and London, UK (2006).
Mayfield, B. "Christa's Lost Lessons", *Space Educator's Handbook*, OMB/NASA Report #S677.
Official NASA biography of Christa McAuliffe, *nasa.gov* (April 2007).
Stahl, L. "Christa McAuliffe's Mother Had Premonition", *The Hour* (quotidiano di Norwalk), January 20, 1996, p. 48.

6

Kathryn D. Sullivan: The First American Spacewalker

Credit: NASA

U. Cavallaro, *Women Spacefarers*, Springer Praxis Books, DOI 10.1007/978-3-319-34048-7_6

Mission	*Launch*	*Return*
STS-41G	October 5, 1984	October 13, 1984
STS-31	April 24, 1990	April 29, 1990
STS-45	March 24, 1992	April 2, 1992

Commemorative cover of mission STS-41G, signed by Kathy Sullivan (from the collection of Umberto Cavallaro)

Kathryn D. Sullivan was one of the first six women selected for the NASA Astronaut Corps in 1978 and holds the record for the first American woman to walk in space. She participated in three Space Shuttle missions, including the one that launched the space telescope Hubble in 1990.

Kathryn D. Sullivan was born in Paterson, New Jersey, on October 3, 1951, but considers Woodland Hills, Los Angeles, California, to be her hometown, since she moved there with her family when she was 6. She was in the first grade at the Havenhurst Elementary School of Los Angeles at the same time as her friend, Sally Ride, the first American woman in space—as they discovered when talking about their childhoods—but neither could really remember the other.

An interest in space is among her childhood memories and intrigued her since she was a little girl, when she remembers running out onto the front lawn with her father, an aerospace engineer working at Lockheed Corporation, to see whether they could watch

the Sputnik. "Sputnik happened when I was six; in fact, one day after my sixth birthday. I was born on October 3rd, so I'm almost a Sputnik baby," she says, recalling the anxiety that the Sputnik going overhead represented for Americans. Given her father's profession and the passion of her brother for flying, space became almost a family affair:

> Every single issue of Life and of National Geographic in the early sixties seemed to have a really entrancing story about who are these guys, what they are going to do, and what is spaceflight like. Issues were arriving every month with these breathtaking, amazing stories about the new space frontier, the seven astronauts, Sputnik, and what it all means. It was fabulous. I ate all of that stuff up. I just was fascinated by all of those things, although nothing in my thinking at that time was oriented towards job or career.

After graduating from Taft High School, Woodland Hills, California, in 1969, she received a Bachelor of Science degree in Earth Sciences (with Honors) from the University of California, Santa Cruz, in 1973, after spending 1 year, in 1971–1972, as an exchange student at the University of Bergen, Norway. There she discovered a gift for languages and learned fluent Norwegian. She also found time to climb some of the highest mountains in Norway. Back in America, she decided to specialize in marine geology:

> I came back from the Norway experience really bent towards earning an advanced degree in oceanography as opposed to general geology, with a bit of a picture in my mind of research oceanography as the thing I wanted to do.

She therefore enrolled in Geology at Dalhousie University in Halifax, Nova Scotia, Canada. During a vacation in California for the Christmas holidays in December 1976, she learned by chance from her brother that NASA was searching for women:

> My head at that time was just on the oceanography side of things, and so I blew him off. 'I'm working in 14,000 feet of water depth. It's already hard enough to understand the bottom of the ocean from a surface ship, and now you want me to go 200 miles above that! This is not what you do to further understand the bottom of the ocean.' I went back to Nova Scotia, dismissing it. Within a week or so I saw one of NASA's own small ads about the recruitment in one of the US science publications

> that the library received. When I read that, a different gear clicked. I recognized a strong parallel between the mission specialist role, as they described it, and the oceanographic expeditions.

Kathy decided to accept the challenge and filled her application. In January 1978, she was selected as a NASA astronaut and, a few months later, she received a doctorate in Geology from Dalhousie University (Halifax, Nova Scotia, Canada). Over the 5 years, she participated in several oceanographic expeditions that studied the floors of the Atlantic and Pacific Oceans and was offered a position for a post-graduate qualification in marine geology of depth, which involved dives in the little Alvin submersible with the opportunity to see the deep-sea floor to perform the volcanology part of marine geology and geophysics:

> I had an Alvin-diving postdoc in my hands, so I wasn't going to lose. Two fabulous things were in front of me, either of which just seemed tremendous things to get to be involved in. It made my mother a little crazy that I was either going 10,000 feet down in the ocean or 200 miles up off of the planet, and there was nothing exciting on the surface, but she quickly got over that!

She joined NASA in 1978 and was initially assigned to the WB-57F high-altitude research aircraft program, which allowed her to be certified by the US Air Force as qualified for pressure suit operations; she subsequently worked in the support group for extravehicular activity (EVA) and spacesuit support crew for several Shuttle missions. For many flights, she also played the role of CapCom (capsule communicator), the Mission Control radio operator in charge of maintaining contact with the astronauts during spaceflight—a role that is traditionally performed by a NASA astronaut.

A veteran of three spaceflights, Dr. Sullivan was a mission specialist on the STS-41G (October 5–13, 1984), STS-31 (April 24–29, 1990), and STS-45 (March 24–April 2, 1992) missions. In October 1984, she participated in the STS-41G mission, which launched from Kennedy Space Center, Florida, on October 5, 1984, for the first time with a crew of seven. And, for the first time, two women astronauts flew together. Sally Ride was in fact flying on her second mission, with Kathy performing an "EVA" and thus becoming the first American woman to do a spacewalk. "The attention of the media," recalls Kathy, "was initially focused on the second flight of Sally Ride. The flight was announced all of about five or six months after her first landing, so there's still a flood of interest surrounding her."

To those who joked that women on board would be useful for keeping the place clean, Kathy—playing the game—retorted that she would be "happy to do the windows, but only from the outside!"

One of the goals of the mission was that of verifying the feasibility of the Orbital Refueling System (ORS), the first ever tried in space to increase a satellite's orbital lifetime. They used a mock-up of the Landsat's cluster of fuel valves. As the rocket fuel hydrazine is extremely instable and can explode if heated or over-pressurized, it was essential to make proper refueling valve connections without leaks or spills. The success of the operation—a key task for satellite servicing—increased the confidence that future satellite problems could be solved by EVAs and satellites would no longer have to be abandoned in space as litter.

To complete this task, Kathy Sullivan, along with her colleague David Leestma, performed an EVA of almost three and a half hours. Before returning, they moved to the front of the Space Shuttle to work on a communication antenna mounted outside the Shuttle and the camera system of radar images—a system that Kathy knew well, having worked on it on the ground. Upon returning aboard, the team talked to President Reagan, connected via satellite, with Kathy telling him: "The spacewalk was one of the most fantastic experiences in my life!"

Kathy was the first woman to wear the Shuttle-era spacesuit: the 225-pound Extravehicular Mobility Unit (EMU), a ready-to-wear suit, not custom-made, with interchangeable arms, legs, and torso units in different sizes. She said she found it pretty comfortable to wear and work in while in zero gravity, although the fit did not quite match where her knees and elbows actually were, making it somewhat harder to move her limbs. It was a great satisfaction that also left a little bitter taste in the mouth when it was decided that the commander of the spacewalk would be Dave Leestma. Kathy pointed out during the interview that she had in NASA's *Oral Histories* program:

> I'm a class senior to Dave. I've been in the program longer than Dave. I've worked in the suits more than Dave. I worked this payload longer than Dave did, and I'm number two to him on the spacewalk. That's really bad optics. It is the norm in the office, and in the culture, that class rank matters, and the senior class guy leads; that's rarely, rarely breached, and I wasn't real comfortable that it was breached in this case.

On the other hand, Kathy found it amusing that so much attention was paid to the precise time she spent on her "spacewalk" in October 1984, and that someone at NASA would have liked it to last a few minutes more than 3 h and 29 min, in order to beat Svetlana Savitskaya, who had lasted 3 h and 34 min:

> The EVA flight team was actually watching the duration clock very carefully and was very mindful of where we were relative to Svetlana's time. They were trying to figure out, how do you compose a flight note to go up on air-to-ground that isn't quite as tacky as "stay out for seven more minutes so you beat her.

"Melodrama around the spacewalk and spacewalking records," she recalled in her *Oral Histories* interview.

In April 1990, Kathy participated as a mission specialist in the Shuttle *Discovery* STS-31 mission, whose main objective was to deploy the space telescope Hubble, which was positioned at an altitude of about 600 km. When one of the solar arrays jammed, Kathy and Bruce McCandless, who had been trained for every foreseeable contingency, suited up and prepared the airlock to go outside and deploy the array manually. However, Mission Control resolved the problem via software—a good but disappointing solution for the two potential rescuers. They also operated a variety of cameras, including both the IMAX in-cabin and the cargo bay cameras, for Earth observations from their record-setting altitude of 380 miles.

Kathy eventually served as Payload Commander—the first woman astronaut assigned to that position—on STS-45, the first Spacelab mission dedicated to NASA's Mission to Planet Earth ATLAS-1 (Atmospheric Laboratory for Applications and Science), focused on measuring atmospheric chemistry and dynamics. During this 9-day mission, the crew operated 12 experiments to obtain a vast array of detailed measurements of atmospheric chemical and physical properties, which significantly contributed to improving our understanding of Earth's climate and atmosphere. During her three missions, she logged more than 532 h in space.

In 2006, with the rank of Captain, she retired from the US Navy, which she had joined in 1998 as an Oceanography Officer Reserve. After leaving NASA in 1992, she accepted from George W. Bush a presidential appointment as Chief Scientist at NOAA (the National Oceanic and Atmospheric Administration). In this position, she oversaw a wide array of research and technology programs ranging from climate change and marine biodiversity to mapping services and satellite instrumentation.

Dr. Sullivan has used her national visibility to promote the public's awareness and understanding of science and education. She served for almost 10 years (1996–2006) as President and CEO of COSI (Center of Science and Industry), Columbus, Ohio—one of the most important scientific museums in the US. She then served for 5 years as the first director of the Battelle Center for Mathematics and Science Education Policy at Ohio State University.

In January 2011, President Obama appointed her as the Commerce Department's Assistant Secretary for Environmental Observation and Prediction—a post that also serves as Deputy Administrator of NOAA. In 2013, the president named her Acting Under Secretary for Oceans and Atmosphere and NOAA Administrator. US Senate confirmed her in that position in 2014.

She speaks five foreign languages and is a member of the US National Academy of Engineering; she has been awarded honorary degrees by Dalhousie University (1985), the State University of New York, Utica (1991), Stevens Institute of Technology (1992), Ohio Dominican University (1998), Kent State University (2002), and Brown University (2015) for her "abundant contributions to science, education and the public good, and her ongoing commitment to improving the state of our planet for future generations."

References

Diaz-Sprague, R. "From the Stars, Down to Earth: A Conversation with Kathy Sullivan", *myhero.com* (March 10, 2005).

Gibson, K.B. *Women in Space: 23 Stories of First Flights, Scientific Missions and Gravity-Breaking Adventures*, pp. 99–104. Chicago Review Press, Inc., Chicago (2014).

Official biography of Kathryn D. Sullivan, *jsc.nasa.gov* (April 2014).

Contacts by e-mail with the Author in May 2016.

Ross-Nazzal, J. "Kathryn Sullivan: Oral History Transcript", *jsc.nasa.gov* (May 10, 2007).
Ross-Nazzal, J. "Kathryn Sullivan: Oral History Transcript", *jsc.nasa.gov* (September 11, 2007).
Ross-Nazzal, J. "Kathryn Sullivan: Oral History Transcript", *jsc.nasa.gov* (March 12, 2008).
Ross-Nazzal, J. "Kathryn Sullivan: Oral History Transcript", *jsc.nasa.gov* (May 28, 2009).
Shayler, D.J.; Moule, I. *Women in Space – Following Valentina*, pp. 219–221, 257, 260. Springer/Praxis Publishing, Chichester, UK (2005).

7

Anna Lee Fisher: The First Mother in Space

Credit: NASA

U. Cavallaro, *Women Spacefarers*, Springer Praxis Books, DOI 10.1007/978-3-319-34048-7_7

Mission	*Launch*	*Return*
STS-51-A	November 8, 1984	November 16, 1984

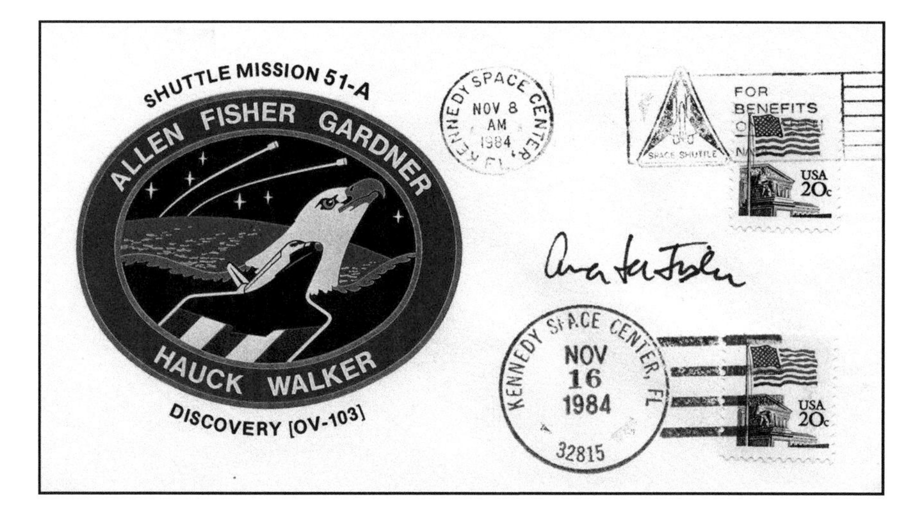

Commemorative cover of mission STS-51A, signed by Anne Lee Fisher (from the collection of Umberto Cavallaro)

Awarded "Mother of the Year" in 1984 when she became the first mother in space, Anna Lee Fisher is one of the longest-serving active astronauts (and the last remaining from the 1978 class) and has been involved in three major historical NASA manned space programs: the Space Shuttle, the International Space Station, and the Orion Project.

Anna Lee Tingle Sims Fisher was born New York City on August 24, 1949, but moved to different locations and considers San Pedro, California, to be her hometown: "My father was in the military. I'm an Army brat, so we moved every 2 or 3 years."

She graduated in 1967 from the San Pedro High School, attracted by science, and received a Bachelor of Science degree in Chemistry in 1971 at the University of California, Los Angeles (UCLA), specializing in X-ray crystallographic studies of metallocarbonanes. She co-authored three publications relating to these studies for the *Journal of Inorganic Chemistry*. She was accepted into the chemistry graduate department in 1971 before entering the UCLA School of Medicine the following year and received her Doctor of Medicine degree in 1976. After completing a 1-year internship at Harbor General Hospital in Torrance, California, she specialized in emergency medicine and worked in several hospitals in the Los Angeles area. Anna later received a Master of Science degree in Chemistry from UCLA in 1987 based on her earlier year in the graduate school. In her memories, her interest in space is connected to the flight of Alan Shepard. She reported in an *Oral History* interview:

We were out at physical education class and our teacher had a little transistor radio. We were listening to Alan Shepard's first flight. That was when I first really thought, 'Wow, I would love to go do something like that.' But of course all the astronauts at that time were male. They were all fighter pilots. For whatever reason, it never even entered my mind to consider trying to go to pilot training. That wasn't something that I had access to or that entered my imagination.

As she was doing her internship at the Harbor General Hospital, she knew, through a friend of her then fiancé and later husband Bill Fisher, that NASA was looking for people. She filled in the complex 11-page form—which at that time had to be filled in by hand—with the help of Bill, who in turn decided to do the same, and sent the application the day before the deadline. Anna and Bill married in August 1977. She remembers that NASA called them to come for an interview when they were about to plan their wedding:

I'm not one of these plan a wedding a year in advance kind of person. We were sitting there and I said, 'It's NASA. They want me to come interview.' He said, 'Say yes, we'll figure it out.' So I said yes. I said, 'Okay, now what?' That began one of the hardest weeks of my life.

The wedding was planned at the last minute:

We went and got dresses, we got a photographer, did all this stuff. We called a couple of our friends that we wanted to come. Bill's family wasn't even able to adjust, but they were very understanding. We did all that. We got married. Then I got on a plane. I didn't really have time to think about the interview process, study, or research.

Anna was selected as an astronaut candidate in January 1978—one of the first six female NASA astronauts. Bill was rejected and was accepted 2 years later into the following NASA astronauts class in 1980.

In August 1979, Anna completed her training and evaluation period and actively participated in the set-up of the Space Shuttle program, contributing to the verification of flight software at the SAIL (Shuttle Avionics Integration Laboratory), where she also reviewed test requirements and procedures for ascent, in-orbit, and Risk Management System software. As the crew representative, she supported development and testing of the Remote Manipulator System (RMS) and of payload bay door contingency spacewalk procedures, of the extra-small EMU (Extravehicular Mobility Unit), and of contingency repair procedures. For STS-5 through STS-7, she supported vehicle integration and payload testing at Kennedy Space Center: "I did the first payload flows that we supported," she recalls, "because the previous flights up to then had not really had real payloads."

Another open problem was that of the spacesuits. Anna Fisher, who was petite, was assigned to help in designing a suit for women: "They simply thought they could take big space suit and proportionally cut them down." Fisher found the unisex stock sizes simply unworkable: "Women are not smaller men. Women are built differently." This had nothing to do with fashion and everything to do with safety. For a while, NASA thought it would be more economical to select a large candidate who would fit the suit, rather than adjust the suit to fit a smaller woman, but tall women are also built differently: another lesson NASA learned.

In addition, Fisher collaborated in the development of procedures for emergency action, paying special attention to the operational aspects and medical intervention, and provided both medical and operational inputs to the development of rescue procedures. As a physician in rescue helicopters, she actually supported the first Orbital flight-test launches and landings (STS-1 to STS-4) at either the primary or backup landing sites, ready to render emergency medical aid. She was then CapCom (capsule communicator) for Shuttle mission STS-9, in which the first Spacelab, the European space laboratory built in Italy, flew. Referring to all these experiences, she explains:

> I got many interesting assignments that really helped me to understand the Shuttle and its procedures and helped me to become a good crew member when it was my turn to fly on the Shuttle. I'm glad I did this experience, because it is not the same thing to train without figuring out what's going on in SAIL, what they do in the Mission Control Center, how they operate at the Cape and without really knowing people for having worked with them, which is something different from having met them in some meeting. All these experiences gave me the opportunity to see up at close hand the main projects handled at the time by my office.

In July 1983, a few weeks before the birth of her first baby, Anna was assigned to the STS-51A mission:

> I'm probably the only person who's been assigned to their flight about two weeks before they deliver. I doubt that's probably ever happened in the history of the space program since, which I thought was really neat that he showed that confidence in me.

After a short leave of absence for the birth of her daughter, Kristin Anne (born on July 29, 1983), Anna started her training and, in November 1984, left for space as a mission specialist on Shuttle STS-51A, thus becoming the first mother to fly into space. The crew's flight patch was designed with six stars: five representing each of the crewmembers and one star for the newborn. She also kept a picture of her baby girl on her Shuttle locker while she was in space, and later gifted each of her girls copies of the necklace she wore on the flight, specified as "flown on my flight" or in Bill's flight. During the mission, the crew successfully deployed two satellites (Canada's Anik D-2 (Telesat H) and Hughes's LEASAT-1 (Syncom IV-1)) and recovered two others (Westar 6 and Palapa B2) whose kick motors failed to ignite, returning them to Earth for refurbishment and later relaunch.

Fig. 7.1 The STS-51 crew's flight patch was designed with six stars: five representing each of the crewmembers and one star for Anna's newborn. Credit: NASA

Fisher was assigned to assist Commander Hauck during the final stages of the complex rendezvous with both satellites (Fig. 7.1).

Anna was eventually assigned to mission STS-61H, scheduled for launch on June 24, 1986, using *Columbia*. However, the mission was cancelled after the *Challenger* disaster in January 1986. She went back to graduate school in chemistry and received a second Master of Science degree in Chemistry from UCLA in 1987. In the same year, she also served on the Astronaut Selection Board for the 1987 class of astronauts: Class 12.

Anna was due to participate in Shuttle mission STS-29, scheduled for the beginning of 1989, but she took a leave of absence from the Astronaut Office to raise her second daughter, Kara Lynne (born on January 10, 1989). She returned to NASA in January 1996 during the early phase of the building of the International Space Station (ISS). It wasn't easy:

> Coming back was the most difficult thing I've ever done in my life because by that time pretty much everyone in my group had flown two to four flights, and most of them were gone; there were only a few people left from my group, and the Office was full of new people whom I didn't even know.

From 1996 through 2002, she was the chief of the ISS branch. In that capacity, she coordinated inputs to the operations of the space station for the Astronaut Office, working closely with all the international partners and supervising assigned astronauts and engineers. She made headlines when, in 2012, speaking during the event of the delivery of the Shuttle *Discovery* to the Smithsonian Museum, she advised those who wanted to become astronauts to "get studying Russian"—a joke that someone, perhaps rightly, interpreted as referring to the new space policy of President Obama.

From January 2011 through August 2013, Fisher served as an ISS CapCom working in the Mission Control Center and was also the lead CapCom for Expedition 33.

Currently, Fisher is a management astronaut, working on display development for the Orion MPCV (Multi-Purpose Crew Vehicle), and supports European payloads for the ISS Integration branch.

References

Evans, B. *Tragedy and Triumph in Orbit: The Eighties and Early Nineties*, pp. 295–297. Springer Praxis Books, New York (2012).

Kevles, T.H. *Almost Heaven: The Story of Women in Space*, p. 74. The MIT Press, Cambridge, MA, and London, UK (2006).

Official biography of Anna Lee Fisher, *jsc.nasa.gov/Bios* (July 2014).

Personal contacts by e-mail in May 2016.

Ross-Nazzal, J. "Anna L. Fisher: Oral History Transcript", *jsc.nasa.gov* (February 17, 2009).

Ross-Nazzal, J. "Anna L. Fisher: Oral History Transcript", *jsc.nasa.gov* (March 3, 2011).

Ross-Nazzal, J. "Anna L. Fisher: Oral History Transcript", *jsc.nasa.gov* (May 3, 2011).

Shayler, D.J.; Moule, I. *Women in Space – Following Valentina*, pp. 221–224, 252. Springer/Praxis Publishing, Chichester, UK (2005).

Woodmansee, L.S. *Women Astronauts*, pp. 58–60. Apogee Books, Burlington, Ontario, Canada (2002).

8

Margaret Rhea Seddon: The First Wedding Among NASA Astronauts

Credit: NASA

U. Cavallaro, *Women Spacefarers*, Springer Praxis Books, DOI 10.1007/978-3-319-34048-7_8

Mission	*Launch*	*Return*
STS-51D	April 12, 1985	April 19, 1985
STS-40	June 5, 1991	June 14, 1991
STS-58	October 18, 1993	November 1, 1993

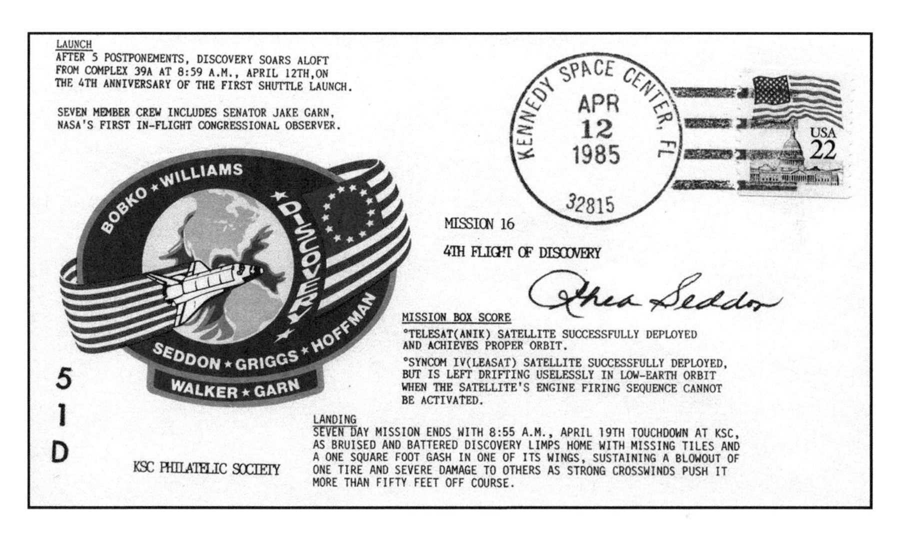

Commemorative cover of mission STS-51D, signed by Rhea Seddon (from the collection of Umberto Cavallaro)

Margaret Rhea Seddon was already a qualified physician and surgeon when she learned that NASA, for the first time in a decade, was selecting new astronauts, opening its arms to women and promising “equal employment opportunities.” Rumors had been circulating for 3 years but evidently it was not easy to reconcile the new “vision” with the needs of test pilots, who had to be safeguarded. Help came from the Space Shuttle program, which opened up new spaces for new positions, such as payload specialist and mission specialist, requiring an in-depth university education in areas of science or medicine or engineering. Rhea was prepared long before and waiting just for this, saying: “Since I knew wherever human explorers went they would have to have doctors along, I decided to be a doctor. When NASA finally accepted women, they wanted scientists for the Space Shuttle and that included doctors.”

Margaret Rhea Seddon was born in Murfreesboro, Tennessee, on November 8, 1947. The daughter of an attorney who was on the board of the local hospital, from a young age, she enjoyed piano and classical ballet and developed an early love for the sciences but, in her small town, there weren't many possibilities at the time: "I can't remember ever having any science before the seventh grade." She was interested in space travel from the very beginning of the space program. "I watched the Russian satellite Sputnik fly overhead when I was a young girl in 1957," she recalls, "and knew that space would be what my generation would explore." Space in those years made headlines and she started to avidly read everything at hand that dealt with space, bringing the space charm into everything she was doing:

> I remember in seventh grade we had to do a science poster for our final grade. This was before Gagarin flew in space, and there was a Life Magazine article about what might happen to humans when they went away from gravity. There were all these swell pictures, pictures of the man going into space, things that might happen, and G-forces. I just thought that was interesting.

Rhea knew that Russia had had a woman as a cosmonaut since the inception of their space program and she figured out that, by the time NASA took on women astronauts, they would also need physicians. However, after graduating from Central High School in Murfreesboro, Tennessee, in 1965, she realized that it was hard to pursue her interests there, left the South, and went to the University of California, Berkeley: "When I was looking for colleges I wanted to find out what was the best college in the country in life sciences, and that's how I ended up at Berkeley." At Berkeley, she first encountered the Free Speech Movement (FSM)—a student protest which was taking place during those years on the campus of the University of California—and became aware for the first time that careers previously barred from women, such as medicine and aviation, were now within reach. But, even at Berkeley, where she did not find gender discrimination or discouragement from professors, she found only a few women on her course when she began, and even fewer as she approached graduation.

After finishing her degree at Berkeley in 1970, Rhea returned to the University of Tennessee, College of Medicine, where she gained her medical doctorate in 1973. Her father helped her to get a job at the local hospital, where they were opening the first intensive-care unit. She says:

> As the unit delayed in its opening they sent me to surgery. That's how I got interested in surgery. I applied then to several surgery programs around the country, but I don't think they were prepared to take women. I did get a slot with the University of Tennessee, where I went to medical school. They knew me; they knew I was interested.

> I'd worked with a number of the surgeons there who knew I was a hard worker and serious about doing this. I was the only woman in the surgery residency program.

There she became interested in the nutrition of patients who have had major surgery:

> Sometimes we'd operate on a patient, and we could fix what was wrong but they couldn't eat for a while. They just basically starved to death; they never healed. We were beginning to be able to feed people by vein. I got interested in that, so did some research in that.

And, progressively, she also started to investigate the effects of radiation therapy on nutrition in cancer patients.

Rhea decided that it was time to do what she had dreamed of for years: learn to fly. A pilot's license would increase her chance of becoming an astronaut. She worked at night in the hospital emergency rooms and took flying lessons during the day. Within a few months, she had her pilot's license.

It was 1977. Rhea heard from a colleague that NASA was offering jobs to women for the Space Shuttle program. She recalls:

> But he didn't have any idea where I should apply. The only thing I remembered was that astronauts trained somewhere in Houston. I addressed an inquiry letter. This was of course back before you could just Google stuff. Sent a letter to NASA, Houston, Texas, and it got to the right person. I got the information back. I wrote for the application and sent it in.

She was thrilled when she was invited to Houston for physical tests and interviews, and had to compare herself with many outstanding candidates; several were MD/PhDs who had done NASA research: "They were just incredible people. I couldn't imagine that I would compete with them. What have I got to offer to compare with that? I don't know why they decided that I should come in and the others shouldn't."

On January 16, 1978, Rhea received a phone call from George Abbey: "Well, we were just calling to see whether or not you were still interested in this job." She recalls: "I asked, 'How many will be in the class?' He said, 'There are 15 pilots and 20 mission specialists.' I said, 'How many women?' He said, 'Six.' I thought, 'That's great. That's really great.'" In this way, she learned that she was one of the TFNGs or Thirty-Five New Guys.

Given her previous professional interest in nutrition, her first technical assignment was designing the food packages for the Shuttle program:

> I was a little disappointed that I didn't get something more general. I think Sally Ride was a CapCom [capsule communicator], Anna Fisher got suits, and I ended up with the cooks. … I was used to, a lot of times, being assigned to follow something that wasn't the big deal, but if you prove yourself in doing whatever job you're given, better jobs come your way.

Rhea was charged with the responsibility of finding a menu suitable for women in space. According to the standard menu, she recalls, "each astronaut had a daily food allowance of 3100 calories a day. But four of the women were under 5'5. That much food would have turned us into blimps." So she replaced puddings and cookies with fruits and vegetables. "One of the problems in space," she explains, "is that you don't have gravity to work against, so you really have to work at it to burn calories."

Eventually, Rhea also worked in the SAIL (Shuttle Avionics Integration Laboratory) where she contributed to the set-up of the Shuttle software: "That was a very valuable experience for me, and it taught me a great deal about how all those facts that I had learned in the lectures would come together and how it really worked." She cooperated in the development of the equipment that was carried in the Orbiter medical kits. As a helicopter physician, she was part of the search-and-rescue team during the launch and landing of Shuttle missions, entered NASA's Aerospace Medical Advisory Committee, and served as CapCom in the Mission Control Center in Houston. Remembering those years, she says: "I think I didn't pass one day without learning at least 10 things."

Rhea was the first active astronaut to wed when, on May 30, 1981, she married Astronaut Robert Lee "Hoot" Gibson, a former navy pilot who had entered the Astronaut Corps in the same TNFG class. The following year, the two had their son, Paul, who became the first son born from two astronauts, soon nicknamed by *Time* magazine "the first U.S. Astrotot," as Rhea recalls in her book, *Go for Orbit* (p. 142). She was the fifth American woman to fly in space and was assigned to three Shuttle flights, as mission specialist of missions STS-51D (1985) and STS-40 (1991) and payload commander of mission STS-58 (1993)—in charge of all science activities on her final flight—totaling altogether more than 722 h (over 30 days) in space.

Rhea had first-hand experience of the uncertainty of the assignment to a flight, which, in a very competitive environment such that of the astronauts, puts one through the wringer. Rhea recalls in her book (p. 188) the moment when she received an unusual phone call from Mr. Abbey's secretary saying "Mr. Abbey would like you to come over to talk to him":

> It was unusual for George to assign flights from his office. It was traditionally done off in a corner at some party or during happy hours. Since the baby, I hadn't been partying much, and my happy hours consisted on collapsing in a heap on the floor to play with Paul on a Friday evening. Still the call to his office caused a nagging fear that I had not measured up or committed some unpardonable sin, and he was going to tell me I wasn't suited for this job.

She should have originally been the *third* American woman to fly in space. When Judy Resnik was selected as one of the first six female astronauts in January 1978, it was widely expected that either she or Sally Ride would be the first to fly. In fact, Rhea remarks in her oral history that Judy and Sally received "the sorts of technical assignments that really prepared them for flight," such as robot arm work and CapCom duties. But delays and cancellations to her scheduled flight pushed her further back, as she reports in *Go For Orbit* (p. 199):

> I went from being the third woman in our ASCAN group to fly, to the fifth. I would also not be the first mother in space because Anna Fisher's daughter Kristin had been born in 1983, a year after Paul, and she would be on Rick Hauck's flight in the fall. In the grand scheme of life, perhaps these weren't major losses, but they were important to me. Everyone wanted to be near the front of the line with their flights and to be the first in something.

Rhea was originally assigned to mission STS-41F, which that was scheduled to fly in August 1984 as the second flight of the *Discovery* and would deploy a pair of communications satellites and the Spartan solar observatory. Unfortunately, on June 26, during the maiden flight of *Discovery*, its three main engines, for the first time in the Shuttle's history, shut down on the pad just a few seconds before lift-off. The entire Shuttle mission was cancelled. Rhea and her crewmates were then reassigned to another mission, STS-51E, in the spring of 1985. That mission would deploy the TDRS-2 (Tracking and Data Relay Satellite), but was itself cancelled, days before launch, with *Challenger* already on the pad, ready to go, when a problem with one cell of the TDRS 24-cell flight battery was discovered and it required replacement. Rhea explains:

> And you go through that frustration of not being named to a flight, being named to a flight, having them cancel the flight, having to retrain for a different payload. We trained on three different payloads. We had Patrick Baudry, the French astronaut, with us for a while. We had Patrick and Jake Garn for a while. We had Greg Jarvis for a while: unfortunately, he flew on Challenger 51-L. Then we had Charlie Walker. We ended up with Garn and Walker. But it was that turmoil of, 'Who's really going to fly with us, what are we really going to fly on, and are we really going to get to fly?' It had its ups and downs. Bo Bobko was just always cool and calm. He just never got upset about stuff like that. That's the kind of commander to have. When they cancel the flight Bo says, 'They'll give us another one. Don't worry about it. We'll get to fly, we'll have fun, let's go take a vacation for a week and come back, and they'll have decided.' That made it really easy. I think if Bo had been absolutely furious at NASA and the world and had led us in that direction we'd all have been furious at the world, but he just handled it with great aplomb, and so did we. We went home and put a fist through the wall, felt sorry for ourselves, and then we came back and started over. We eventually got to fly.

Eventually, Rhea's crew made it into orbit on mission STS-51D on April 12, 1985. They deployed two communications satellites: Telesat-I (Anik C1) and Leasat-3. Their flight made history when Leasat failed to ignite its perigee kick motor upon deployment. After Rhea unsuccessfully attempted to activate a deployment switch with a makeshift flyswatter, a contingency extravehicular activity (EVA) was arranged. The mission was consequently extended by 2 days to ensure that the satellite's spacecraft sequencer start lever was in its proper position. Griggs and Hoffman performed the unscheduled EVA, assisted by Rhea who, with caution and great skill, was maneuvering the robotic arm of the Shuttle, to try to spin up the satellite that contained six tons of very dangerous hydrazine propellant. Despite several attempts, they failed to revive the satellite. Ironically, the person whom Senator Garn had displaced on this crew was Greg Jarvis, a payload specialist who had worked for the Hughes Aircraft Company in the Leasat program.

The spokesman of Mission Control commented with acclaim: "She has the skill of a good seamstress!" Sally Ride, who at that time was CapCom, corrected him promptly: "… of a good surgeon." There was sometimes rivalry among the first women astronauts but, if necessary, they defended each other.

During the mission, Rhea also performed several medical experiments to study the cardiovascular system in microgravity. Given her medical background, Seddon was then assigned to the Spacelab-4 mission scheduled for March 1987. The mission was cancelled due to the *Challenger* disaster and the subsequent grounding of the Shuttle fleet. On the day of the tragedy, she was in training, sitting in the conference room at a contractor facility, and remembers that they turned on the TV to see the launch of *Challenger*:

> We watched it, and it was like, 'What was that?' Then I realized, maybe before the others, that this was something really bad. When you began to see these big chunks of stuff falling into the ocean it was clear to me that this was not just an early separation of the boosters, this was a catastrophic failure.

The tragedy of the *Challenger* was gone through as something horrible not only for the loss of close friends, but also because it called into question the very future of the US space program. She recalls:

> We started raising the issue: 'Should we do away with human spaceflight? Is my job going to go away? Even if my job doesn't go away, are we ever going to fly again, and if so, how many years is it going to be?'

The mission was eventually renamed Spacelab Life Sciences (SLS) and split into SLS-1 and SLS-2. Rhea was assigned to both missions that flew respectively with Shuttle STS-40 (SLS-1 in 1991) and STS-58 (SLS-2 in 1993).

During mission SLS-1, the crew performed experiments which explored how humans, animals, and cells respond to microgravity and readapt to Earth's gravity on return. STS-40 was the first mission on which three women flew together in space: besides Rhea, Tamara E. Jernigan and Millie Hughes-Fulford were also part of the crew, both on their first flights.

The countdown for the launch of Shuttle STS-58 (SLS-2), scheduled for October 14, 1993, was halted 31 s before the lift-off due to a malfunction in the on-board computer. After a further attempt the following day, it was rescheduled for October 18 and this time launched without glitches. Dr. Seddon was responsible for the laboratory as the payload commander on this life science research mission, which received NASA management recognition as the most successful and efficient Spacelab flown to date. During the 14-day flight, the seven-person crew performed some 650 neurovestibular, cardiovascular, cardiopulmonary, metabolic, and musculoskeletal medical experiments on themselves and 48 rats, expanding our knowledge of human and animal physiology both on Earth and in spaceflight. Rhea recalls:

> I think the experience I had on SLS-1 was really valuable in helping things to work so smoothly on SLS-2. Everything went well. I think they told us we got 115 percent of the data that we were supposed to get, because the hardware worked well, we had timelined it right, we had a great crew, and we worked really well and we worked really hard and we were willing to stay a little bit late at night. And we did a lot of what I call preplanning. In other words, we took a little bit of time at the end of the day to say, 'How can we get ready for tomorrow?'

Between her second and third flights, Rhea served as CapCom:

> It was one of the best jobs that you could have. It really kept you in the loop of what was going on currently with the flying world. As a CapCom, you really got involved. You were representing the crew, and you had to pull in resources. you were in charge of explaining everything to the crew or explaining the evolution of what was going on. So I gained a lot of appreciation for what went on on the ground, and I really enjoyed the job. It gave me a greater appreciation of what went on on my first flight, where they had to do all the re-planning and lots and lots of people had to work on lots and lots.

After her return, she worked for a while on the development of the NASA-Mir payloads, in preparation for the Shuttle/Mir STS-71 flight of Norman Thagard and his Mir-18 expedition in 1995. She was the liaison with Norm and Bonnie Dunbar and verified the difficulties of going to live in Russia, especially living in Star City. "I felt so bad for Bonnie," she remarks, "because she got sent over there at the last minute, and it was tough on her." She therefore declined the offer to participate in a Mir long-duration mission:

> You were going to have to be willing to learn fluent Russian and you had to be willing to uproot your family and go live there for two years. The food was different; the daily life was different. They had no washer and dryer when they got there. They scrubbed things in the sink. The electrical connections were not good, so that they'd blow up your computers. They had trouble with the phone lines.

Rhea was largely baffled by the Russian command structure, which was completely different. While the Space Shuttle commander was the final authority on everything, Russians waited instead for permission from the ground before proceeding: they received a bonus for doing what they were supposed to do during the mission and, if they did something they weren't supposed to do, their bonus was cut; this is why they were not going to do anything without asking the ground. Also different was their approach to scientific experiments: their attitude was passive. Rhea says:

> At NASA I had worked on the Bioethics Task Force and it surprised me about the different ways different cultures addressed doing experiments on humans, and so we had to factor that in as we wrote the NASA rules about informed consent. Flying with the Russians and eventually flying on the International Space Station opened up a lot of other issues and questions and how are we going to ethically perform human experiments. It wouldn't necessarily be the way we would do it in our country, but the way other countries insist upon doing it. The Russians had to be paid to be subjects, and that's against the law in the U.S. You had to work those things out. With Japanese it would not occur to them to refuse to do anything. You could have proposed all kinds of horrible things, because in their culture you bow to authority. They didn't care about informed consent. It was their duty, it was part of what they did for the good of the mission and the team.

With the arrival of her third baby, Rhea decided to leave NASA. As she needed one more year with NASA to qualify for a pension, she agreed with NASA to return to Tennessee and work from the Vanderbilt University, Nashville. Vanderbilt had been working with the Neurolab people for the last Spacelab mission and there was a Neurolab experiment coming out of Vanderbilt. After retiring from NASA in 1997, she was appointed

Assistant Chief Medical Officer in the Vanderbilt Medical Group of Nashville until 2008. There she also led an initiative aimed at improving patient safety, quality of care, and team effectiveness by the use of an aviation-based model of Crew Resource Management. "After NASA," she says, "I applied the aviation-based lessons I learned about teamwork and communication to the healthcare world to help avoid errors that harmed patients." Now with LifeWings Partners, LLC, she teaches this concept to healthcare institutions across the US.

In 2015, Rhea Seddon was the eighth woman inducted into the Astronaut Hall of Fame on Saturday May 30 and in the Tennessee Women's Hall of Fame® (TWHOF) on October 26. One week prior to her induction into the Astronaut Hall of Fame, she released a book entitled *Go For Orbit* about her career at NASA. As she mentions on her website, she had the privilege of presenting a copy of her memoirs to President Obama for his daughters the week following the ceremony during a visit to the White House with husband, fellow inductee Captain Robert Lee "Hoot" Gibson.

She is now an author, speaker, and consultant. "I hope," she says, "I will inspire other young people, especially women, to pursue exciting careers – possibly with the space program."

References

astronautrheaseddon.com

Briggs, C.S. *Women Space Pioneers*, pp. 49–56. Lerner Publications, Minneapolis (2005).

Evans, B. "'For All Womankind': America's First Female Astronauts', *americaspace.com* (February 10, 2012).

Evans, B. *Tragedy and Triumph in Orbit: The Eighties and Early Nineties*. Springer/Praxis, New York (2012).

Kevles, T.H. *Almost Heaven: The Story of Women in Space*, pp. 55–56, 66–67, 73, 76–77. The MIT Press, Cambridge, MA, and London, UK (2006).

Official NASA biography of Margaret Rhea Seddon, *jsc.nasa.gov/Bios* (November 1998).

Personal communication through e-mail in April/May 2016.

Ross-Nazzal, J. "Margaret Rhea Seddon: Oral History Transcript", *jsc.nasa.gov* (May 20, 2010).

Ross-Nazzal, J. "Margaret Rhea Seddon: Oral History Transcript", *jsc.nasa.gov* (May 10, 2010).

Ross-Nazzal, J. "Margaret Rhea Seddon: Oral History Transcript", *jsc.nasa.gov* (May 10, 2011).

Seddon, R. *Go for Orbit*. Your Space Press, Murfreesboro, Tennessee (2015).

Woodmansee, L.W. *Women Astronauts*, pp. 60–61. Apogee Books, Burlington, Ontario, Canada (2002).

9

Shannon Lucid: The First American Woman to Participate in a Long-Duration Mission

Credit: NASA

U. Cavallaro, *Women Spacefarers*, Springer Praxis Books, DOI 10.1007/978-3-319-34048-7_9

Mission	*Launch*	*Return*
STS-51G	June 17, 1985	June 24, 1985
STS-34	October 18, 1989	October 23, 1989
STS-43	August 2, 1991	August 11, 1991
STS-58	October 18, 1993	November 1, 1993
STS-76	March 22, 1996	
STS-79		September 26, 1996

Commemorative cover of mission STS-34, signed by Shannon Lucid (from the collection of Umberto Cavallaro)

Shannon Lucid was the oldest and tallest in the class of the first six American female astronauts and was the last to fly in space. She is one of six women who rocketed into space five times, but she made history as she was the first and only American female astronaut to fly on a Russian space station in the "NASA-2 increment" to Mir—breaking the record for endurance in space by a woman, held until then by Elena Kondakova—and the second American after Norm Thagard. "My arrival on MIR, eight months after the end of the mission of Thagard," she said, "was the beginning of a continuous American presence in space, which has lasted for more than two years." She lived and worked there for more than 188 days, the longest stay of any American on that space station. The record she set as a female in space was exceeded only in 2007 by Sunita Williams on the International Space Station (ISS).

Shannon Matilda Wells Lucid was born in Shangai, China, on January 14, 1943, to Baptist missionary parents. When she was 6 weeks old, her family became Japanese prisoners of war. After a year in an internment camp, they were released in a prisoner exchange and returned to the US. After the war was over, they returned to China. But they had to leave again when the Communists took control.

Those who know Shannon describe her as a person of innate patience, faith, and good humor: qualities she likely learned from her parents at a very young age, but she also developed herself by working hard at them as she grew. The dream of exploring space began when she was a little girl:

> It was a long time before there ever was such a thing as a space program. I was just real interested in the American West. I was interested in exploring. When I was a child, then I thought by the time I grew up that the world would be explored, so what would be left for me to do when I grew up? Then I read about Robert Goddard and the rockets that he'd been doing out in New Mexico. I started reading a little bit of science fiction, and it just sort of clicked. Well, you can go explore the universe and that wouldn't get used up before you grew up.

After graduating from Bethany High School, Oklahoma, she attended the University of Oklahoma and earned a Bachelor of Science degree in Chemistry in 1963 and then a Master of Science and a PhD in Biochemistry from the same school in 1973. She then devoted herself to academic work. She was selected for the NASA Astronaut Corps in 1978—the only applicant to already have children when she was selected. She had three children when selected.

Her first spaceflight was in June 1985 on STS-51G. She then flew missions STS-34 (1989), STS-43 (1991), and STS-58 (1993), and finally visited Mir in 1996.

During her first mission (STS-51G), three communication satellites were released into space for Mexico (Morelos), the Arab League (Arabsat), and the US (Telstar). Then, *Atlantis* STS-34 was a 5-day mission that deployed the Galileo spacecraft toward the planet Jupiter and carried out a mapping of atmospheric and radiation measurements. *Atlantis* STS-43 released the fifth TDRS (Tracking and Data Relay Satellite). During the mission, 32 scientific experiments were conducted, mostly relating to the Extended Duration Orbiter and Space Station Freedom, then in its design phase. *Columbia* STS-58 was a record-duration 14-day mission that was recognized by NASA management as the most successful and efficient Spacelab mission.

But Shannon Lucid made history mainly for her fifth and final flight when, in 1996, she spent 188 days in space, including 179 days on board the Russian space station, Mir, so setting the record for endurance in space of an American astronaut in one single mission and, more generally, of a non-Russian spacefarer and the world record for a woman in space. She flew on the Space Shuttle both for launch (STS-76) and for return (STS-79). She recounts:

> I had been a NASA astronaut for 15 years and had flown on four shuttle missions when one Friday afternoon, I received a phone call from my boss, Robert 'Hoot' Gibson, then the head of NASA's astronaut office. He asked if I was interested in starting full-time Russian-language instruction with the possibility of going to Russia to train for a Mir mission. My immediate answer was yes. Hoot tempered my enthusiasm by saying I was only being assigned to study Russian. This did not necessarily mean I would be going to Russia, much less flying on Mir.

Shannon viewed the Mir mission as a perfect opportunity to combine two of her passions: flying and working in laboratories. "For a scientist who loves flying," she says, "what could be more exciting than working in a laboratory that hurtles around the earth at 17,000 miles (27,000 kilometers) per hour?" She was also happy to do the training with her friend and colleague, John Blaha, with whom she had already flown twice on the STS-43 and STS 58 missions. After 3 months of intensive language study, she got the go-ahead to start her training at Star City, the cosmonaut training center outside Moscow. She arrived there in January 1995, in the depths of a Russian winter—the phones didn't work:

> You'd try to make a phone connection and you might try forty-five minutes to an hour to get a connection, and then many times it was so poor, you couldn't hear. So it was very, very unsatisfactory. But my daughter sent me over CompuServe, which she bought, and I installed it on my computer. With CompuServe, I was able to make a connection back home every day and to pick up e-mail messages.

She had to work harder during that year than at any other time in her life:

> Every morning I woke at five o'clock to begin studying. As I walked to class I was always aware that one misstep on the ice might result in a broken leg, ending my dreams of a flight on Mir. I spent most of my day in classrooms listening to Mir and Soyuz system lectures – all in Russian, of course. In the evenings I continued to study the language and struggled with workbooks written in technical Russian. At midnight I finally fell exhausted into bed.

She would have to learn the systems and operations for Mir and Soyuz in case of an emergency and a need to evacuate from the station. "It's very different from training in the U.S.," she said. "Basically you're in the Soyuz, you put on your 'skafonder' you sat in the Soyuz and my job was to be quiet and not interfere." She never met other cosmonauts:

> All my training in Russia was with John Blaha. John Blaha and I trained together. We sat in a classroom together. It was just the two of us and the instructor for whatever classroom it was. We just talked to each other all day long and that was it. I mean, we didn't interface with anybody else. Only toward the end did we do just a

very few sims with the Russian crew. I mean, we got in the Soyuz and went through a simulation. But it was very, very minimal. There wasn't training with a crew like you think in the U.S.

A few weeks before launch, she realized that the Russians were nervous about sending another woman. They had sent four into space before her: Tereshkova, whom they considered a failure; Savitskaya, whom they had not liked very much; Sharman, who - they said - did not count because she was more of a tourist than a cosmonaut; and Kondakova, whose temperament some found trying. Suddenly, the media assaulted her with pedantic and sometimes heavy questions. Luckily, it didn't last long:

At last, in February 1996, after I had passed all the required medical and technical exams, the Russian spaceflight commission certified me as a Mir crew member. I traveled to Baikonur, Kazakhstan, to watch the launch of the Soyuz carrying my crewmates-Commander Yuri Onufriyenko, a Russian air force officer, and flight engineer Yury Usachev, a Russian civilian. Then I headed back to the U.S. for three weeks of training with the crew of shuttle mission STS-76. On March 22, 1996, we lifted off from the Kennedy Space Center on the shuttle Atlantis.

Her long mission began on Saturday evening, March 23, 1996, when the *Atlantis* STS-76 crew delivered Lucid to Mir to begin the increment known as NASA-2. She later said:

It was just pretty neat to look out the window and see Mir, and know that it was going to be your home .... It was great to see Yuri and Yury. They'd been up there a month before I got there. They acted very happy to see me. I believe that they really were. So, as soon as the hatch opened, I moved over and became part of the Mir-21 crew.

Although there was some concern about the legendary Russian chauvinism, Lucid's crew—Commander Yuri Onufriyenko and flight engineer Yury Usachev, whom the Russian media had affectionately dubbed "the two Yuris"—turned out to be particularly harmonious. Shannon says:

They're both very, very nice people and I've enjoyed working with them very, very much. They have different personalities .... I think the personalities mesh quite well together. Yuri the commander tends to be a little more quiet, and Yury the flight engineer always has something to say. We really had a good time together. We really enjoyed being there together. Yuri and Yuri were absolutely fantastic to work with. I mean, I could not have picked better people to spend a long period of time with. We just lived every day as it came. We enjoyed every day. We enjoyed working together and joking around together. It was just a very good experience, I think, for all of us.

Soon after arriving, Lucid had to soften a polemic in the media that had been started by a comment from a Russian space official. General Yuri Glaskov, Deputy Commander of the Gagarin Cosmonaut Training Center at Star City, had earlier hinted that the two male cosmonauts would welcome Lucid "because we know that women love to clean." He tried then to rectify it by saying: "The simple presence of a lady onboard the Mir station helps ... because our crewmembers simply pay more attention to the way they behave, they act, they speak, and so on." Lucid had to return to the subject in a press conference from space:

"That kind of thinking doesn't bother me. We all work together to keep the place pretty tidy." Commander Onufriyenko said that all three crewmates on Mir would share housekeeping and that Lucid would improve the "cultural level" on board the station.

The three returned to this topic during the long Mir expedition. One evening, she was talking with Yuri and Yury, who repeated the usual arguments that it was right that there weren't more females in the Russian Space Program, etc., and Shannon objected: "Now, you guys, that's not the way that you are, and that's not the way you're treating me." They said: "Oh, but you're an American, so it's okay for American women to do!"

Usachev had welcomed her and offered her the Spektr module as her private apartment where Shannon could sleep and take her personal belongings, including her Bible ("I have a small Bible I always carry when I travel. I brought it up on the Shuttle, so I had it with me the whole time I was there"), and she found a Gideon New Testament that was there. Shannon brought many books. Because she was in Russia and could not buy her own books to bring, before launch, she said to her daughter: "Go to Half-Price Bookstore and just get some things that you think I haven't read. The only requirement is they had to have a lot of words per page." Her daughter, who was an English major, picked out a lot of books by English authors of the seventeenth century. One of these was *David Copperfield* by Charles Dickens. She recounts:

> I thought 'Wow, here was this guy that lived in a totally different era than we're living, and it had never ever crossed his mind that his book would be being read by an American on a Russian space station'. I just thought about that a lot, about the power that authors have and his ideas and his story was transcending the centuries.

In fact, on Mir, there was "the first library in space." Shannon says:

> I left most of my books up there. Obviously, I couldn't bring them back, and I thought the other guys would enjoy reading them. The Russians have a fairly extensive library up on Mir, because over the years, there's lots and lots of Russian books that have come up. Behind one panel was just full of books. There were even a few English in there, because the foreign astronauts have brought up.

Her positive dealings with the Russians and her humor helped to create huge warmth between two distant partners in what was, at times, a difficult relationship. Kevin, the Commander of STS-76, had presented her, as a surprise, with two pink socks that she took with her over to Mir and she decided to wear them on Sundays and to make Sunday a special day during her stay on Mir, where she established funny traditions such as sharing a bag of Jello every Sunday night. But the key to the success of the mission involved the few rules that the team chose to adopt, such as, for example, meeting together for meals:

> We ate all our meals together and spent a lot of time talking to each other over mealtimes. Then many times we did take a small break in the middle of the late afternoon where we'd have tea and cookies together. Lots of times we would have tea and cookies together just before we went to bed at night.

Every day, before lunch, they had 45 min of exercise developed by Russian physiologists to prevent muscles from atrophying in the weightless environment. "I'll be honest: the daily exercise was what I disliked most about living on MIR," says Shannon.

On April 26, the seventh and final Mir module, the Priroda Laboratory, arrived in which Shannon had placed many of the US experiments. Priroda's launch was supposed to have taken place 6 weeks earlier, on March 10, so that the laboratory would be ready for Shannon when she arrived, but it was delayed. Shannon spent a lot of time conducting most of her 28 scientific experiments there. "As a graduate student years ago, I fantasized about having my own laboratory," she said. One experiment determined that the average radiation exposure on Mir was equivalent to undergoing eight chest X-ray radiographs per day. NASA scientists believe, however, that an astronaut would have to spend at least several years in orbit to suffer any appreciable risk of developing cancer.

Shannon was also involved in a long-running experiment to grow wheat in a greenhouse on the Kristall module and to study the chemical, biochemical, and structural changes in plant tissues and their contribution to spaceflight life support, as they produce oxygen and food while eliminating carbon dioxide and excess humidity from the environment:

> John Blaha, the American astronaut who succeeded me on Mir, harvested the mature plants a few months later and brought more than 300 seed heads back to the earth. But scientists at Utah State University discovered that all the seed heads were empty.

(Unfortunately, precisely because of this mission and this handover—which Blaha thought sketchy—there was some tension between Shannon and John. But misunderstandings were resolved and the old friendship was restored.)

Shannon discussed her experiments at least once per day with colleagues Bill Gerstenmaier and Gaylen Johnson at the Mission Control Center in Moscow. Unlike the Space Shuttle, which transmits messages via a pair of communications satellites, Mir was not in constant contact with the ground. The cosmonauts could talk to Mission Control only when the space station passed over one of the communications ground sites in Russia. These "comm passes" occurred once per orbit, about every 90 min, and generally lasted about 10 min. She explains:

> Commander Onufriyenko wanted each of us to be 'on comm' every time it was available, in case the ground needed to talk to us. This routine worked out well because it gave us short breaks throughout the day. Often, when we started a new experiment, Gerstenmaier made sure that our conversations were heard by interested researchers, who in case could answer all my questions. Our work schedule was detailed in a daily timeline that the Russians called the 'Form 24'. The cosmonauts typically spent most of their day maintaining Mir's systems, while I conducted experiments for NASA.

Many investigations aimed at evaluating the effect of microgravity on the human immune system, while others were aimed at providing useful data to the designers of the ventilation systems and firefighting equipment on the ISS:

> We found out lots of interesting things. What I thought was most interesting was that it didn't answer all the questions. All it did was pose lots more questions that they need to have answered in the future. But this is important work, not only from the science perspective, but also because we need to know how things propagate in space, in case there ever is a fire or something like that so that we know how to put

> them out, because flames are different. Fire is a little bit different and propagates differently in a microgravity environment than it does down here on Earth.

Shannon was involved in many Earth observation projects from the Kvant-2 module and she also documented some unusual events on Earth's surface. On one occasion, she took photographs as part of her Earth observation work of giant plumes of smoke over Mongolia, as if the whole country was burning. Shannon noted with concern that she had never seen fires of such ferocity on any of her previous four Shuttle missions. She immediately informed the Control Center. A few days later, the crew was informed that news was arriving from Mongolia that massive forest fires were devastating parts of the country. She also photographed the devastation left by a volcanic eruption on the Caribbean Island of Montserrat. Dormant for centuries, the Sufriére Hills volcano erupted in July 1995, quickly destroying and burying the Georgian-era capital, Plymouth. Only one of her 28 scheduled experiments failed to yield results because of equipment breakdown. The crew also deployed an aluminum and nylon pup-up model of a Pepsi Cola® can, which they then filmed against the backdrop of Earth. The soft-drink company paid for the procedure and planned to use the film in a TV commercial, but it was never used, reportedly because Pepsi® later changed the design of the can.

Aboard the Mir, Shannon and her Mir-21 crewmates spoke exclusively in Russian. She says:

> At the beginning communication was at times adventurous. Yuri and Yuri at that time didn't speak English. We used Russian. We made a lot of jokes about it. A lot of times I would just take an English word and pronounce it like a Russian word, and put an ending on it, and that worked. You'd be surprised the number of times that worked. If they didn't understand it, then I would tell them that they ought to go to Russian class. Little by little we developed a new language, a 'cosmic language.' If a Russian teacher had been listening to me, they would have stuck their hands over their ears. I know that a lot of the times I wasn't saying anything correctly. But Yuri Usachev had really a great ability, what a linguistic would call restorative ability. He always knew what I was saying

On April 12, Lucid and her crewmates observed Russia's Cosmonautics Day and celebrated the 35th anniversary of the first human in space, Yuri Gagarin in 1961, and the 15th anniversary of the first US Space Shuttle launch, *Columbia* in 1981. She recounts:

> One evening Onufriyenko, Usachev were very curious about my childhood in Texas and Oklahoma. Onufriyenko talked about the Ukrainian village where he grew up, and Usachev reminisced about his own Russian village. After a while we realized we had all grown up with the same fear: an atomic war between our two countries. I had spent my grade school years living in terror of the Soviet Union. We practiced bomb drills in our classes, all of us crouching under our desks, never questioning why. Similarly, Onufriyenko and Usachev had grown up with the knowledge that U.S. bombers or missiles might zero in on their villages. After talking about our childhoods some more, we marveled at what an unlikely scenario had unfolded. Here we were, from countries that were sworn enemies a few years earlier. I was living on a Russian space station working and socializing with a Russian air force officer and a

Russian engineer. Just 10 years ago such a plot line would have been deemed too implausible for anything but a science-fiction novel.

After her 150th day in space, on September 19, Soyuz TM-24 arrived with the crew Valeri Korzun, Alekandr Kaleri, and French colleague Claudie André-Deshais Haigneré. After the departure of the "two Juris," Shannon spent the last few days with Valeri Korzun and Alekandr Kaleri. "I had seen Valeri quite a bit in Star City," Shannon recalls, "but I had never seen Sasha at all. I met him for the first time when he came out of the hatch."

Her stay on Mir lasted beyond all expectations, first because of a problem on the Shuttle solid rocket boosters and then—when the technical problems were solved—because of Hurricane Fran that further delayed the launch of Shuttle STS-79. Her return was delayed by 6 weeks and one of her first thoughts upon hearing of the extension was: "Oh, no! Not another month-and-a-half of treadmill running!" The delayed landing would mean that she had to miss both her son's 21st birthday in August and her daughter's 28th birthday in mid-September. She said:

> That's life, you know. And, we'll just go on and I'll continue to have an enjoyable time. My family would be surprised at the patience I've developed in space. I hope I can bring some of that back with me.

On September 7, Shannon broke Elena Kondakova's 169-day record for the longest stay in space by a woman. During a NASA news conference at about this time, Yuri Glaskov, Deputy Commander of Russia's Gagarin Cosmonaut Training Center, was asked what he thought about Lucid's taking the Russian record. Glaskov said: "I don't think you've taken the record from us. We have offered this record to you." He also said: "As far as Dr. Shannon Lucid is concerned, I would like to extend my sincerest thanks to the management of the program for making such a selection. Because everybody is fond of her. … Everybody loves her."

Lucid prepared for the end of her own stay and the arrival of *Atlantis* by conducting a thorough inventory of experiment supplies and equipment in the Spektr and Priroda modules of Mir for her handover to John Blaha. Finally, on September 26, 1996, she was able to return home after 188 days in space. President Bill Clinton awarded Lucid the Congressional Space Medal of Honor after her Shuttle/Mir mission. She is the only woman (apart from the 4 women astronauts died in the two Shuttle accidents) to have received this award. Russian President Boris Yeltsin awarded her the Order of Friendship Medal, the highest Russian award that can be presented to a non-citizen.

After this mission, from February 2002 until September 2003, she was NASA's Chief Scientist at the agency's headquarters located in Washington, where she oversaw the development and implementation of projects and programs to communicate NASA's scientific goals to the wider world. Her suggestions after this experience were important for the next setting of the work and life of the ISS. On the one hand had to be avoided stressful rhythms such as those that, in the mid-seventies, had led to the rebellion of the Skylab crew, harassed by a Mission Control that did not understand that operations in a lack of gravity become slower. On the other hand, her concern was to ensure that, during long-duration missions, teams had enough productive things to do so that they were using their time wisely, avoiding make-work from the ground or a low level of productive activity which could lead to depression. The workload on the Mir was too low—she had run out of all her duties for a while. When the announcement of the delay of the Shuttle came, she knew she had to find a way to occupy her time. Luckily, she was prone to talking, despite the language barriers, and was an avid reader.

The following year, she returned to Johnson Space Center in Houston and resumed technical assignments in the Astronaut Office and was CapCom (capsule communicator) in Mission Control during some 15 Shuttle and ISS missions, from talking to the STS-114 mission until the end of the Shuttle program (STS-135).

After working as an astronaut for more the three decades, Shannon Lucid retired from NASA on January 31, 2012. She had spent a combined total of 246 days in space.

References

Briggs, C.S. *Women Space Pioneers*, pp. 70–78. Lerner Publications, Minneapolis (2005).
Davison, M. "Interview with Shannon Lucid", *history.nasa.gov* (June 17, 1998).
Evans, B. "'A Job? But You're a Girl!' The Triumphant Career of Shannon Lucid", *americaspace.com* (February 2, 2012).
Gibson, K.B. *Women in Space: 23 Stories of First Flights, Scientific Missions and Gravity-Breaking Adventures*, pp. 105–111. Chicago Review Press, Inc., Chicago (2014).
Haven, K. *Women at the Edge of Discovery: 40 True Science Adventures*, pp. 150–156. Libraries Unlimited, Westport, London (2003).
Kevles, T.H. *Almost Heaven: The Story of Women in Space*, pp. 157–163. The MIT Press, Cambridge, MA, and London, UK (2006).
Lucid, S. Six months on MIR. *Scientific American*, **278**(5), 46–55 (1998).
Morgan, C. NASA-2 Shannon Lucid: Enduring Qualities. In: Morgan, C. *Shuttle-MIR: The United States and Russia Share History's Highest Stage*, pp. 54ff. NASA SP-4225, Houston (2001).
Official NASA biography of Shannon Lucid, *jsc.nasa.gov/Bios* (February 2012).
Communication through e-mail with the Author in May 2016.

10

Bonnie J. Dunbar: The First Female Astrocosmonaut in History

Credit: NASA

U. Cavallaro, *Women Spacefarers*, Springer Praxis Books, DOI 10.1007/978-3-319-34048-7_10

Mission	*Launch*	*Return*
STS-61A	October 30, 1985	November 6, 1985
STS-32	January 9, 1990	January 20, 1990
STS-50	June 25, 1992	July 9, 1992
STS-71	June 27, 1995	July 7, 1995
STS-89	January 22, 1998	January 31, 1998

Bonnie J. Dunbar was the first astrocosmonaut in history and is one of the six women astronauts who have been in space five times (Fig. 10.1).

Fig. 10.1 Bonnie J. Dunbar's astrocosmonaut patch. Credit: spacepatchdatabase.com

She was one of the first NASA astronauts to move to Star City in the framework of the agreement for the Shuttle–Mir program. Russians, pressed by their financial constraints, had invited Americans, their former enemies, to enter in an extraordinary partnership and America, in exchange for its investment of US$400 million, would send seven American astronauts to the Russian space station to experience long-duration missions, in preparation for the International Space Station (ISS). But, after signing the agreement, NASA had a hard time in finding, among its 120 active astronauts, enough people willing to train in Russia for about a year and a half. Russia and Star City did not have a good reputation in Houston.

Bonnie was one of the first to accept, and initially with great enthusiasm. She had already studied some Russian and she had been to the Soviet Union, specifically Tashkent, Uzbekistan, for a scientific meeting in 1991 and afterward she had accepted an invitation to visit Star City and had liked what she had seen. She didn't mind the spartan conditions in Star City, to which she had been accustomed with her family since childhood and, as the daughter of a Marine, she had a great spirit of duty: the country was looking for volunteers and she felt compelled to make herself available. In 1994, she arrived in Russia as the backup for Norm Thagard, who would inaugurate the new Shuttle–Mir program. Her assumption was that she would be assigned to the next long-duration Shuttle–Mir mission according to a non-written "rotation" rule—that was the way it usually worked. However, she was never assigned to a long-duration mission. Nevertheless, she visited Mir twice as a crewmember of two Shuttle–Mir missions.

Roberta ("Bonnie") J. Dunbar was born in Sunnyside, Washington, on March 3, 1949, and—the first of four children—grew up in an isolated ranch near Outlook, five miles from Sunnyside, where her father, a Marine Corps veteran, had bought 40 acres of rocky, savage land in a government raffle opened to veterans after World War II. It was years before they had electricity or running water.

In this environment, Bonnie developed tenacity and a strong work ethic: "I just don't see obstacles," she says. "I see challenges." She was driving a tractor by age nine and, with her two younger brothers and sister, helped to round up cattle, milked cows, and worked in the fields. Both her parents were self-sufficient, inventive, entrepreneurial people who placed a high value on education. They did not treat the girls any differently from the boys. They encouraged all their children to reach for their dreams. They also set high standards, and expected their children to do their best at whatever they chose to do: "My parents started with nothing," she says. "They built something. They are also very good people. All my parents ever expected is that we be good productive people, not that we be famous. They encouraged us to take whatever God-given talents we had and use them. I came out of that environment thinking I could do just as well as anyone else." Also, she recalls: "My father used to say 'I fought for my sons and my daughters to be able to become what they want to become, if they wanted to work hard enough to do it'."

Bonnie began dreaming about space in 1957, when *Sputnik* was launched. She remembers going outside with her parents when she was eight and looking up at the starry sky for traces of the tiny satellite. She also remained impressed when she visited the "Century 21 Exposition," the 1962 Seattle World's Fair, where a major theme was to show that the US was not really "behind" the Soviet Union in the realms of science and space: "That was the only vacation we ever took when I was a kid," she says, "and I was very interested in these futuristic things."

When, in the eighth grade, her principal asked her what she wanted to do or become, she was a little too embarrassed to tell him she wanted to be an astronaut, so she just told him she wanted to build spaceships. "Well, you'll have to know algebra," he said. She signed up for algebra in the ninth grade and later discovered that it was the key.

After graduating at the Sunnyside High School in 1967, Bonnie decided she wanted to go to college. She recalls:

> I really wanted to go to college. And I promised my parents that if I had that opportunity, I would. They believed that education was a gift and they didn't have that opportunity. I was very interested in going to college. I applied to be an astronaut the first time when I was 18. I received a very nice letter back from NASA—not sure where it is, I'll have to look for it—which basically said that I needed a college education and they would post the announcements in trade journals of some kind.

Her physics teacher suggested that she should study engineering as a way to reach her goals and so she did: "I was very, very fortunate. People gave me good advice at the right time. I had a very supportive family."

Dunbar was just finishing her junior year when she suffered the first, and most searing, loss of her life. Her younger brother Robert, who had followed in their father's footsteps and joined the Marines, was killed in Vietnam on May 31, 1970. Only 16 months apart in age, she and "Bobby" were very close as children, always doing things together: "My brothers and I were very close," she says, "but we would always one-upmanship, whether it be in athletics or anything. It was a healthy, friendly competition, but it made me a better person."

In June 1971, she earned her *cum laude* Bachelor of Science and master's degrees in Ceramic Engineering from the University of Washington and, in 1975, obtained a Ph.D. in Mechanical and Biomedical Engineering from the University of Washington:

> There was the hippie environment, but I was not part of that. One of the first things I did when I arrived on campus, was try to join Air Force ROTC (Reserve Officers' Training Corps). But they weren't yet accepting women. I joined the auxiliary at that time called Angel Flight. And there were 50 of us selected from across the campus. It was a community development, community service, and also an Air Force mission-related organization, which still exists as Silver Wings right now. It gave me an opportunity to meet people of like interests, people who loved flight.

It was not an auspicious time for anyone to be looking for work, especially for women, who, in a male-dominated field, represented less than 1 % of the workforce. She applied to any company that even remotely had anything to do with aviation and also applied for an engineering position at Boeing: "They offered me a position in a new organization called Boeing Computer Services. Since I had studied Fortran IV programming as an engineer, they knew I could program computers." So she became a Common Business-Oriented Language (COBOL) systems analyst and programmer. After 2 years, sponsored by her former professor and adviser in ceramic engineering, Dr. Suren Sarian, she succeeded in taking a position as *visiting scientist* at the AERE (Atomic Energy Research Establishment) in Harwell, near Oxford, UK. This was one of the few places overseas that were doing research in the field of ceramic engineering. After a few months, she learned that the Space Division of Rockwell International was offering her a job, so she returned to the US and was immediately sent to Palmdale, California, to help in providing support and overseeing the startup of the tile-production facility for the Space Shuttle program. In 1978, she won a company-wide award as the "Rockwell International Engineer of the Year" for her solution to a problem involving waterproofing the tiles.

The first application that she sent to NASA in 1977 to become an astronaut was not accepted. "I was not a part of the class," she says, "but I was extremely honored to become a finalist. There were several thousand who had applied." Although she had not been selected, she was interviewed for some positions at NASA's Johnson Space Center (JSC). Since all the successful candidates (except those who were pilots) had doctoral degrees, she decided to go back to graduate school.

In July of 1978, Bonnie was invited to join the Payload Operations at JSC. She took the job and, at the same time, began working toward a doctorate in Mechanical and Biomedical Engineering at the University of Houston (which she completed in 1983).

All of a sudden, they discovered a problem with Skylab re-entry. Bonnie was trained as a Guidance and Navigation Controller for Skylab and sat on the console for 9 months until July 1979. After that, she went back to her Payload Officer job and started preparing for STS-1.

She applied again in 1980 and this time she was accepted as a Candidate Astronaut. This was another lesson she had learned as a child: "If you fall off the horse, you get back on. Failure is a part of life; success is getting up and trying again."

Her very first assignment was to SAIL, the Shuttle Avionics Integration Lab. In 1983, she obtained a doctorate in Mechanical/Biomedical Engineering from the University of Houston, Texas. Her multi-disciplinary dissertation on materials science and physiology involved evaluating the effects of simulated spaceflight on bone strength and fracture toughness.

She flew in space five times. In 1985, she was *mission specialist* in the STS-61A mission, with Spacelab D-1, the first Spacelab entirely managed and controlled by West Germany. The media were very interested in that mission, especially in Germany, "but not always for the right reasons," Bonnie says. A German reporter once asked her what she thought of *Dallas*: "Well, you know, I've only visited it once or twice, but I hear it's a pretty city," she replied. She did not understand that the reporter was referring to the TV show with J.R. Ewing that at the time was quite popular in Germany, but of course Bonnie had never seen or known about.

It was the first time the Shuttle flew eight astronauts, and the first time that an activity carried out on the Shuttle was fully controlled by a non-American country. During the mission, 75 experiments were conducted in various scientific disciplines including physiology, biology, materials science, and navigation. Bonnie Dunbar was responsible for Spacelab operations and was involved in several experiments that she had contributed to prepare, traveling for 7 months in Germany, Switzerland, France, and the Netherlands to become familiar with the protocols of the different experiments. That was the last mission of the *Challenger*, which disintegrated a couple of months later.

In 1988, Bonnie was assigned to the STS-32 mission and, that same year, she married Ronald M. Sega, an engineering professor at the University of Colorado and Officer of the Air Force Reserve. The STS-32 mission, which deployed the telecommunications satellite Syncom 4 into space, was launched in January 1990. During the mission, the news arrived that her husband had been selected for the NASA Astronaut Corps. Bonnie, maneuvering the robotic arm, recovered the LDEF (Long-Duration Exposure Facility) platform which, with its burden of long-term experiments, had been placed in low orbit by the STS-41C more than 5 years before, in April 1984. Its recovery had been delayed because of the *Challenger* accident in 1986. During these first two missions, she had been trained for a possible contingency spacewalk:

> There still is a sizing problem for some of the females. It's driven by economics more than anything. Back during Gemini, Mercury, and Apollo the astronauts wore custom suits. When we started the shuttle program, they decided so many more astronauts were going to fly it would be economical to build suits that had interchangeable arms and lower waists. On my first flight, I flew a medium suit. Then Challenger happened and there were budget cuts. On my second flight they forgot to certify my configuration and I had to get into a small hard upper torso. So, I could hardly breathe because it was small. And the other was the arms were too long; my

> fingers didn't reach the end of the gloves. And they said, 'Well, we don't have enough money to change it.' You need to be able to function. It's like putting on your father's gloves and being asked to do brain surgery. But these decisions were being made in Washington D.C. while the training is happening in Houston. Your suit tech and trainer are not in the position of changing the budget.

In 1992, she was the Payload Commander during the STS-50 USML-1 Spacelab mission (United States Microgravity Lab-1). For almost 14 days, the astronauts, led by Bonnie Dunbar, in four shifts a day, conducted 30 experiments in materials science and life science in microgravity, with the participation of over 100 American scientists on the ground.

As mentioned before, Bonnie arrived at Star City in 1994 to train as a backup for Norman Thagard, who had to participate in the MIR-18 expedition. She says:

> I'd actually been taking Russian lessons before I went to Russia, but I hadn't been speaking it. All of the training was there being delivered in Russian. It's all technical. So you're still learning the language, how to speak it, but you're also expected to grasp all of the technical concepts. It was like being in first grade and grad school at the same time.

She felt herself suddenly plunged into a culture far away from what she was used to, with Russian colleagues who saw nothing wrong with putting their arms around her and paying her gallant compliments. She wasn't allowed to use the cosmonauts' gym because no women's dressing room was available there—the Cosmonaut Training Center was indeed a military base. She experienced at first hand the aversion of the space environment for Russian women. She remembers that a Russian instructor told her—with no scientific data to support it—that it was unlikely that she would ever fly on Mir because women need special toilets, as "the female urine has a different chemistry than males' urine and MIR toilets are not able to treat them." She felt very alone. What hurt her was that, even if Norm Thagard was there and heard such comments or saw this sexist behavior, he would never come to her support and the rift resulted in both stopping talking to each other. She realized that, despite the primates of Tereshkova and Savitskaya, women in the Russian program had made only sporadic appearances, dictated for propaganda purposes.

In the end, she wasn't assigned to a Mir long-duration expedition, but was part of the team in the *Atlantis* STS-71 mission (1995), the first Space Shuttle mission to dock with the Russian space station Mir. The *Atlantis* had been modified to carry a docking system compatible with Mir and this was the first time that a Shuttle crew switched members with the crew of a station.

Dunbar served as MS-3 on this flight, which carried in the payload bay the Spacelab, the European laboratory built in Turin, Italy, by Alenia Spazio. In the Spacelab, the crew performed medical evaluations on the returning Mir crew, including investigations on the effects of weightlessness on the cardiovascular system, the bone/muscle system, the immune system, and the cardiopulmonary system.

For the first time, ten crewmembers were in space at the same time in the same orbital complex that, until then, had had a crew of three people, who at most became six during crew exchange, usually carried out using the small Soyuz spacecraft—a record for Mir and for space in general.

In January 1998, Bonnie participated, for the second time as Payload Commander, in the STS-84 mission, the eighth Shuttle–Mir docking mission (that also was the fifth and last exchange of US astronauts via the Shuttle). Bonnie was responsible for delivery of more than four tons of scientific equipment, supplies, water, scientific equipment, and logistical hardware to Mir. Besides all the payload activities, Bonnie was also responsible for conducting 23 technology and science experiments.

After her last mission, she stayed at NASA for another 7 years, working in senior management in a number of different positions, including Associate Director of Technology Integration and Risk Management at the JSC, and 5 years as the assistant director for University Research at Johnson.

She left NASA in September 2005 and began a new chapter of her life, serving as President and CEO of the Seattle Museum of Flight, Seattle, Washington, where she established a new Space Gallery and expanded its K12 STEM educational offerings. The position appealed to her, she said, because it involved "trying to inspire the future by preserving the past." She says:

> History shows us when you invest in technology and lead, it translates right back into the quality of life for every American, and their influence in the geopolitical environment. We're risking our nation if we don't have scientists and engineers—the infrastructure of the nation. Where are the engines of the economy? Where are the solutions to the environmental problems, the energy problems? Those are scientists and engineers. It's not just about appropriating the money out of your government. It's what you do with it. The pyramids were not materialized out of ether. They were built by engineers and mathematicians. Navigation came from astronomers and the stars, and the mathematics that came along with that.

Since she was at NASA, Bonnie is strongly convinced that:

> … to address the grand challenges of the future, we need the new ideas, new companies, and new industries created by STEM careers. This has been historically, and will be in the future, the key to great progress in the United States.

She seeks to stimulate the general public in Congress and make everyone understand that, on the one hand, we must invest more in scientific education and, on the other, we must ask for the students:

> When we went to the moon in the '60s, the whole world watched that, and we ought to be proud that we were leaders doing it because this has benefited our society in many ways. Not just the quality of life and the technology. The quality of life has freed us for other pursuits and learning. Philosophy, political leadership, democracy.

So we should never underestimate the role that science and technology has had in every major civilization since the dawn of time. Do away with science and technology and you start to undermine the very form of government.

The fundamental question is: "Are we a spacefaring nation or are we not? If you're saying no, then let the Sun set on our civilization. We'll sit back and be a nation of watchers and not doers. Our quality of life will not quite be the same and we could even become another Third World country. But if the answer is yes, you have to ask yourself how do we move forward? We invest in education. We invest in research. We invest in technology."

In April 2010, Bonnie undertook with passion the fourth career of her life and led, until the end of 2015, the new STEM Center (science, technology, engineering and math) of the University of Houston.

Currently, she is a professor of aerospace engineering at Texas A&M University and serves as Director of the Institute for Engineering Education and Innovation (IEEI)—a joint entity in the Texas A&M Engineering Experiment Station (TEES) and the Dwight Look College of Engineering at Texas A&M University.

A veteran of five space missions, she has logged 1208 h in orbit—more than 50 days—aboard the shuttles *Atlantis*, *Challenger*, *Columbia*, and *Endeavour*.

In 2012, Bonnie was elected to the Executive Committee of the International Association of Space Explorers (ASE), thus becoming the first women in that committee's 25-year history.

On April 20, 2013, Bonnie was inducted, together with Curt Brown and Eileen Collins, into the prestigious "Astronaut Hall of Fame" (AHoF) group—the elite group of American space explorers which includes some 30 famous space legends such as Neil Armstrong, Alan Shepard, John Glenn, Sally Ride, Jim Lovell, Tom Stafford, and Gus Grissom. The 2013 class was the first to include more women than men, incidentally on the 30th anniversary of the first American woman flying in space, the late Sally Ride, who was inducted in 2003.

References

Briggs, C.S. *Women Space Pioneers*, pp. 64–70. Lerner Publications, Minneapolis (2005).

Heffernan, T. "Bonnie Dunbar: An Adventurous Mind", *sos.wa.gov/legacyproject/oralhistories* (June 18, 2009).

Kevles, T.H. *Almost Heaven: The Story of Women in Space*, pp. 154–157. The MIT Press, Cambridge, MA, and London, UK (2006).

Official biography of Bonnie J. Dunbar, *jsc.nasa.gov/Bios* (September 2005).

Ross-Nazzal, J. "Bonnie Dunbar: Oral History Transcript", *jsc.nasa.gov* (September 14, 2004).

Ross-Nazzal, J. "Bonnie Dunbar. Oral History Transcript", *jsc.nasa.gov* (December 22, 2004).

Ross-Nazzal, J. "Bonnie Dunbar. Oral History Transcript", *jsc.nasa.gov* (January 20, 2005).

Shayler, D.J.; Moule, I. *Women in Space—Following Valentina*, pp. 245–246, 321. Springer/Praxis Publishing, Chichester, UK (2005).
Tate, C. "Dunbar, Bonnie J.", HistoryLink.org Essay 9865, *HistoryLink.org* (June 27, 2011).
Wright, R. "Interview with Bonnie Dunbar", *spaceflight.nasa.gov/history/shuttle-mir* (June 16, 1998).

11

Mary L. Cleave: Flying at Age 14

Credit: NASA

U. Cavallaro, *Women Spacefarers*, Springer Praxis Books, DOI 10.1007/978-3-319-34048-7_11

Mission	*Launch*	*Return*
STS-61B	November 26, 1985	December 3, 1985
STS-30	May 4, 1989	May 8, 1989

Mary Louise Cleave was born in Southampton, New York, on February 4, 1947. The eldest of three sisters, she soon proved very athletic and strong for sport, and started to practice field hockey, lacrosse, volleyball, and basketball early on, until she was told she was too short. She also dreamed about flying and her mother drove her to the local airport so she could take flying lessons. She says:

> No one's really sure why I was so crazy about airplanes when I was a little kid but I was crazy about airplanes. Nobody else in my family flew except for my mother's brother, who was a pilot that was killed in World War II, so I didn't even know him.

Her instructor, an old retired Army Air Corps test pilot, confirmed to her parents that she had an ability to do this: "This is the first person I've ever had in an airplane that doesn't have anything to forget about driving mechanics; her whole interaction is with the airplane." Mary, working in her spare time as a baby-sitter to earn the money for the lessons, soloed when she was aged 16 years and received her pilot's license at the age of 17, but her parents still had to drive her to the airport so she could fly. She recalls:

> Since I lived in New York City, you had to be eighteen to drive a car, so there was a period of my time where I was legal to fly people before I was legal to drive, which is a very bizarre thing.

People would take her to the airport in the car and then she would take them for a ride in the airplane! In 1969, when she graduated from college, Mary applied to become a commercial pilot but was told that women couldn't become pilots. Still wanting to fly, she applied to become a flight attendant! But, this time, it wasn't her sex that was the problem; it was her size and, once again, she was told that at 5′2″ she was too short: in those days, women were required to be at least 5′4″ tall to be an airline stewardess.

As Mary liked animals, she started a pre-veterinary medicine program. That year, the Faculty of Veterinary accepted only two women and, after realizing that she preferred dissecting plants rather than animals, she changed her major to Biological Sciences and, in 1969, she received a Bachelor of Science degree in Biological Sciences from Colorado State University. After practicing teaching in Denver for a few months, she obtained a job on a "floating campus" tutoring college kids in biology as the ship tramped from port to port. Wherever she went, Mary took samples of the water. The memory of Lake Champlain never left her. In 1971, she decided to go back to school, to Utah State University. She specialized in phycology, which is the study of algae, and worked in the Utah Water Research Laboratory (UWRL) until 1980, carrying out research on a number of microbial, environmental, and engineering projects, including the productivity of algal components of cold desert soil crust in the Great Basin desert south of Snowville, the prediction of the minimum river flow necessary to maintain certain game fish, and the effects of increased salinity and oil shale leachates on freshwater phytoplankton productivity.

In 1975, Mary earned a Master of Science degree in Microbial Ecology from Utah State University. "My co-workers in the lab," she says, "encouraged me to move to engineering because they saw I was more of a problem solver than a scientist." In 1979, she received a doctorate in Civil and Environmental Engineering, so becoming the first woman to earn a Ph.D. in engineering at Utah State University.

When Mary was working at the UWRL, she accidentally learned from a colleague research engineer working at the lab that, in the local post office, there was an advertisement from NASA saying that they were looking for scientists and engineers to work in space: "My colleague handed me the flyer because he said I was the only one crazy enough to pursue it." And she applied:

> Space flight was great, but my interest was airplanes, and, in fact, when I found out there was a possibility of flying in T-38s, I really thought that was neat because I never thought I'd get to fly in a high-performance jet.

This time, her size was not a barrier; instead, it was an asset—she was the only astronaut small enough to carry out narrow repairs, such as fixing the Shuttle's commode when it malfunctioned! And, in 1980, Mary was invited to Houston for an interview: "I was too short to serve coffee in flight, but tall enough to fly into space."

Although she applied for the astronaut selection in 1978, she was not selected. But she succeeded in 1980. The news that she had made the first cut did not come as a totally unexpected surprise. When federal investigators dropped by Wellsville to conduct a "discreet" background check, everybody in town kept her apprised of what the feds were asking:

> The first inkling I had was they start doing a background check on you. Although nobody was supposed to know they were doing this and they asked everybody not to talk about it, if you live in Wellsville, Utah, which is a town of 2,000 people, I mean, forget it. This stranger shows up, and I was getting phone calls saying, 'All right. He's just left the

> building. It looks like he's driving down to 23rd Street. I bet he's going to see Maude'. So those poor guys, I knew where they were the whole time they were in the valley.

Finally, in the last week of March 1980, Mary received an invitation to come to Houston for an interview. She knew her chances of selection were remote. This time, there were 2880 applicants for mission specialist (396 of them women) and 583 applicants for pilot (18 women applied to be pilots, but none of them qualified). Interviews for each group lasted from Monday until Friday. Some applicants, who might have forgotten that there is always the prospect of mortal danger in the space program, emerged from an excursion in the Shuttle simulator feeling quite shaken. During the claustrophobia test, while some of the applicants panicked, Mary fell asleep! She was selected as an astronaut in May 1980. She was the first in her class to be assigned a real job. Based on her background of microbial ecology and water sewerage and pollutants, John Young, the moonwalker (Apollo 14) who at that time was the Chief of the Astronaut Office, assigned Mary as "sanitary engineer" to fix the Shuttle waste-management systems which hadn't worked properly on the first flight of the Shuttle. She was then assigned to the SAIL (Shuttle Avionics Integration Laboratory) to test flight software and helped to develop crew equipment, including the personal-hygiene kit and flight documentation such as the *Malfunction Procedures Data Book*. She also served as CapCom (capsule communicator) in five missions. She was CapCom in the first female-to-female link, during mission STS-7 when Sally Ride flew. In November 1985, Mary was assigned as mission specialist to mission STS-61B that deployed three communication satellites: MORELOS-B, AUSSAT II, and SATCOM K-2. She also conducted a few experiments for 3M and several for the Mexican government. The mission had the heaviest payload so far carried into orbit by the Space Shuttle. During that mission, Mary, who was the only civilian astronaut and the only non-test pilot in the group, was responsible for operating the Space Shuttle Robot Arm, or SRMS. She explains:

> They couldn't fit me in a spacesuit because I was too small, and they decided not to buy these small hard upper torsos to save money. They were going to fit the midrange of the astronaut corps, so they ended up having to train me as a flight engineer. So I was a flight engineer on that flight, and I flew the arm.

She was a bit disappointed at first but, in the end, she was happy with how it turned out. "It seemed like," she explained later to the NASA oral historian, "they assigned women to fly the arm [Shuttle Remote Manipulator System or Canadarm] more often than guys, and the rumor on the street was because they thought women did that better." Mary had experienced quite a difference between using the training model on the ground or in the water tank and using the real arm in space:

> We were facing the Earth and I was working the arm. We had the Earth going at Mach 25 underneath, and it was a moving target and really distracting. Somebody pointed out, in the Gemini Program, they almost missed a rendezvous because of the same phenomena. What they did was, they always did it against deep space, and I went, 'Well okay, so we made the same mistake twice'.

The recommendations she provided in the debriefing after this experience helped to improve SRMS operations.

The view of our planet from space reinforced Mary's passion for the environment:

> You get an appreciation for just how small the atmosphere is. Just a very thin layer that separates us and our planet from a really nasty vacuum out there. So you get real respectful for how we should treat the atmosphere, how we should treat the oceans, how we should treat the land. Really, being in space changes a person's attitude. Our planet is quite finite and is 70% ocean. Whereas we know these things intellectually from our Earth-bound field trips and from science books, the finite nature of our planet and the small portion that we actually inhabit, instantly becomes real when you look down from low Earth orbit.

But mainly she enjoyed the weightlessness: "Weighting nothing is great fun. You don't need a stool to reach things!"

In 1989, Mary took part in the STS-30 mission and she was the first woman to fly after the *Challenger* accident. During the ascent phase, as there was a crew of five and one person had to be alone downstairs, she accepted to fly by herself: "I was by myself and I got to enjoy the ride," taking pictures and having a good time.

The mission successfully deployed the Magellan Venus-exploration spacecraft. It was the first time that a spacecraft that was going to another planet had been deployed from the Shuttle. Magellan has been one of NASA's most successful scientific missions, which arrived at Venus in August 1990 and mapped over 95 % of its surface, providing valuable information about its atmosphere and magnetic field. Mary commented:

> It was great work. We ended up working real hard until we deployed it: the thing is that as it gets it out of the payload bay, then it's not JSC's [NASA Johnson Space Center, Houston, Texas] problem anymore, it belongs to JPL [NASA Jet Propulsion Laboratory, Pasadena, California]. Working with the guys from JPL was great.

For the rest of the mission, Mary had a good time, taking pictures. This was 4 years after her first spaceflight and, as an environmental engineer, looking down, she realized that there were huge changes and she was bothered by how quickly Earth was changing. She therefore decided to take care of the environment and to leave Houston's JSC and move to Greenbelt, Maryland, working in the NASA Goddard Space Flight Center. When she told colleagues of her plans, they tried to discourage and dissuade her. She comments:

> The amount of pressure that goes in to people when you decide you don't want to do what most people consider to be the best job on the planet was really pretty tough: 'What do you want to go to Goddard for? God, work with those unmanned spacecraft. Don't do this. It's a one-way gate.' It was standard military practice, don't do any traumatic decisions until you think about it for a year.

She left JSC in May 1991 to join NASA's Goddard Space Flight Center, where she worked in the Laboratory for Hydrospheric Processes as the Project Manager for SeaWiFS (Sea-viewing, Wide-Field-of-view-Sensor), an ocean color sensor aimed at monitoring vegetation globally and providing global measurements of all of Chlorophyll a on Earth every 48 h. The spacecraft launched in August 1997 and operated successfully for 13 years. The data from the spacecraft helped to increase understanding of climate change, oceanography, and atmospheric science.

From 2004 onwards, Mary served as Deputy Associate Administrator for Earth Science at NASA Headquarters in Washington, DC, where she was responsible for numerous spacecraft exploring space from the edge of our Solar System to the Sun's orbit. After retiring from NASA in October 2007, Mary joined Sigma Space, a company based in Lanham, Maryland, which provides optoelectronic aerospace instrumentation for NASA, DOD, and commercial customers.

Mary has logged a total of 262 h in space, orbited Earth 172 times, and traveled 3.94 million miles.

References

Gibson, K.B. *Women in Space: 23 Stories of First Flights, Scientific Missions and Gravity-Breaking Adventures*, pp. 67–68. Chicago Review Press, Inc., Chicago (2014).

Kevles, T.H. *Almost Heaven: The Story of Women in Space*, pp. 68–69. The MIT Press, Cambridge, MA, and London, UK (2006).

Official biography of Mary L. Cleave, *jsc.nasa.gov/Bios* (2007).

Shayler, D.J.; Moule, I. *Women in Space—Following Valentina*, pp. 184–185, 246–248. Springer/Praxis Publishing, Chichester, UK (2005).

Woodmansee, L.S. *Women Astronauts*, pp. 67–68. Apogee Books, Burlington, Ontario, Canada (2002).

Wright, L. "Space Cadet", Texas Monthly (July, 1981), 114–119.

Wright, R. "Interview with Mary Cleave", *jsc.nasa.gov/history/oral_histories* (March 5, 2002).

12

Ellen Baker: An Internist Physician on the Shuttle

Credit: NASA

U. Cavallaro, *Women Spacefarers*, Springer Praxis Books, DOI 10.1007/978-3-319-34048-7_12

Mission	*Launch*	*Return*
STS-34	October 18, 1989	October 23, 1989
STS-50	June 25, 1992	July 9, 1992
STS-71	June 27, 1995	July 7, 1995

Commemorative cover of mission STS-34, signed by Ellen S Baker (from the collection of Umberto Cavallaro)

Growing up in New York, Ellen Baker dreamed of playing baseball for the Yankees but, at that time, girls were not accepted to play in major-league baseball. She says:

> I've always been interested in the space program and I thought it was exciting and challenging, but they didn't let girls go into space when I was little. So I didn't really think about it as a possibility for me until I was out of medical school. The first group of women were selected in 1978 and that's the year I graduated from medical school.

Ellen Louise Shulman Baker was born on April 27, 1953, in Fayetteville, North Carolina, but considers New York City to be her hometown. She graduated from Bayside High School, New York, in 1970 and received a Bachelor of Arts degree in Geology from the State University of New York at Buffalo in 1974 and a Doctorate of Medicine from Cornell University Medical College in 1978. She recalls:

> When I was finishing medical school, I saw an article in the newspaper that NASA was accepting astronaut applications and women were urged to apply. I started to think about the possibility. Of course, it seemed an unreachable goal.

After training in internal medicine at the University of Texas Health Science Center, San Antonio, Texas, Ellen was certified by the American Board of Internal Medicine and, in 1981, she joined NASA as medical officer at the Johnson Space Center (JSC) Flight Medicine Clinic.

As one of the 4934 applicants, Ellen was selected as a NASA astronaut in May 1984 and became part of the Astronauts Corps Group 10 "The Maggots" together with two other women: Kathryn Thornton and Marsha Ivins. Ellen once said to me:

> What I like best are the people. I work with a wonderful group of people and it's hard to imagine that there's a better group of people anywhere. And of course I like the program, the excitement, the challenge and I like working for something that I think is important. I like to make a difference.

She received much encouragement from the women who went before her, including close friend Shannon Lucid.

Ellen participated in STS-34 (1989), STS-50 (1992), and STS-71 (1995). She first flew in space as a mission specialist on STS-34 *Atlantis* in October 1989. The crew successfully deployed the Galileo probe to explore Jupiter. They also mapped atmospheric ozone on Earth and conducted medical and scientific experiments.

Ellen was accompanied in this flight by Shannon Lucid, who was flying her second mission and whom she described as being "completely at home" in space. Having set Galileo on its way, the primary mission of STS-34 was over. Several secondary experiments were performed, including the first flight of the Shuttle Solar Backscatter Ultraviolet (SSBUV) instrument in the payload bay. This was part of an ongoing NASA effort to calibrate ozone on free-flying satellites and verify the accuracy of atmospheric ozone and solar irradiance data. During the rest of the mission, Ellen took stunning photographs documenting the impact of human activity upon the world's ecosystem: "We took pictures of everything and brought them back to the experts to analyze," she said. The images from space eloquently convey our planet's fragile beauty. She was also responsible for some experiments, including the GHCD (Growth Hormone Concentration and Distribution) that studied the effect of microgravity on the distribution and concentration of growth hormones in plants.

Ellen next flew in June 1992, together with Bonnie Dunbar, on Space Shuttle *Columbia* STS-50, which was the first flight of the USML-1 (United States Microgravity Laboratory) and the first Extended Duration Orbiter flight. Over a 2-week period, as mission specialist, Ellen was the flight engineer and monitored and operated vehicle systems during the flight while the payload crew conducted scientific experiments involving crystal growth, fluid physics, fluid dynamics, biological science, and combustion science. This mission set a new record as the longest US manned spaceflight, which surpassed the Gemini 7 mission (behind the three manned Skylab missions). Describing some side effects of weightlessness, she pointed out that fluids tend to flow toward the head, meaning that "your legs look skinny, which is good, but your face looks fatter, which isn't."

Ellen flew for the third time in 1995, with the first Shuttle–Mir missions, flown by *Atlantis* STS-71. This was the 100th manned space launch by the US and was indeed the first Space Shuttle mission to dock with the Russian space station Mir. The Orbital Docking System (ODS) of *Atlantis* was equipped for this purpose with a special Androgenous Peripheral Docking System developed by Russians to link the Orbiter to a similar "Krystall" docking module aboard Mir. Everything performed flawlessly. The mission involved an exchange of crews: launched from the Kennedy Space Center with a seven-member crew, *Atlantis* returned with an eight-member crew. The crew also performed various life-sciences experiments and data collections. In this mission, Ellen flew again with Bonnie Dunbar and was responsible for the operation of the Spacelab-Mir module and several of the biological experiments. She referred to this experience with the Russian colleagues as "a fascinating but linguistically challenging information exchange."

A veteran of three Space Shuttle missions, Ellen logged 686 h (over 28.5 days) in space. She then served for many years in different positions in JSC and became Chief of the Education/Medical Branch of the NASA Astronaut Office. After more than 30 years with NASA, she retired in December 2011 to pursue other interests, planning to return to work in the medical field.

References

AA.VV. "Astronaut Ellen Baker's Post-NASA Career", *collectspace.com* (January 2012).

Anon. "Women Shaping History 2014: Ellen Baker", *educationupdate.com* (March/April 2014).

Official NASA biography of Ellen Baker, *jsc.nasa.gov/bios* (January 2012).

Communication with the Author through e-mail in May 2016.
Roa, G. "Astronaut Baker's Talk Inspires NIH Audience", *nihrecord.nih.gov* (April 21, 1988).
Shayler, D.J.; Moule, I. *Women in Space – Following Valentina*, p. 255. Springer/Praxis Publishing, Chichester, UK (2005).
Woodmansee, L.S. *Women Astronauts*, pp. 70-71. Apogee Books, Burlington, Ontario, Canada (2002).

13

Kathryn Thornton: The "Space Walker Mom"

Credit: NASA

U. Cavallaro, *Women Spacefarers*, Springer Praxis Books, DOI 10.1007/978-3-319-34048-7_13

Mission	*Launch*	*Return*
STS-33	November 22, 1989	November 27, 1989
STS-49	May 7, 1992	May 16, 1992
STS-61	December 2, 1993	December 13, 1993
STS-73	October 20, 1995	November 5, 1995

Commemorative cover of mission STS-33, signed by Kathryn Thornton (from the collection of Umberto Cavallaro)

Kathryn Cordell Ryan "K.T." Thornton was born in Montgomery, Alabama, on August 17, 1952, and graduated in 1970 from Sidney Lanier High School. She later received a Bachelor of Science degree in Physics at Auburn University in 1974. She went on to earn a Master of Science in Physics and a Ph.D. in Physics at the University of Virginia in 1977 and 1979, respectively.

While in graduate school, Kathryn participated in the nuclear physics research programs at the Oak Ridge National Laboratory, the Brookhaven National Laboratory, the Indiana University Cyclotron Facility, and the Space Radiation Effects Laboratory. Her work included statistical analyses of heavy-ion nuclear reactions and light-ion production from bombardment of various nuclei with high-energy ion beams. As an expert in nuclear physics, she became a member of the main professional associations and oriented her career on a more academic track. After earning her doctoral degree in 1979, Kathryn was awarded a NATO Postdoctoral Fellowship, allowing her to continue her research at the Max Planck Institute for Nuclear Physics in Heidelberg, West Germany.

In 1980, Kathryn returned to Charlottesville, Virginia, and began working as a physicist for the US Army Foreign Science and Technology Center. One day, she ran across an advertisement about NASA's search for astronauts. That ad not only caught her eye, but changed her life: "It was a notice from NASA. I hadn't really thought about becoming an astronaut before, but I decided to send in the application. I figured all they could do was say, 'no'." She was one of just 120 to make the cut from a pool of 5000 applicants and was invited to Houston for an interview. She said:

> It was a week-long process of mostly medical tests, but there was also an interview with the selection board and visits with astronauts to find out what the jobs were like and that sort of thing. I met so many interesting people and I had a great time, but I just knew when I left there that I would never see Houston again.

Two months later, Kathryn received instead another call announcing that she had been selected as one of the 17 astronaut candidates of Group 10:

> There were about 5,000 applications in that round and they interviewed about a 120 of us about 20 at a time for the next six weeks. Then they selected 17…to this day I believe they called the wrong number when they called me.

Leaving Virginia and moving to Houston was a tough decision, as her husband was a professor of physics at the University of Virginia, and their daughter was only 2 years old. They decided, however, that it was an opportunity that could not be passed up so, in July of 1984, she packed her bags and moved to Houston with her daughter. She confesses:

> When I first went down there, I didn't know which end of the shuttle was up; I mean I knew absolutely nothing, I didn't really know what I had gotten myself into, to tell the truth, but it sounded like a good deal.

And so she started what she calls "the greatest job in the world" that brought her four times into space.

After the basic astronaut training, Kathryn worked in NASA's Mission Development Branch and the SAIL (Shuttle Avionics Integration Laboratory). The *Challenger* tragedy, on February 1, 1986, delayed all space activities for more than two and a half years. Kathryn served meanwhile as a team member of the VITT (Vehicle Integration Test Team) at Kennedy Space Center, taking care of the payloads' integration and—when, towards the end of 1988, the Shuttle flights resumed—served in the Mission Control Center as a spacecraft communicator (CapCom).

Her first flight assignment came in November 1989—when she was already a mum of two daughters—as a mission specialist aboard the Space Shuttle *Discovery* STS-33, carrying a secret Department of Defense (DoD) payload. STS-33 was the first crew to return to orbit following the 1986 *Challenger* disaster.

After waiting a long time, Kathryn was anxious to fly:

> By the time you board the shuttle you have done a lot of training and you are just anxious to get on board. You know something could go wrong, but waiting another day or two, a week or even five years isn't going to change that.

The mission was launched at night from the Kennedy Space Center in Florida on November 22, 1989. Kathryn was the first woman to fly on a secret military mission, and she and her crewmate Story Musgrave were the first civilians assigned to a military Shuttle flight. Sources said that the DoD protested the use of civilians on such a mission, but NASA refused to make a change.

Despite details being classified, because of the DoD's reluctance in divulging flight and cargo information, speculation abounded on what was inside *Discovery*'s payload bay, pointing out that the Soviets would know exactly what was aboard the Shuttle within hours of launch, and the only people who would not know would be the Americans. The *Washington Post* was denounced for its front-page story. It turned then out that—besides conducting a variety of routine engineering and scientific experiments—the mission launched a 6000-pound "Magnum" ELINT (ELectronic INTtelligence) spy satellite to ferret out voice, telemetry, and other broadcast signals from Soviet military installations, and overhear military and diplomatic communications from the Soviet Union, China, and other Communist states.

Apart from a bit of motion sickness on the first day, Kathryn found the flight absolutely amazing:

> I had a broad grin on my face for a month and I just couldn't get it off, I was so happy. I was ready to go and would have climbed aboard another shuttle the next day. The sad part about coming back is that you have to wait a long time before you get another shuttle assignment.

It would be 3 years before Kathryn's next assignment, during which time she worked on the ground in the simulators, with Mission Control, and trained continuously while also raising two daughters and having a third baby:

> Being an astronaut is a full-time job. If you are not training for your own mission, then you are supporting another mission. There are lots of ground jobs to do, such as working with mission control, participating in tests, and working in the simulator. So basically you stay in training all the time.

In 1992, Kathryn participated in the maiden flight of the Space Shuttle *Endeavour*, mission STS-49, whose main goal was to retrieve, repair, and re-launch Intelsat (the International Telecommunications Satellite). During this mission, the crew completed a record four extravehicular activities (EVAs) and also set another EVA record: the first three-person EVA. NASA learned a great deal about spacewalking from this mission. Also, Kathryn took her first spacewalk, afloat in a 350-pound suit for 7 h and 45 min, so becoming the second American woman to perform a spacewalk:

> When you climb into the suit you immediately add an extra 350 pounds to your body weight and it restricts your ability to control it somewhat, but after you have been out there a while, your body begins to adjust.

During the EVA, Kathryn—with Tom Akers—also demonstrated construction techniques for the Freedom Space Station (eventually implemented as the International Space Station (ISS)) and tested the ASEM (Assembly of Station by EVA Method) approach. For this purpose, they constructed a truss segment with both astronauts floating freely. It was a very hard task: "Every minute," she commented, "was a working minute."

In 1993, Kathryn was assigned as a mission specialist EVA crewmember to her third mission, *Endeavour* STS-61, whose goal was to capture the Hubble Space Telescope, the world's first space-based optical telescope, and install corrective optics on it. Once more, the four-astronaut EVA crew had to complete a record five space walks to restore the Hubble to full capacity.

Even Kathryn, who served as a mission specialist, made two spacewalks. Both the EVAs were performed with Thomas Akers. The first, lasting 6 h and 36 min, was taken on December 6 and the second was made on December 8, lasting 7 h and 21 min. Putting together all three of her spacewalks, Kathryn scored a total of 21 h spent in open space, thus becoming the third woman to perform extravehicular activities (the second American, after Kathryn D. Sullivan) and the first to make more than one. Her record was surpassed only in 2007 by Sunita Williams, during the ISS Expedition-14.

In her fourth and final flight on *Columbia* STS-73 in 1995, Kathryn served as payload commander of USML-2, the second mission of Spacelab, configured as the United States Microgravity Laboratory. Another sort of first occurred when Kathryn, during the EVA, used a power screwdriver from the Shuttle's toolkit. The video clip of her impressively working in open space was used by NBC in an episode of the movie *Home Improvements* in which astronaut Tim Taylor (actor Tim Allen) manages to create his first "intergalactic screw-up." Kathryn, then mother of three girls, was dubbed "Space Walker Mom." But her daughters, she said, weren't all that impressed that their mum was an astronaut mom: "It's all they'd ever known." Kathryn logged over 975 h in space (over 40.6 days), including 21 h of EVA.

In 1996, after delivering a graduation speech at the University of Virginia, Kathryn was asked by the university to return to the academic community. She decided to seize the opportunity and left NASA on August 1, 1996: "I loved every minute of it," she said, "but it was time to return to Charlottesville and have more time with my family."

That same year, Kathryn joined the Engineering Faculty and, 3 years later, became associate dean for graduate programs. In addition to managing the engineering graduate programs and mentoring undergrads, she is also a professor of mechanical and aerospace engineering.

Since leaving NASA, Dr. Thornton has served on several review committees and task groups, including the National Research Council Study, Science Opportunities Enabled by Constellation (2007) as co-chair, and the NASA Return to Flight Task Group (2006), which evaluated NASA's work in meeting goals set by the Columbia Accident Investigation Board prior to resumption of Space Shuttle flights. In 2010, she was inducted into the Astronaut Hall of Fame—the only third woman to hold this honor at that time.

When she is able to find a moment, Kathryn enjoys a quick flight: "I have my pilot's license, so I like to go out sometimes and just defy gravity."

References

Barker, M. "Kathryn Thornton '74: A Leader Among Women in Space", *Journey & Spectrum* (2006), 20–22.

Bronstein, N. "Prestigious Awards Acknowledge Kathryn Thornton's Stellar Accomplishments as a Leader of Women in Space", *seas.virginia.edu* (May 24, 2010).

Cavallaro, U. NASA-ESA Hubble Space Telescope: The Greatest Leap Forward in Astronomy since Galileo. *Astrophile*, **57**(1, #317), 13–19 (2015).

Harland, D.M. *The Story of the Space Shuttle*, pp. 181–182. Springer Praxis Books, New York (2004).

Kevles, T.H. *Almost Heaven: The Story of Women in Space*, pp. 187–188. The MIT Press, Cambridge, MA, and London, UK (2006).

Official biography of Kathryn Thornton, *jsc.nasa.gov/bios* (January 1996).

Woodmansee, L.S. *Women Astronauts*, pp. 71–72. Apogee Books, Burlington, Ontario, Canada (2002).

14

Marsha Ivins: Aspiring Astronaut at 19

Credit: NASA

U. Cavallaro, *Women Spacefarers*, Springer Praxis Books, DOI 10.1007/978-3-319-34048-7_14

Mission	*Launch*	*Return*
STS-32	January 9, 1990	January 20, 1990
STS-46	July 31, 1992	August 8, 1992
STS-62	March 4, 1994	March 18, 1994
STS-81	January 12, 1997	January 22, 1997
STS-98	February 7, 2001	February 20, 2001

Commemorative cover of mission STS-46, signed by Marsha Ivins (from the collection of Umberto Cavallaro)

When Marsha was 19 years old—those were the days of Apollo—she wrote a letter to Deke Slayton, then boss of the Astronaut Office, to inquire whether she could become an astronaut and what career path would be best for her:

> I was in my second year of engineering school, engineering selected because that's what astronauts seemed to be after military pilots. I wrote the letter because I was looking for guidance perhaps for specific kinds of engineering to study.

Marsha Ivins
Farrand Hall Box 77
Boulder, Colorado 80302

Mr. Richard Slayton
Director of Astronaut Training
Manned Spacecraft Center
Houston, Texas

Dear Sir:

I am a freshman Engineer at the University of Colorado, in the Aerospace Engineering program. I am also an instrument rated commercial pilot presently working on my instructor's rating. My ambition is to become an astronaut (or whatever one would call a female astronaut) or, if that is not possible, to work somewhere in the astronaut training program. What I would like to know is if you think there will be a future for women in the space program, and if so, will Aerospace Engineering provide the best background or is there a better subject to study. Any information you can give me would be greatly appreciated.

Thank you very much.
Sincerely,

Marsha Ivins

Fig. 14.1 This letter by Marsha Ivins was found in the Johnson Space Center (JSC) History Collection, UHCL Archives (University of Houston-Clear Lake), and was first published in neumannlib.blogspot.it in October 2011. This letter is not dated, but Deke Slayton's response (reproduced in page xxii), encouraging her to continue her education because female astronaut selection was "inevitable", is dated May 4, 1970

Deke Slayton answered by return, saying that he didn't expect any selection of new astronauts in the near future (see Fig. 14.1), and in this he was right: the next selection would come 8 years later, in 1978. Marsha applied to the astronaut program three times before she was selected in 1984.

She is one of the six women astronauts who have flown in space five times:

> I wanted to be an astronaut since I was 10. In 1961, Alan Shepard made history by becoming the first American to go to the space. Over the next decade US astronauts ventured on their moon missions. I remember my entire family would gather in front of our little black and white television set and watch the broadcasts absolutely transfixed. I was captivated by the whole idea of going to space. And from then on, one thing I wanted to do with my life was to work for the space program.

Marsha Sue Ivins was born in Baltimore, Maryland, on April 15, 1951. She started to fly very early, when she was 15 years old: "In fact I learned to fly before I learned to drive a car." After graduating at the Nether Providence High School in Wallingford, Pennsylvania, in 1969, she continued her study and, in 1973, earned a Bachelor of Science degree in Aerospace Engineering from the University of Colorado in Boulder. In July 1974, she went to work for NASA's JSC as an engineer for Orbiter displays and controls and man machine engineering. It was during the time at which they were designing the Space Shuttle. The group she worked for designed all of the cockpit, and the displays and controls. Her major assignment in 1978 was to participate in the development of the Orbiter Heads-Up Display (HUD):

> Our job was to make sure that the 1,800 switches and circuit breakers that were in the shuttle could be reached, could be seen by astronauts while they were launching, while they were in orbit, and when they were coming back to land.

In 1980, Marsha was hired as a flight engineer for the specially modified Gulfstream II aircraft—the NASA jet trainer used by Aircraft Operations to train astronauts in the approach and landing techniques required to land the Shuttle Orbiter. She also served as a co-pilot on Gulfstream I, the NASA administrative aircraft.

She holds a multi-engine Airline Transport Pilot License with Gulfstream I type rating, as well as commercial single-engine airplane, land, sea, and glider licenses, and flight instructor ratings. She has logged over 7000 h in civilian and NASA aircraft.

Marsha applied to be an astronaut in 1978 but was rejected. She applied again in 1980 and was interviewed but again not accepted. She tried again in 1984 and was selected for Astronaut Group 10, "The Maggots."

During her 26-year career in the Astronaut Office, Marsha reviewed Orbiter safety and reliability issues and participated in the improvement of avionics and Orbiter cockpit layout. She also worked in the SAIL (Shuttle Avionics Integration Laboratory) for software verification and was CapCom (capsule communicator) in Mission Control and then "Cape Crusader," leading the Astronaut Support Personnel team in Kennedy Space Center that supported Space Shuttle launches and landings. She was the lead for crew equipment, stowage and habitability, and imagery review. Prior to her retirement in 2010, she led the Constellation Branch, supporting crew issues during the development of the Constellation Program and early commercial crew and cargo development. A veteran of five spaceflights, she has logged over 1318 h (55 days) in space.

Marsha's first experience in space was in 1990 with mission STS-32, in which she served as the ascent and entry flight engineer:

> We accept flying in planes as part of daily life. But flying in space is still a wow moment, even if you've worked in the space programme, supported other space flights, and spoken to everyone who has flown before you.

During the mission, she was primarily responsible for the LDEF's retrieval support, photographing the condition of the "Long Duration Exposure Facility" to get information vital to the development of the future space station and other spacecraft, prior to berthing it back into the payload bay. LDEF had been placed in low orbit by the STS-41C in April 1984 and had been exposed in outer space for more than 5.5 years, as its recovery had been delayed because of the accident of the *Challenger* in 1986. The crew on this mission successfully deployed a Syncom satellite.

In 1992, Marsha was flight engineer on the STS-46 mission on which Franco Malerba, the first Italian astronaut, also flew. Malerba describes her as follows in his book, *The Summit*:

> Marsha, the Flight Engineer, is like a little bird; small and speedy, perfectly at ease, she floats from one corner to another as a bird would fly from a branch to another, she can fit in the most incredibly small holes, she seems born in this environment. When she happens to untie her braid, her tiny face is soon encompassed by a halo of long, thick flowing hair due to the lack of gravity.

This sense of lightness and freedom also surfaces in a recent interview in which Marsha says:

> Space takes me away from the individual problems and stresses. You are disconnected from what's going on in the planet. I don't have to look at the phone and worry about things like have I left the tap running. When you look at the earth from the space, you don't see any natural borders or barriers between countries and continents ···. Some of the best days of my vacations were in space.

The first experiment with the Italian Tethered Satellite System (TSS, designed by Alenia Spazio in Turin, Italy), the strangest Shuttle experiment of all time, was performed during this mission—an unusual experiment that Marsha described as "weird science." The experiment unfortunately failed, as it refused to unroll more than about 256 m, instead of its planned 12 miles. NASA later identified a fault in the release mechanism provided by its partner Martin Marietta and since, in NASA, there was a strong interest in the results of the experiment, a re-flight under TSS-1R (STS-75 in January 1996) was scheduled. The mission also deployed the ESA-EURECA (EUropean REtrievable CArrier), the first European reusable satellite and dedicated microgravity free-flyer that—also built in Italy by Alenia Spazio—would be retrieved by a later mission after 11 months.

During the STS-62 mission in 1994, Marsha was responsible for the Dexterous End Effector (DEE), a new and improved end effecter of the Shuttle robotic arm. It was her longest flight, which lasted for 14 days. She also contributed to several experiments in the USMP-2 "Microgravity Experiments Package" and, as in the previous flights, was responsible for all photographic and TV set equipment procedures and objectives in the mission.

In 1997, Marsha participated in the fifth of nine planned Shuttle–Mir missions (STS-81) which was mainly devoted to logistics transfer and—using the Spacehab "double module," built in Italy by Alenia Space for Spacehab Inc. Austin, Texas, now Astrotech Corporation—delivered 2710 kg (5970 lb) of supplies, including food, water, scientific materials, and space parts, to Mir. Marsha had primary responsibilities for the hand-held laser used for distance measurement during rendezvous and docking operations with Mir and was tasked with photographing the separated external tanks. She also was the loadmaster, with primary responsibility for logistic and scientific transfer and supervising cargo movements.

Marsha's last mission (STS-98, February 2001) was an assembly flight of the International Space Station (ISS). During that mission, as the primary remote manipulator system (RMS) operator, she was in charge of taking the US Laboratory Destiny out of the cargo bay and integrating it on the ISS:

> This was probably the hardest thing I have ever done. And the scariest because it was a one of a kind, 1.4 billion dollar laboratory module without which there would be no Space Station science, so the pressure was on me to actually do this job.

She was also responsible for the mission's TV and photographic documentation tasks.

Marsha left NASA on December 31, 2010. That year, with the closure of the Shuttle program, the drastic decrease in the probability of flying, and no prospect of an American spacecraft for many years, many other astronauts stepped down, looking for new opportunities outside, as had happened in the early 1970s with the end of the Apollo program. Marsha said:

> I was very fortunate to be selected and assigned to flights in an era where you could actually fly every two or three years. That's never going to happen again. The guys that are in the office now realize that they'll be lucky to get one flight in the 10 or 15 years they spend in the office, maybe two. And so there's a huge level of frustration, and it's not just with astronauts, but with the NASA administrator, with the program managers. You look at the moon every night like a big giant tease out there because it's not any closer.

Marsha works now as an independent engineering consultant.

References

AA.VV. "Astronaut Marsha Ivins Leaves NASA", *collectspace.com* (January 4, 2011).
Briggs, C.S. *Women Space Pioneers*, pp. 92–95. Lerner Publications, Minneapolis (2005).
Glass, I. "Nice Work If You Can Get It", *thisamericanlife.org* (April 6, 2007).
Interview with Marsha Ivins. I Learned to Fly before I Learned to Drive. *Zoom in America*, **IX**(98) (2012), *usinfo.pl*.
Malerba, F. *The Summit*, p. 28. Tormena, Genova (1993).
Mooney, C. "What It's Like to Spend 55 days in Space", *motherjones.com* (September 20, 2013).
Official biography of Marsha Ivins, *jsc.nasa.gov* (January 2011).
Communication with the Author through e-mail in May 2016.

Paul, R. "Some of My Best Vacations Were in Space, Says Marsha Ivins", *dnaindia.com* (April 24, 2012).
Roper, C. "An Astronaut Reveals What Life in Space Is Really Like", *wired.com* (November 19, 2014).
Shayler, D.J.; Moule, I. *Women in Space – Following Valentina*, pp. 256, 261, 269, 277, 294. Springer/Praxis Publishing, Chichester, UK (2005).
Woodmansee, L.S. *Women Astronauts*, pp. 73–74. Apogee Books, Burlington, Ontario, Canada (2002).

15

Linda M. Godwin: Physics and Astronomy

Credit: NASA

U. Cavallaro, *Women Spacefarers*, Springer Praxis Books, DOI 10.1007/978-3-319-34048-7_15

Mission	*Launch*	*Return*
STS-37	April 5, 1991	April 11, 1991
STS-59	April 9, 1994	April 20, 1994
STS-76	March 22, 1996	March 31, 1996
STS-108	December 5, 2001	December 17, 2001

"Astronauts Land MU Faculty Positions" read the headlines in newspapers in 2010 when Linda Godwin decided, with her husband, astronaut Steven Nagel, to leave NASA after 30 years and return to Columbia, Missouri, to teach at the university's Department of Physics and Astronomy. Linda explains:

> It was a tough decision because we had been in Houston for a long time, but I knew it would be rewarding to come back and work with students. I hope to be able to add my efforts to those of others here at MU to interest students in math and science. I hope to use my experiences at NASA in that effort and educate students about opportunities for their futures.

Like her husband (who prematurely died in August 2014), Linda has flown four times on Space Shuttle missions (once together, before becoming a couple).

Linda was born in the hospital of Cape Girardeau, Missouri, on July 2, 1952, but her hometown is Jackson, Missouri. She graduated from Jackson High School in 1970:

> I grew up with a really big interest in math and science; I liked it. I grew up watching a lot of the coverage of the early US space program, all the way back, starting with Mercury and then through Gemini and Apollo, and of course to the Moon as main part of the Apollo program, and that fueled an interest in the science that I had already. So I was very interested in working for NASA, but I didn't see a path

available to me until I was in graduate school working on a Ph.D. in Physics, and NASA began hiring astronauts for the shuttle program, and for the first time including women.

Linda obtained a Bachelor of Science degree in Mathematics and Physics from Southeast Missouri State University in 1974. She earned a Master of Science degree from the University of Missouri in 1976 and a doctorate in Physics in 1980 from the same university with research on "low temperature solid state physics, including studies in electron tunneling and vibrational modes of absorbed molecular species on metallic substrates at liquid helium temperatures;" she published the results of her research in several scientific journals:

> When I was already two-thirds of my way through a Ph.D., NASA announced they were hiring shuttle astronauts, kind of a new category of science astronauts called Mission Specialists, and the first time they were hiring women. And I was fortunate enough to have chosen some educational routes that put me in a position where I could at least have a shot at it, so that's when I really began to think about applying.

Linda's first application for becoming an astronaut in 1978 was not successful. After trying again in 1980, she was invited to Houston for an interview and was offered a job as an engineer at the Johnson Space Center (JSC) in Houston, Texas, where she worked in payload integration (Spacelab and attached payloads) and as a flight controller and payloads officer on several Shuttle missions. Five years later, she tried again and, in June 1985, she became one of the 13 astronaut candidates selected by NASA for Group 11. Her first technical assignment was in the SAIL (Shuttle Avionics Integration Laboratory) working with flight software verification. She then coordinated mission development activities for the Inertial Upper Stage (IUS), deployable payloads, and Spacelab missions.

As Assistant to the Director of Flight Crew Operations, Linda supported several Shuttle missions. She held different positions in the Astronauts Office until, in 1993, she became, for almost 7 years, Deputy Chief of the Astronaut Office (the first woman to assume that role in the Office) and then Chief of the Astronaut Office's CapCom Branch and Assistant to the Director for Exploration at the JSC. A veteran of four spaceflights, Dr. Godwin has logged over 38 days in space, including over ten extravehicular activity (EVA) hours in two spacewalks.

During her first mission (STS-37 in 1991), Linda was mission specialist. The mission deployed the Compton Gamma Ray Observatory (GRO) to study gamma ray sources in the universe. At 17.5 tons, it was the heaviest payload ever carried by the Shuttle. It was deployed by Linda using the remote manipulator system (RMS). Linda flew in this mission with Commander Steven R. Nagel, whom she married a few years later in 1995.

In her second mission (STS-59 in 1994), Linda was payload commander in charge of the Space Radar Laboratory (SRL), which consisted of three large imaging radars operated in three frequencies and four polarizations that provided information about Earth's surface over a wide range of scales not discernible with previous single-frequency experiments. One of the goals was to map the changing global environment and to distinguish human-induced environmental changes against natural changes. This was the first of two SRL missions flown in 1994. While in training for this mission, in March 1993, she was named Deputy Chief of the Astronaut Office.

In 1996, Linda flew aboard *Atlantis* STS-76, the third docking mission to the Russian space station Mir. One of the goals of the mission was to transfer to Mir her colleague Shannon Lucid, who began a continuous presence of US astronauts aboard Mir for the next 2 years. Using the Spacehab pressurized logistic module, built in Italy by Alenia Spazio, the crew also delivered 2200 kg of science and mission hardware, food, water, and air to Mir, and repacked over 500 kg of items no longer needed on the station for return to Earth, together with US and European Space Agency (ESA) science and Russian hardware, and the "Biorack," a small multi-purpose laboratory used during this mission for research of plant and animal cellular function. Linda recalls:

> Spacelab module had a suite of experiments developed by the European Space Agency. I and Ron Sega were the operators for the experiments; we had spent time training at the facility in Florida and met many of the Principal Investigators. During the mission we operated the experiments, even while we were docked to the MIR. I enjoyed working with that program.

An important assignment for Linda was also the first flight of Kidsat, the program initiated by Sally Ride—at that time professor at the UCSD (University of California, San Diego)—to allow middle-school students to use a digital camera aboard the Space Shuttle to capture photographs of specific places on Earth for science and education. In her third mission, Linda not only flew in, but also stepped out of, the Space Shuttle, becoming the fourth woman to walk in space when she performed a 6-h EVA to mount, together with her colleague Richard Clifford, experiment packages on the Mir docking module to detect and record contamination in the space environment around the station over the next 18 months. Describing her spacewalk, Linda says:

> When you have to make yourself pause and you actually look around, you can see some part of the space station with Earth in the background. You can see that from inside the station or the shuttle, but to be out there in your own space suit and looking at it, it's pretty awesome.

This was a Shuttle-based EVA, since the docking module of the Russian segment for the EVA purposes fell under the jurisdiction of the Russian cosmonauts. And it was the first American EVA while docked to an orbiting space station since the end of the Skylab program in 1974.

With *Endeavour* STS-108, officially nicknamed International Space Station (ISS) UF-1 "Utilization Flight-1," 5 years after her previous space mission, Linda visited the ISS. She was one of the few women astronauts who visited both Mir and the ISS. She recalls:

> They were both quite interesting to visit – of course! Mir felt older. The ISS is larger, even when I visited, and more familiar as we spent more time in the ISS similar. Both were fantastic experiences to rendezvous with an outpost in Earth orbit and get to stay a few days.

One of the goals of the mission was to deliver the Expedition-4 crew and return to Earth the Expedition-3 crew. The mission used the Raffaello Multi-Purpose Logistics Module (MPLM), built in Turin, Italy, by Alenia Spazio, for the second time and unloaded over 2700 kg of supplies, logistics, and science experiments. Linda was primary RMS operator for the mission, responsible for the transfer of supplies and equipment to the station. Using the Shuttle's robotic arm, she picked up the Raffaello MPLM from the shuttle cargo bay and attached it to the ISS for unloading, supported by Mark Kelly, who finally, with the help of Linda, relocated it into the Shuttle payload bay for return to Earth, once filled with the experiment results and unwanted material. She was also loadmaster in charge of keeping track of the unloading and reloading for return. During this mission, she also performed her second 4-h, 12-min session of extravehicular activity (EVA). The main objective of the spacewalk was to install thermal blankets on mechanisms that rotate the ISS's main solar arrays. She explained in an interview just before that flight:

> UF-1 is still kind of the focal point for saying this is now the transition time between enough major construction that we have a self-sustaining, self-sufficient station up there, and so we're kind of ready to bridge on to the next stage of using it for the purpose of why we built it. So we are taking up science, we are taking up a lot of supplies. And so not only are our astronaut crews up there on the ISS maintainers and builders of station, now they've become researchers as well.

Before undocking from the station on December 16, the team deployed the satellite STARSHINE-2 (Student Tracked Atmospheric Research Satellite Heuristic International Networking Experiment). "It's a small satellite," Linda explains, "but it's made up of many mirrors that have been worked on literally by thousands of students who have polished these mirrors to a very perfect, smooth surface, and then they've been assembled on this satellite." This was the second of three small, optically reflective spherical satellites, constructed largely from spare flight hardware and fitted with almost 900 small aluminum mirrors that had been machined by technology students in Utah, and eventually polished by 25,000 students of 660 schools in 26 countries around the world. Linda explains:

> It stays in orbit for a period of time and because it's so reflective, and it also has a very slow spin on it so it kind of seems to shine as it rotates, the students on the ground can see it, and they'll be able to track it. And they can do their measurements, make calculations on where it is and how high it is and how its orbit is changing, so it gives them a lot of experience in using mathematics to look at orbital.

In her last few years at NASA, she was the representative of the Astronaut Office in the Constellation program—"the program," she says, "which had a suite of vehicles: a new rocket, a crew module, and potentially a lunar lander although that was never very developed." Her tasks involved following the requirements, the budget, and the design.

Linda left NASA, as mentioned, in 2010 and is now a professor of Physics and Astronomy at the University of Missouri.

References

Anon. "Astronauts Deploy Satellite to Be Tracked by Students", *spaceflightnow.com* (December 16, 2011).

Lindsey, L. "Two Former Astronauts Make MU Their Home", *Missouri.edu* (August 22, 2011).

Official biography of Linda M. Godwin, *jsc.nasa.gov/Bios* (August 2010).

Personal communication of the Author with Linda Godwin in April and May 2016.

Pojmann, K. "To Mizzou and Beyond", *mizzouwire.missouri.edu* (October 13, 2011).

"Preflight Interview: Linda Godwin", STS-108, *spaceflight.nasa.gov* (July 4, 2002).

Shayler, D.J.; Moule, I. *Women in Space – Following Valentina*, pp. 257, 269, 275–276, 295. Springer/Praxis Publishing, Chichester, UK (2005).

Woodmansee, L.S. *Women Astronauts*, pp. 75–76. Apogee Books, Burlington, Ontario, Canada (2002).

16

Helen Sharman: The First Briton in Space

Credit: ESA

U. Cavallaro, *Women Spacefarers*, Springer Praxis Books, DOI 10.1007/978-3-319-34048-7_16

Launch		*Return*	
Soyuz TM -12	May 18, 1991	Soyuz TM -11	May 26, 1991

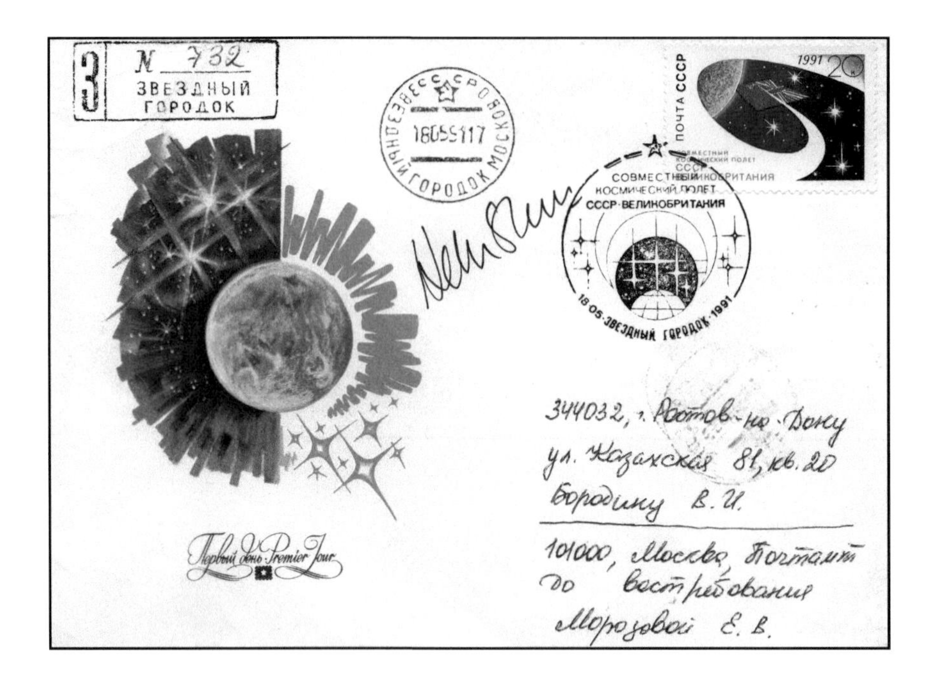

Commemorative cover of mission Soyus TM-12, signed by Helen Sharman (from the collection of Umberto Cavallaro)

Helen Sharman was the first Briton to fly in space, and the first European, first non-Russian, and first non-American woman. She was also the first to break the gender barrier on the Soviet Mir space station and the first non-Soviet woman to visit a space station ever, as well as the last foreign visitor of the Soviet era, since the Soviet Union disintegrated shortly after her return to Earth.

For the record, it must be added that she made the UK the first of only three nations up to that date to have had a woman as its first astronaut. (Helen has since been joined by Iran's Anousheh Ansari in 2006 and by South Korea's Yi So-yeon in 2008.)

Recently, feminists flared up when Tim Peake (first British man to travel into space as a European astronaut as well as the first to live aboard the International Space Station (ISS)) was mistakenly called "the first Briton in space."

Helen doesn't like to be classified as "the first space tourist": "I was there on a commercial mission, I was paid to do it as a job, doing experiments on someone else's programme."

Helen Patricia "Lenochka" Sharman was born in a hospital in Sheffield, England, on May 30, 1963, the daughter of a college lecturer father and a nurse mother, and she lived with her family in Grenoside, a suburb of the industrial town of Sheffield, where she received much of her early education. In her autobiography, *Seize the Moment*, she described her upbringing as "decidedly unremarkable." At school, she was good at French and German, and loved science. She also played the piano.

Helen graduated in Chemistry from the University of Sheffield in 1984. She was hired as an engineer for the General Electric Company Plc in London. "Even though the salary on offer was the lowest," she says in her autobiography, "the work they were offering was varied" and her role encompassed solving production problems, organizing schedules, and doing research on and development of cathode ray tube components. Her employer eventually offered her the opportunity to further her studies with a part-time Ph.D. to explore the luminescence of rare-Earth ions in crystals and, in 1985, Helen enrolled at Birkbeck College in London. In 1987, while still studying at Birkbeck, she began working as a research technologist for MARS—an American global manufacturer of confectionery and other food products that was planning new ice-cream products:

> "The company wanted to make a new and delicious ice cream and I was to be part of the team that scaled up production from a few bars in the laboratory to tons of ice cream every day in the factory."

Once she had completed this project, Helen worked in the Chocolate Department, "investigating the properties of chocolate and using different ingredients and machinery to make chocolate more quickly, more cheaply, and with the same flavor." She found the work fascinating, not least for the fact that it enabled her to incorporate many facets of everyday life into her work.

One evening in June 1989, the 26-year-old Helen, while sitting in a traffic jam heading back from MARS in Slough, Berkshire, to her flat in Surbiton, south-west London, casually heard a radio announcement: "Astronaut wanted. No experience necessary." In her autobiography, she describes in hindsight the events of that evening as "the crucial, pivotal moment in my life." A career path that she had never considered possible was fairly straightforward.

The advertisement was looking for astronaut volunteers under the project "Juno," a private British space program, since a national space program did not exist. No experience was required. The only requirements were British citizenship, age between 21 and 40 years, formal scientific training, proven ability to learn a foreign language, and a good standard of physical fitness.

In fact, the plan of Project Juno was to send a British astronaut to the Mir space station, under a cooperative arrangement between the Soviet Union (which was seeking to cement relations with Britain as part of President Mikhail Gorbachev's policy of *glasnost* and to raise hard currency) and a consortium of private British sponsors led by Antequera, Ltd. Soviet promotional operations had already succeeded in selling a ticket for space to Japan, Cuba, Austria, Syria, Afghanistan, Mongolia, and many more.

Although the UK had become the third country in the world (after the USSR and the US) to launch a satellite when, on April 26, 1962, it put into orbit Ariel 1, constructed in the US by NASA Goddard Space Flight Center on behalf of the British Empire, and launched by NASA from Cape Canaveral, its aspirations in space exploration almost exclusively focused upon unmanned research, with particular emphasis upon Earth observation, and there has not been government policy to create a British astronaut corps. Since the 1980s, despite the setting-up of the British National Space Centre (BNSC), the Conservative government of Margaret Thatcher virtually gutted any chance of a human space program. Four military Britons were training with NASA before the *Challenger* accident. By the time NASA was flying Shuttles again, Britain had decided not to participate.

Helen recounts:

> "Immediately that small dream that I had had when I was younger, ... that maybe—just maybe—space travel was possible. Suddenly that was there in front of me. What would it be like to sit on top on a rocket and wait for that launch? What would it be like for me, as a chemist, to be able to grow crystals, crystals that you cannot grow on Earth? But probably best of all, what was it going to be like floating about feeling weightless, and for those reasons, very selfish reasons, when I got home that night I applied for that job of astronaut."

After applying by telephone and a short interview, Helen received a long application form which she completed and posted, becoming one of over 13,000 applicants for that job. Rigorous physiological and medical tests followed, after which only 22 candidates made the cut through the selection process; and three of these were women. A final medical review produced the final four, who were then "sequestered" for 2 weeks in a Russian hospital, where medical tests were performed. Helen and her backup, British Army Major Timothy Mace, then underwent 18 months of grueling training in Soviet Russia at the Juri Gagarin Center in Star City, 30 kilometers north-east of Moscow, where Soviet cosmonauts used to train, enduring challenges such as the centrifuge (to experience G-forces) and hydro tank (for spacewalk training). Helen was measured in 54 different places to ensure the suit fitted perfectly. "That's the only suit I have ever had that was made to measure," Helen jokes.

Once it was announced that she could fly, the press started to call her the "Girl of Mars." Although quite tiresome, the sobriquet was thought out as a means of helping to sell the project to the public at large. At the time, Soviets requested US$12 million (£7.5 million) to cover the expenses of the Soyuz taxi and 1 week on the space station. The government of Margaret Thatcher, who had ended Britain's manned space program in 1986, refused to contribute towards the cost and made it clear that the funding would have to be raised by private finance. Antequera, Ltd—headed by Gregory Pattie, a member of the British Parliament and a former space minister—was responsible for the selection and started to raise the funds needed: US$12 million to pay Energia for the fly to Mir and another US$8 million to cover the expenses of 18 months of training. However, only one of the 500 British companies connected with the Aerospace industry bought a share: British Aerospace. Memorex and Interflora eventually joined the pool of corporate sponsors and ITV bought the television rights, but the consortium as a whole failed to raise the entire sum. At one point, rumors spread that the British astronaut was selected by lottery. What actually happened was that, given the difficulty in finding a commercial sponsor, a lottery was invented to try (indeed without much success) to raise funds: only a paltry US$1.7 million was generated. Helen and Mace had been training in Star City for over a year when they learned that their contact in London had left, as Antequera had not been able to raise sufficient funds. It looked as if the enterprise would be cancelled. Indeed, there were many problems.

Besides financial problems, there was the Soviet reluctance that had started to surface since Antequera had announced that the two finalists were Helen Sharman and Timothy Mace, the military pilot and a member of the national team of British skydivers. It seemed clear to almost everyone that some "technical" Soviet circles were not very happy to send to Mir a woman—perhaps out of fear that their own female cosmonauts who had been trained, but had never flown and were grounded in Star City, would complain. Reportedly, Mikhail Gorbarchev personally stepped in to save the day in the interests of international co-operation. The Soviets arranged for the Moscow Narodny Bank—a British commercial bank subsidized by the Soviet state—to take over the program.

On May 18, 1991, at the age of 27, the cosmonaut researcher Helen Sharman, accompanied to the launch pad by Alexei Leonov—then deputy Director of the Cosmonaut Training Center in charge of crew training, who offered her a number of "unofficial" items to carry aboard Mir—lifted off aboard Soyuz TM-12, thus gaining a place in both space and history. She lifted off from Baikonur—from which, 30 years earlier, Gagarin and Tereskova had left—accompanied by two Soviet cosmonauts: the commander Anatoly Artsebarsky and flight engineer Sergey Krikalev.

When it was confirmed that Helen would fly, Artsebarsky—who had counted on flying with the famous British pilot, Tim Mace—was unable to hide his disappointment, although he later declared that Helen's performance during the mission was impeccable.

At their arrival to Mir, Helen was honored by Viktor Afanayev and Musa Manarov, who invited her to enter the station first and greeted her with the traditional bread and salt that Russian used to welcome a new guest, and then offered her a bedroom to herself.

With the demise of much British commercial participation in the mission, the British element of the Project Juno mission was Sharman herself. "The mission was purely commercial," Helen explained, "but having said that the funding was not readily available so a deal was done with the Soviets that I would do their experiments in return for a seat on their flight." Rather than being a representative of a Soviet–capitalist partnership, as anticipated, she became a guest of the Soviets and her experiment program was almost exclusively designed by the Soviets, with a primary emphasis upon the life sciences: she operated 17 Soviet biotechnological, medical, and technical experiments. She also wore electrodes to track her heart rate, monitored her mental co-ordination and reaction speed, and took blood samples from the tips of her fingers 12 times a day. Additionally, Sharman took air samples throughout Mir to assess the prevalence of dust in the station: dust and crumbs floated everywhere, causing frequent sneezing and also disturbing sleep. She brought back paper filters with samples to be studied on Earth. Helen also had carried on board half of the 125,000 pansy seeds given to her by Suttons, a British company providing quality seeds, and placed them in the Kvant-2 airlock, the portion of the station least shielded from cosmic radiation. She then brought them back to Earth and distributed a sample of seeds flown in space and one of those left behind to British schoolchildren, in order for them to sprout the seeds and check whether the space exposure had affected the growth of the seeds.

From Mir, Helen also participated in a televised advert for Interflora by "ordering" flowers for delivery to her mother: the first order ever received from outer space! "Just imagine how Helen's mum must have felt when we delivered them into her arms!" advertised Interflora on its website.

Finally, a significant portion of Helen's time was devoted to photography of the UK and watching at the Blue Planet from space. She told *The Independent*:

> "It's something no astronaut ever gets tired of doing. You get this constantly changing image of the Earth spinning below. You get these fabulous views and you get time to think about that."

After eight days on Mir, Helen returned to Earth in the Soyuz TM-11 descent module, with fellow cosmonauts Viktor Afanayev and Musa Manarov on May 26, 1991. The "soft" landing of Soyuz took place in a very windy day that caused the capsule to laterally collide during the impact with the ground and to roll several times before stopping. In the end, everything worked out and Helen only reported a few bruises to her face due to an impact with the built-in microphone in the helmet.

Following her flight in 1991, Helen became a TV presenter and a public speaker on space and science, hoping to inspire a new generation of men and women into joining the fields of science. A brilliant ambassador for science and keen advocate of human space travel, she pointed out that Britain was alone among the major industrialized nations in not having a human space program:

> "We need to be pushing our human boundaries. We were a sea-faring nation and that exploration made us the country that we have become. The fact that we've stopped human space exploration has become a real problem. It has stopped us from being proud of being part of this international community."

Many people may think that Britain has a manned program because there have been a handful of astronauts over that past three decades with joint US citizenship who have flown on the Shuttle. However, they have all done it under the American flag:

> "I am sure this has led the public to believe that we as a country have had quite a long interest in human space flight as a nation, and we haven't. I think that is something that the Government has been very happy to allow us to continue to think."

Helen never returned to space, but was a shortlisted candidate for the European Space Agency (ESA) in 1992, when she applied with other two British candidates. She unsuccessfully tried again in 1998. She eventually reconnected with science as Group Leader of the Surface and Nanoanalysis Group at the National Physical Laboratory. In 2015, she joined Imperial College London as Operations Manager in the Chemistry Department.

References

Cavallaro, U. "Helen Sharman 25 years ago, the first Briton in space", *Ad*Astra, ASITAF Quarterly Journal*, 28 (June 2016), pp. 31–34.

Connor, S. "Don't Forget to Look out of the Window", *independent.co.uk* (May 27, 2013).

Evans, B. "'No Experience Necessary': The Story of Project Juno", *americaspace.com* (May 2013).

Gibson, K.B. *Women in Space: 23 Stories of First Flights, Scientific Missions and Gravity-Breaking Adventures*, p. 167. Chicago Review Press, Inc., Chicago (2014).

Hamill, J. "Sexists Have Forgotten Helen Sharman Was First British Person in Space, Angry Feminists Claim", *mirror.co.uk* (December 17, 2015). "Interflora Reaches Out", *interflora.co.uk* (1991).

"Helen Sharman and Project Juno", *citizensinspace.org* (January 27, 2013).

Kevles, T.H. *Almost Heaven: The Story of Women in Space*, pp. 138–144. The MIT Press, Cambridge, MA, and London, UK (2006).

"Moon over Helen Sharman", *squeamishbikini.com* (March 15, 2012). Personal contacts by e-mail in May 2016.

Personal communication with the Author by email in April/May 2016.

Shayler, D.; Hall, R. *Soyuz: A Universal Spacecraft*, p. 341. Springer, London (2003).

Shayler, D.J.; Moule, I. *Women in Space – Following Valentina*, pp. 198, 314–318. Springer/Praxis Publishing, Chichester, UK (2005).

Woodmansee, L.S. *Women Astronauts*, pp. 76–77. Apogee Books, Burlington, Ontario, Canada (2002).

17

Tamara Jernigan: An Astrophysicist Out Among the Stars

Credit: NASA

U. Cavallaro, *Women Spacefarers*, Springer Praxis Books, DOI 10.1007/978-3-319-34048-7_17

Mission	Launch	Return
STS-40	June 5, 1991	June 14, 1991
STS-52	October 22, 1992	November 1, 1992
STS-67	March 2, 1995	March 18, 1995
STS-80	November 19, 1996	December 7, 1996
STS-96	May 27, 1999	June 6, 1999

Tamara "Tammy" Elizabeth Jernigan was born in Chattanooga, Tennessee, on May 7, 1959. She is one of the six women spacefarers with five Shuttle flights to their credit. She graduated in 1977 from the Santa Fe High School, in Santa Fe, California, where she was also awarded a trophy as the best female athlete in Santa Fe Springs' Lakeview School and was volleyball "Player of the Year" in 1977. She enjoys volleyball, racquetball, softball, and flying, and she has been a successful student and athlete as far back as she can remember. As with all the kids, she was fascinated by the Moon landing:

> "I think my first memory of aspiring to the astronaut program was when I was a kid and we put the first man on the moon. I remember going out the front door and wanting to look at the moon knowing that there were humans on the moon."

But, at that time, she didn't think seriously at it:

> "I think my more serious interest came when I was a sophomore at Stanford. I was a physics major. In 1978, they started taking astronauts from the scientific community in earnest and also taking women. I think it was in 1978 that I thought I might have a chance to be selected."

College-acceptance letters arrived from Princeton, the Air Force Academy, Stanford, and University of California-Berkeley. After a semester at Princeton, Tamara transferred to Stanford: "Stanford had excellent physics and athletic programs, especially volleyball." After receiving a Bachelor of Science degree in Physics (with honors) from Stanford University, in June 1981, she joined, as a research scientist, the Theoretical Studies Branch at NASA Ames Research Center, where she contributed to the study of bipolar outflows in regions of star formation, gamma ray busters, and shock-wave phenomena in the interstellar medium.

In 1983, Tamara earned a Master of Science degree in Engineering Science at Stanford University and, in 1985, a Master of Science degree in Astronomy from the University of

California-Berkeley. The same year, she was selected from among 792 women and 4,142 men and entered as mission specialist in Group 11 of NASA Astronaut Corps, thus becoming, at only 26, the youngest NASA astronaut. Her first assignment was software verification in the SAIL (Shuttle Avionics Integration Laboratory); she was then in charge of operations coordination on secondary payloads and spacecraft communicator (CapCom) in Mission Control, Houston, for five Shuttle missions: STS-30, STS-28, STS-34, STS-33, and STS-32. In 1988, she was awarded a doctorate in Space Physics and Astronomy from the Rice University. Over the years, she held different positions in the Astronaut Office, as Chief of the Mission Development Branch and then Deputy Chief of the Astronaut Office.

A veteran of five Shuttle missions, Tamara logged 1512 h (approximately 62 days) in space and one extravehicular activity (EVA) of 7 h and 55 min. She was a mission specialist on Space Shuttle *Columbia* STS-40 (June 5–14, 1991), where she flew with the two colleagues Millie Hughes-Fulford and Margaret Rhea Seddon. This mission took into orbit Spacelab for the first time since October 1985 and, for the first time, the research was entirely dedicated to life science with SLS-1 (Spacelab Life Sciences). During the 9-day flight, crewmembers performed experiments which explored how humans, animals, and cells respond to microgravity and readapt to Earth's gravity on return. SLS-1 performed the most detailed and interrelated physiological measurements since the Skylab, two decades before. It investigated materials science, plant biology, and cosmic radiation. In the small amount of spare time, she enjoyed watching Earth: "The stars provide a beautiful view but the most beautiful of all is the Earth rim and going over the various continents and over the vast oceans and enjoying the beauty of our own planet."

Even during her following mission, the 10-day flight of the Space Shuttle *Columbia* STS-52 mission, Tamara flew as mission specialist primarily responsible for the deployment of the Italian satellite LAGEOS-II (Laser Geodynamic Satellite) which—launched from the Shuttle, using the innovative IRIS launcher (Italian Research Interim Stage) built in Turin, Italy by Alenia Spazio—would be used to measure movement of Earth's crust. Tammy was the first female professional astronomer to fly in space. The mission also carried into space US Microgravity Payload 1 (USMP-1). During this mission, research for the future implementation of the *Freedom* space station, that later evolved into the International Space Station (ISS), were also conducted and—using a small target assembly which was released from the remote manipulator system—the Space Vision System (SVS), developed by the Canadian Space Agency, was tested. This mission demonstrated the Shuttle's flexibility in being a satellite launcher, a space laboratory, and a technology test bed, all in the same mission.

During the record-setting 16-day Space Shuttle *Endeavour* STS-67 mission, Tamara, as astronomer, was the Payload Commander, primarily responsible for Astro-2, the second flight of the Astro observatory—a unique complement of three telescopes that was aimed at exploring 23 different science programs. Working intensively in three shifts, the crew conducted observations around the clock to study the far ultraviolet spectra of faint astronomical objects and the polarization of ultraviolet light coming from hot stars and distant galaxies. "We want to bring home a wealth of information on the universe we all live in," Tamara said. On the recommendation of the astronomers who, just before the launch, had discovered the spectacular explosion of a binary star system, the crew, without the shield of the atmosphere, could perform ultraviolet observations of the event still in progress, and also could photograph the effects of a recent eruption that had taken place in the previous days on the surface of Io, one of Jupiter's moons.

This Shuttle flight is also remembered because, for the first time, 13 astronauts and cosmonauts were in the space at the same time: in addition to the crew of the Shuttle, in those days, the three occupants of Mir and the three crew of Soyuz TM-21, which had visited Mir with the new crew of the Expedition MIR-18, also were flying in Earth orbit.

Tamara was mission specialist even during the mission *Columbia* STS-80, which successfully deployed and retrieved the Wake Shield Facility (WSF) and the Orbiting Retrievable Far and Extreme Ultraviolet Spectrometer (ORFEUS) satellites. The free-flying WSF created a super vacuum in its wake and grew thin film wafers for use in semiconductors and other high-tech electrical components. The ORFEUS instruments, mounted on the reusable Shuttle Pallet Satellite, studied the origin and makeup of stars. Her two planned spacewalks were cancelled in flight due to a jammed hatch on the airlock preventing access to open space:

> "It was frustrating to have done all the preparation for the two space walks and not be able to execute them in flight. The crew wasn't the only group of folks who were disappointed because there had been many people, many engineers and technicians on the ground who had worked very hard to prepare that hardware for flight, so the whole community was pretty disappointed."

This was a record-breaking mission. Launched on November 19, 1996, it was scheduled to return on December 5. Weather conditions, however, forced a delay in the return by 2 days.

Tamara performed her first spacewalk during her next mission, Space Shuttle *Discovery* STS-96 (May 27 to June 6, 1999), where she flew with Ellen Ochoa and Julie Payette and was mission specialist. This was a 10-day mission during which the crew performed the first docking to the ISS, then consisting of only two modules: the Russian Zarya and the American module Unity. During the rendezvous, Tammy was responsible, with Ochoa, for looking at the program RPOP that gave information about the trajectory and controlled approaching and docking to the station, for opening the ISS and for carrying out pressure checks and air sampling before entering. She was then in charge of delivering and storing in the Zarya module (or "Functional Cargo Block") four tons of logistics and supplies in preparation for the arrival of the first crew to live on the station at the beginning of the next year. The cargo was contained in the pressurized module Spacehab, built in Italy. To install some devices outside the ISS, Tamara finally participated, with Daniel Barry, in her first EVA, which lasted for 7 h and 55 min. This was a dream come true, anticipated in what was perhaps her first interview ever, released 20 years before to the *Los Angeles Times* when, at the beginning of her career, she said: "In 20 years, I think I'll be part of a space station." She was supported by Ellen Ochoa, who was inside the Shuttle, operating *Discovery*'s robot arm, and Canada's Julie Payette, who coordinated the effort from the aft flight deck:

> "We're taking up two cranes—she told before the mission—We have a U.S. crane and we also have a Russian crane that we're going to take up with us and install externally during the space walk on the International Space Station. We're using the arm to transport large pieces of structure from the Space Shuttle to the International Space Station and, in some instances, operating at the upper limit of the reach of the arm."

Before returning to Earth on June 5, *Discovery* deployed the small STARSHINE-1 (Student Tracked Atmospheric Research Satellite Heuristic International Networking Experiment) that would be the first of three small, optically reflective spherical satellites, fitted with almost 900 small aluminum mirrors that had been machined by technology students in Utah and eventually polished by 25,000 students of 660 schools in 26 countries around the world. The US co-operative program was an experiment sponsored by NRL (United States Naval Research Laboratory). The students could track the spacecraft they had helped to "build" and use it to measure upper atmospheric density and the response of that region of the atmosphere to solar storms.

Married to the astronaut Peter J.K. "Jeff" Wisoff in 1999, Tamara left NASA in 2001 and served until 2007 as Deputy Director at the Lawrence Livermore National Security, California, dealing with advanced technologies, management of strategic human resources, and workforce planning.

References

Briggs, C.S. *Women Space Pioneers*, pp. 95–98. Lerner Publications, Minneapolis (2005).

Dreyfuss, J. "Two Adventures in Outer Space Are Heading for the Launching Pad: All Systems Go for Tamara Jernigan, Nation's Youngest Astronaut Candidate", *latimes.com* (July 1, 1985).

Official NASA biography of Tamara Jernigan, *jsc.nasa.gov* (November 2001).
"Preflight Interview: Tamara Jernigan", STS-96, *spaceflight.nasa.gov* (July 4, 2002).
Shayler, D.J.; Moule, I. *Women in Space – Following Valentina*, pp. 257–258, 263, 272, 276, 290–291. Springer/Praxis Publishing, Chichester, UK (2005).
Woodmansee, L.S. *Women Astronauts*, pp. 79–80. Apogee Books, Burlington, Ontario, Canada (2002).

18

Millie Hughes-Fulford: The First Female Payload Specialist

Credit: NASA

U. Cavallaro, *Women Spacefarers*, Springer Praxis Books, DOI 10.1007/978-3-319-34048-7_18

Mission	*Launch*	*Return*
STS-40	June 5, 1991	June 14, 1991

Biochemist and molecular biologist, Millie Hughes-Fulford made history in 1991 when she became the first American woman astronaut, leaving aside Christa McAuliffe, to fly on the Space Shuttle as a working expert, or payload specialist. "You have the shuttle pilots," she explains in an interview, "who get you there and they get you home. Then you have mission specialists, who are career astronauts who train for a specific mission. And then you have payload specialists who know the topic."

Millie Elizabeth Hughes-Fulford was born on December 21, 1945, in Mineral Wells, a small city 80 miles west of Dallas, Texas, where she grew up. She wanted to be an astronaut since the age of five. She explained in one interview:

> "I watched Buck Roger on channel 8, out of Dallas that we could get in Mineral Wells. And their pilot was a woman named Wilma Deering. I wanted to be Wilma Deering because she could wear pants. At that time a little girl could not go around in pants. She flew a spaceship and was a professional woman. I watched as many sci-fi programs I could see and fell in love with space."

At age 11, she helped her father, who was crippled with arthritis, in his grocery store alongside a male cousin. And she recalls that she climbed ladders and carried cartoons of Great Giant peas:

> "No one told me I couldn't do it. I was changing oil in the car. I was doing everything the boys were doing. I didn't see that there were limits. It wasn't until I was sixteen that it occurred to me that perhaps women were not treated the same. And I realized that all the real astronauts were men and were all fighter pilots, and women couldn't be fighter pilots. So I decided that I just would be a government scientist. Scientists always save people and they were good and honorable."

But, as she explained later, "it ended up coming full circle, and my science got me into space."

In 1962, at age 16, Millie graduated from Mineral Wells High School and received her Bachelor of Science degree in Chemistry and Biology from Tarleton State University in 1968 and her Ph.D. in Medicine from Texas Woman's University in 1972, specializing in plasma chemistry. She then entered the US Army Reserve Medical Corps, studying cancer growth regulation and contributing to over 120 scientific papers.

In 1976, Millie applied as a NASA astronaut, but didn't pass the penultimate cut and wasn't among the 35 astronauts selected in 1978. She succeeded 5 years later, however, and, in 1983, recently married, she took leave from her job in San Francisco and—with her daughter and her husband, Captain George Fulford, pilot for United Airlines—moved to Houston to start her training at the Astronaut Office as payload specialist. She was the first woman to hold that position. Payload specialists are scientists who are experts on a particular mission's experiments or "payload." Although they undergo rigorous training, payload specialists aren't career astronauts who expect to make space travel their life's work. Instead, their goal is to carry out a group of experiments for other scientists who remain on Earth.

Millie expected to stay in Houston for 2 years, but her mission, Spacelab-4, the first Spacelab mission entirely devoted to biomedical investigations and originally scheduled for launch in late 1985, was delayed several times: first because constructing cages for the rats, jellyfish, and test animals turned out to be more difficult than expected, then because of the *Challenger* tragedy, and finally because of other organizational problems with the complex mission. In 1987, her family decided to return to California.

The Spacelab-4 mission was finally split into two "Spacelab Life Sciences" missions: SLS-1 and SLS-2, and Millie flew in 1991 on *Columbia* STS-40. At the beginning, she experienced space motion sickness—a discomfort suffered by 80 % of astronauts, although few admit to it. She recovered within few hours, however, and, in only 9 days, under the direction of Professor Hughes-Fulford, this long-planned mission completed 18 experiments and brought back more medical data than any previous NASA mission: medical experiments that used the caged animals and fish as well as, mostly, the crew, herself included: "Normally," she explained, "if there is any medical part, the crew is the guinea pig. Whom else are you going to test? All the medical data from the flight come from the crew." It would take several years to analyze all the data that they collected on the human skeleton, heart, and calcium loss, and compare with baseline pre- and post-flight-test data collected on Earth. There were three women on this flight, including Rhea Seddon, the surgeon who had already flown in 1984, and Tammy Jernigan, the rookie astrophysicist who had joined in the class of 1985.

One of the experiments that Millie carried out during her 9 days on board Space Shuttle *Columbia* had been designed by Switzerland-based researcher Augusto Cogoli to examine how the immune system fares in space. The weakening of the immune system in microgravity had been first observed in returning Apollo astronauts. "During the Apollo era," Millie says, "over 15 of the 29 astronauts had an infection either during flight or during the first week of return." Up until then, scientists had believed that perhaps hormones were to blame for space travelers' weakened immune systems. Cogoli's experiment showed that the lack of gravity in space affected the immune system: zero gravity, not hormones, was to blame: "Our research," Millie says, "on the one hand demonstrates that permutations in the space take place with a lower frequency; on the other hand, it confirms that the efficiency of the immune system also depends on the gravity."

After completing her mission, Millie served for 3 years as Scientific Advisor to the Under Secretary of the Department of Veterans Affairs, and now she continues her research at the University of California Medical Center at San Francisco. She cooperates in preparing biomedical experiments to investigate the root causes of osteoporosis that occur in astronauts during spaceflight: "One of our goals is to use microgravity as a novel model system of aging to investigate the molecular mechanisms of immune suppression commonly seen in the elderly population."

Millie retired from the US Army Medical Corps in 1995. In 1996–1997, she sent several experiments into space. She was indeed the Principal Investigator (PI) on a series of SpaceHab/Biorack experiments, which examined the regulation in the growth of osteoblasts (bone cells). These experiments flew on three different missions: STS-76 in March 1996, STS-81 in January 1997, and STS-84 in May 1997. These studies examined the root causes of the osteoporosis that occurs in astronauts during spaceflight. She found changes in anabolic signal transduction in microgravity. Eventually, in collaboration with Dr. Augusto Cogoli of Zurich, Switzerland, she prepared an experiment that was lost in the STS-107 disaster. An ESA/NASA joint experiment to examine changes in T-cell gene induction in spaceflight was sent to the International Space Station (ISS) via the Soyuz TMA-9 in September 2006.

Now, the studies of Professor Hughes-Fulford have verified that T-cell activation is affected by zero gravity or, in other words, that microgravity as such is at the origin of the diminished T-cell activation. With no gravity, T-cells—specialized immune-system cells—recognize infections within the body and initiate a defensive response. Several key changes in the immune systems of the elderly, such as diminished immune-system function and impaired T-cell activation, are the same as those found in astronauts returning from spaceflight. Millie explains:

> "It is difficult to study the genetic and molecular changes associated with aging-related immune suppression because the condition develops over decades, and the elderly often have illnesses that can complicate research studies. Instead, changes in the immune system, including T-cell behavior, quickly occur in space."

The results of this research will hopefully help to better understand the biochemical mechanisms that underlie diminished T-cell activation and develop treatments for a range of auto-immune diseases ranging from suppression to hyper-reaction of the immune-system function that are the origin of disorders such as arthritis, thyroiditis, and diabetes for the general population.

Research continues even now that the Shuttle is no longer available. Millie's most recent experiment to study T-cell activation in aging has been sent to the ISS aboard SpaceX-3 launched on April 18, 2014 (Fig. 18.1). It was the first to go up on a privately operated rocket. The second experiment is a two-part investigation launched on January 10, 2015.

Fig. 18.1 Patch from the "T-cells Activation in Aging" experiment, the first to go up to the International Space Station (ISS) on a privately operated rocket. Credit: NASA

References

Anon., "Our T-Cells Came Back from Space!", *hughesfulfordlab.com.*

Anon., "T-Cell Activation in Aging", *hughesfulfordlab.com.*

Figliozzi, G. "T-Cell Activation in Aging, (SpaceX-3)", *nasa.gov* (April 21, 2014).Official NASA biography of Millie Hughes-Fulford, *jsc.nasa.gov* (March 2014).

Kevles, T.H. *Almost Heaven: The Story of Women in Space*, pp. 118–122. The MIT Press, Cambridge, MA, and London, UK (2006).

Mendonca, K.B. "VA's First Astronaut: Millie Hughes-Fulford, PhD", *sanfrancisco.va.gov* (March 13, 2014).

Nimon, J. "NIH Grant Recipient Dr. Millie Hughes-Fulford", *nasa.gov* (October 29, 2010).

Quirós, G. "Millie Hughes-Fulford: Scientist in Space", Video Story for QUEST Northern California (November 26, 2014).

Shayler, D.J.; Moule, I. *Women in Space – Following Valentina*, pp. 233, 257. Springer/Praxis Publishing, Chichester, UK (2005).

Woodmansee, L.S. *Women Astronauts*, pp. 77–79. Apogee Books, Burlington, Ontario, Canada (2002).

19

Roberta Bondar: The Pioneer of Space-Medicine Research

Credit: NASA

U. Cavallaro, *Women Spacefarers*, Springer Praxis Books, DOI 10.1007/978-3-319-34048-7_19

Mission	*Launch*	*Return*
STS-42	January 22, 1992	January 30, 1992

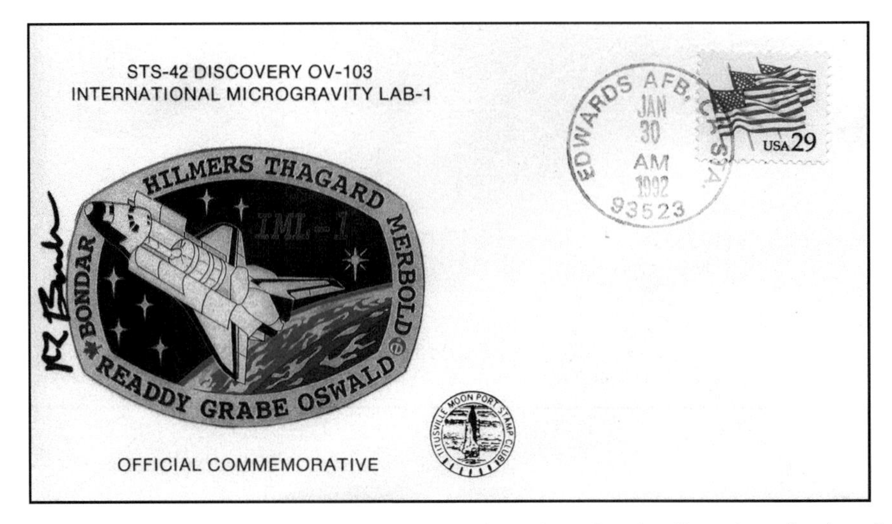

Commemorativecover of mission STS-42, signed by Roberta Bondar (from the collection of Umberto Cavallaro)

"When I was 8 years old to be a spacemen was the most exciting thing I could imagine." But when she grew older, Roberta Bondar realized that her dream was twice blocked: once for gender and then for nationality, as Canada had no space program at the time. Her dream came true at the age of 47, and not only did she become the first Canadian woman to fly in space, but also she was the first non-American woman to fly on the Space Shuttle and the world's first neurologist in space. She is also the only astronaut to use fine-art photography to explore and reveal Earth's natural environment from the surface.

Roberta Lynn Bondar was born to a father of Ukrainian descent on December 4, 1945, in Sault-Sainte-Marie, Ontario, where she completed both her elementary and secondary schooling. At age five, she recalls, helped by her mom, she saved all of her Double Bubble chewing-gum wrappers to win a space helmet which she wore to go around exploring the outdoor surroundings of her house. She was keenly interested in the idea of space travel and avidly read science-fiction comics and books about space: "I longed to soar into space. I wanted to reach out to adventure with my body as well as with my imagination."

Since she was a young girl, Roberta's parents encouraged her to be goal-oriented and she was involved in several activities including Girl Guides, the YMCA, Anglican Church groups, and athletics. She loved sports, such as canoeing and biking, fishing, and hot-air ballooning, and she also competed in basketball and tennis. At school, she excelled in science. Instead of dolls, she preferred to play with science equipment, chemistry sets, and medical bags, and her father built her a laboratory with a microscope in the basement. In grade 13, she recalls that, despite her guidance counselor trying to talk her out of science because she was a girl, she undertook a science project on the Moon.

After graduating from Sir James Dunn Collegiate & Vocational School in Sault-Sainte-Marie, Roberta earned a Bachelor of Science degree in Zoology and Agriculture from the University of Guelph (1968), a Master of Science degree in Experimental Pathology from the University of Western Ontario (1971), a Doctor of Philosophy degree in Neuroscience from the University of Toronto (1974), and a Doctor of Medicine degree from McMaster University (1977). Along the way, she got a pilot's license.

In 1981, Roberta became a fellow of the Royal College of Physicians and Surgeons of Canada as a specialist in neurology. She was teaching medicine at the McMaster University and directing the McMaster Medical Centre of Hamilton clinic for multiple sclerosis when, in 1983, she learned about the agreement between the Canadian Space Agency (CSA) and NASA, and applied to become an astronaut, together with 4,300 other candidates. She was chosen as one of only 19 finalist candidates and the only woman and, in December 1983, she became one of the six astronauts of the first CSA class. Her particular interest in the nervous system and the inner-ear balancing system, especially as it related to the functioning of the eye, had immediate relevance to experiments being planned for the first Canadian spaceflight.

Roberta began training as an astronaut in February 1984. Like Hughes-Fulford, she assumed at first that she would wait for only few months, and continued to train in Houston, although, after a while, she started commuting to Toronto, where she had left her patients and returned to her research using a new kind of trans-cranial Doppler ultrasound to measure blood flow in the brain. Later, she would use a similar instrument in space, in microgravity. She was scheduled to fly in early 1986. In January, while she was in Houston, she learned of her father's sudden death, almost immediately followed by the *Challenger* accident.

Both events seriously challenged Roberta's stay in the space program. She finally decided to continue, but only flew in 1992, as payload specialist in Mission IML-1, the first International Microgravity Laboratory mission, on board the Space Shuttle STS-42. This was a very ambitious scientific mission, with 54 experiments conducted during the 8-day mission. The space laboratory was connected with the Marshall Space Center in Huntsville where the scientists involved in the experiments could participate and make the necessary adjustments. To ensure minimum disturbance to the laboratory, the Shuttle was gravity-gradient stabilized by orienting it vertically, with the tail continuously pointing toward the center of Earth.

Roberta, the first neurologist in space, was involved in over 40 scientific experiments conducted for scientists in 14 countries. Her research was aimed at better understanding the adaptation and recovery mechanisms of the human body after their return from space and possible connections with neurological disorders suffered by many patients on Earth.

Following her spaceflight, Roberta left the space agency to return to research and pursue her interests. For over a decade at NASA, she headed an international research team carrying out scientific studies on the physiological change that occurs in humans in space and continuing to find new connections between astronauts recovering from the microgravity of space and neurological illnesses on Earth, and contributed to studying the data collected during 24 missions on both the Shuttle and Mir. Her contribution to space medicine is universally recognized. "Astronauts have been our guinea pigs in space," she explains. "The longer they stayed in orbit, the more parts of their bodies, cells and functions bent beyond normal." Besides being used by NASA for the preparation of future interplanetary travel, the results of this research also contribute to the preparation of more effective treatments of diseases such as diabetes, high blood pressure, Parkinson's Disease, and Shy Drager's Syndrome.

Roberta's techniques have been used in clinical studies at the B.I. Deaconess Medical Center, a teaching hospital of Harvard Medical School, and at the University of New Mexico. A respected advisor to industry and government, Roberta participates in meetings and conferences on the environment and to conferences and events with students. she wrote in the introduction to the Report of the Working Group of the Canadian Environmental Education:

> "After observing the planet for eight days from space, I have a deeper interest and respect for the forces that shape our world. Each particle of soil, each plant and animal is special. I also marvel at the creativity and ingenuity of our own species, but at the same time, I wonder why we all cannot see that we create our future each day, and that our local actions affect the global community, today as well as for generations to come."

Among other things, Roberta has supported UNEP (United National Environmental Program) to launch various initiatives, including a talk on "Oceans" at the Rio Summit, tree-planting, and promotion of Global Environment Outlook (GEO) reports. She was Chair of a distinguished panel at the 2010 IWF Cornerstone Conference in Guayaquil, Ecuador, and Chair of the Province of Ontario's Working Group on Environmental Education. All recommendations in their report, *Shaping Our Schools—Shaping Our*

Future, were accepted by the Ontario Ministry of Education, thus strengthening environmental education for elementary- and secondary-school students: "Without knowledge, the world is bereft of culture. And so we must be educators and students both. At some point, an educator must broaden the net to include all issues relevant to humanity's challenges." Dr. Bondar is a member of the Governing Council of *icipe*, a pan-African institute researching the influence of climate change on the spread of insect-borne diseases.

An avid photographer, Roberta studied nature photography at the Brooks Institute in California and published several books of photographs. *Touching the Earth*, which she issued in 1994, tells the chronicle of her experiences in space after her 8-day STS-42 mission, during which she was also tasked with taking photographs of Earth. After her space mission, she continued her photographic explorations and published three other books of photographs, including *Passionate Vision* documenting Canada's national parks in 2000. Several exhibitions of her landscape photographs were shown in galleries in London, Vancouver, Toronto, and Calgary. She is co-founder and president of the Roberta Bondar Foundation, which, as her website highlights, "responds to the recognized need within society to educate and improve knowledge of the environment in a way that stimulates interest, excitement, creativity, responsibility, and for some, the desire for study in this area."

Roberta is academically one of the most distinguished astronauts to have flown in space, and was awarded with 24 honorary doctorates from Canadian and American universities, and served as Chancellor of Trent University for 6 years. She has also been inducted into the Canadian Medical Hall of Fame and into the International Women's Forum's Hall of Fame for her pioneering space-medicine research. Several schools and parks are named after her in Canada and her country has featured her on a stamp.

References

"About Roberta – the Inspiration: a Great Canadian, a Great Cause", *therobertabondarfoundation.org*. Biography of Roberta Bondar, *cityssm.on.ca* (2008).

Bondar, R. *Shaping our School, Shaping Our Future: Environmental Education in Ontario Schools*, Report of the Working Group of the Environmental Education, Giugno (2007).

Communication with the Author through e-mail in May 2016.

Gibson, K.B. *Women in Space: 23 Stories of First Flights, Scientific Missions and Gravity-Breaking Adventures*, pp. 171–175. Chicago Review Press, Inc., Chicago (2014).

Gueldenpfenning, S. *Women in Space Who Changed the World*, pp. 49–57. The Rosen Publishing Group, New York (2012).

Kevles, T.H. *Almost Heaven: The Story of Women in Space*, pp. 122–126. The MIT Press, Cambridge, MA, and London, UK (2006).

Muir, E.G. *Canadian Women in the Sky: 100 Years of Flight*, p. 132ss. Dundurn Group, Toronto, Canada (2015).
Official NASA biography of Roberta Bondar, *jsc.nasa.gov* (July 2007).
"Roberta Bondar", *Our Planet: The Planet We Share* (September 2011), *www.unep.org/pdf/op_sept/EN/OP-2011-09-EN-FULLVERSION.pdf*, pp. 34–35.
Woodmansee, L.S. *Women Astronauts*, pp. 81–82. Apogee Books, Burlington, Ontario, Canada (2002).

20

Nancy Jan Davis: Growing Up with the Saturn Rockets

Credit: NASA

U. Cavallaro, *Women Spacefarers*, Springer Praxis Books, DOI 10.1007/978-3-319-34048-7_20

Mission	*Launch*	*Return*
STS-47	September 12, 1992	September 20, 1992
STS-60	February 3, 1994	February 11, 1994
STS-85	August 7, 1997	August 19, 1997

Commemorative cover of mission STS-47, signed by Nancy Jan Davis (from the collection of Umberto Cavallaro)

When Jan Davis was growing up in Huntsville in the 1960s, there simply were no women astronauts. So it wasn't something she was ever considered becoming. But, in 1978, when the first female candidates were selected to be part of NASA's Astronaut Corps, Dr. Davis began to rethink her career path. And, in 1984, while working as an aerospace engineer at NASA Marshall Space Flight Center (MSFC) and attending the University of Alabama in Huntsville (UAH), she applied to be an astronaut: she was not chosen from among the 5000 people who applied that year, but she was selected 3 years later.

Nancy Jan Davis was born in Cocoa Beach, Florida, on 1 November, 1953. She received much of her schooling in Huntsville, where she moved with her family when she was a young girl and that she considers as her hometown:

> "I grew up in Huntsville, which is where Von Braun and the rocket team was. I went to school with a lot of their kids. I moved here in the early'60s, when we were going to the Moon, and I was here during all of the testing they were doing in Huntsville. They tested all of the engines, and the whole town would vibrate. It was really a big deal in Huntsville. When I say the whole town, I mean the windows would shake and you just knew we were doing something to help us to go to the Moon. It was really exciting. With every mission, and every launch, they would make a big deal about it at school. I guess I became interested in space when everybody else in the country did, but it was just a very real thing for me here."

Jan enjoyed math and science:

> "I had a science teacher who really influenced me and made it very interesting and fun for me. I think that made the difference. With math I just enjoyed the challenge of it. I ended up taking all the courses I could, and they didn't have any courses left for me when I was a senior in high school, I started taking calculus at the university when I was in high school. Just because I really enjoyed it."

In 1975, Jan earned a Bachelor of Science degree in Applied Biology from the Georgia Institute of Technology, and a second Bachelor of Science degree in Mechanical Engineering from Auburn University in 1977: "I liked the idea of using engineering to do things that help people and I was going to go into bioengineering but there were no jobs there, So I went where the jobs were." And that was Bellaire, Texas, where, in 1977, she took a job with Texaco as a petroleum engineer in tertiary oil recovery.

While she had a lot of interest in space, Jan had never dreamed of becoming an astronaut herself, but began to rethink her career path in 1978: "When I was growing up," she explains in an interview, "there weren't any women astronauts so I didn't even think it was possible until they started selecting women in 1978. So that's when I first got the idea."

In 1979, Jan left Texaco and joined NASA's MSFC in Huntsville as an aerospace engineer, working in the team responsible for the structural analysis and verification of the Hubble Space Telescope (HST) and then of the other great space observatory, the Chandra Telescope (or Advanced X-Ray Astrophysics Facility); in 1986, she was then named team leader of the HST maintenance mission. While working at NASA, in 1983, she received a Master of Science degree in Mechanical Engineering at the UAH and, in 1985, a doctorate at the same university, studying the long-term strength of pressure vessels due to the viscoelastic characteristics of filament-wound composites: "It was convenient that they had night classes and I could work full time," she says.

After the *Challenger* tragedy, Jan served as chief engineer of the team redesigning the attach ring linking the Shuttle's solid rocket boosters with the external tank. She holds her own patent and authored several technical papers.

After finishing her master's and beginning her doctorate, Jan felt more competitive and applied to be an astronaut again. She says:

> "I think having a graduate degree and my pilot's license, it just seemed like I might have the qualifications, so I might as well give it a shot. I knew it was a long shot but I realized it could happen. It was a possibility."

The application was declined twice and was accepted in 1987 when she became an astronaut as a member of Group 12. Jan's experience at NASA was key to her selection.

Her initial technical assignment was in the Astronaut Office's "Mission Development Branch," where she provided technical support for Space Shuttle payloads, including the Italian Tethered Satellite Systems, deployed by the Colleague Tamara Jernigan during the STS-46 mission. She also served as a CapCom (capsule communicator) in Mission Control for seven Space Shuttle missions. She was CapCom during mission STS-48 when, in September 1991, the first collision-avoidance maneuver ever was performed in the then 10-year-old Shuttle history when, minutes before midnight, thrusters were burned for seven seconds to avoid debris from the Russian Kosmos 955. "I think we scored a space first," Jan told the crew before they went to sleep for the day.

Jan flew three times in space. In 1992, she participated in the STS-47 mission, the 50th mission of the Space Shuttle program and second mission of Space Shuttle *Endeavour*, which brought into space for the third time Spacelab LM-2, built in Italy. During the eight-day mission—a co-operative venture between the US and Japan—she was mission specialist, responsible for operating Spacelab-J and its subsystems, together with Mae Jemison, the first African American woman to fly in space, and performed a variety of experiments in life sciences and materials processing, under the direction of Mark C. Lee, who was payload commander.

After her first spaceflight, Jan served as the Astronaut Office representative for the Remote Manipulator System (RMS), with responsibility for RMS operations, training, and payloads. In 1994, she flew with STS-60, the first Shuttle–Mir mission, which was the first to include a Russian cosmonaut—Sergei Krikalev—in the crew. During the eight-day mission, her prime responsibility was to maneuver the WSF (Wake Shield Facility) on the RMS; she also conducted experiments on thin-film crystal growth and was responsible for performing scientific experiments in the Spacehab, the pressurized module built in Italy for Spacehab Inc. Austin, Texas, now Astrotech Corporation. "It was hard work!" she said, "but we were up there for a reason and we didn't have a lot of time to just look out the window."

After this mission, Jan served as the Chairperson of the NASA Education Working Group and as Chief for the Payloads Branch, which provided Astronaut Office support for all Shuttle and International Space Station (ISS) payloads. In 1997, she flew in space for the third time, as payload commander of mission STS-85. During this 12-day mission, Davis deployed and retrieved the CRISTA-SPAS payload, and operated the Japanese Manipulator Flight Demonstration (MFD) robotic arm.

After returning from this mission, Jan worked at NASA Headquarters as the Director of the Human Exploration and Development of Space (HEDS), Independent Assurance Office for the Office of Safety and Mission Assurance. In that position, Davis managed and directed independent assessments for the programs and projects assigned to the HEDS enterprise. In July 1999, she moved to the MSFC as Director of the Flight Projects

Directorate and was responsible for the ISS Payload Operations Center, ISS hardware, and the Chandra X-Ray Observatory Programme. After the *Columbia* accident, she was named head of Safety and Mission Assurance at Marshall, where she assured the safe return to flight of the Space Shuttle.

Jan retired from NASA in 2005: "I was eligible to retire and it seemed like a good opportunity to learn how private industry worked and to expand my knowledge," she said. She had logged a total of 673 h in space. In October 2005, she accepted a Vice President and Deputy General Manager position at Jacobs Engineering Group, an international technical professional services firm based in Huntsville, Alabama, with over 53,500 employees globally, in 200 offices around the world. She is not very far from her original home: "I oversee the almost 900 employees who support the MSFC programs and projects," she says. "So it's a good way to still stay with NASA!"

References

Evans, B. "Of Marriage, Medaka Fish and Multiculturalism: The Legacy of STS-47", *americaspace.com* (October 28, 2012).

La Chance, D. "College of Engineering Alumna Dr. Jan Davis on UAH", *www.uah.edu* (November 14, 2013).

Official biography of Nancy Jan Davis, *jsc.nasa.gov* (January 2006).

Personal contacts with Author by e-mail in March 2016.

Shayler, D.J.; Moule, I. *Women in Space – Following Valentina*, pp. 261–263. Springer/Praxis Publishing, Chichester, UK (2005).

Woodmansee, L.S. *Women Astronauts*, pp. 82–84. Apogee Books, Burlington, Ontario, Canada (2002).

21

Mae Jemison: Our Limits Are the Stars

Credit: NASA

U. Cavallaro, *Women Spacefarers*, Springer Praxis Books, DOI 10.1007/978-3-319-34048-7_21

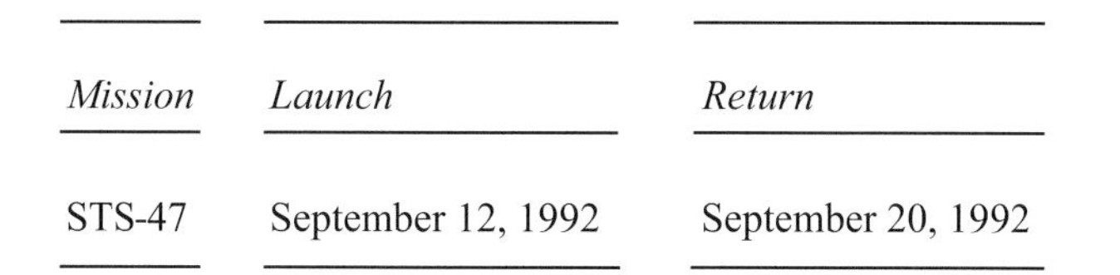

Mission	*Launch*	*Return*
STS-47	September 12, 1992	September 20, 1992

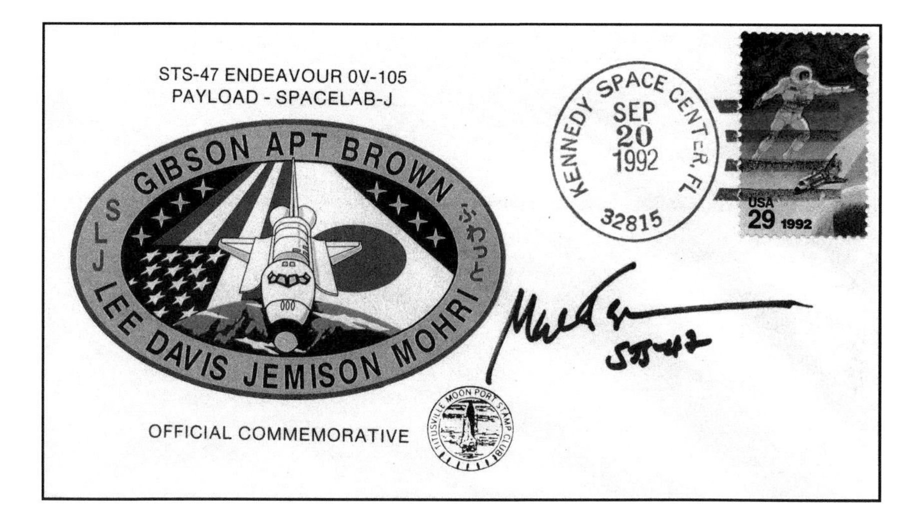

Commemorative cover of mission STS-47, signed by Mae Jamison (from the collection of Umberto Cavallaro)

Mae Jemison was the first Afro American astronaut. A resourceful personality, she is a determined breaker of boundaries and dispels commonplaces: "The best way to make dreams come true is to wake up!" I met her in Turin in July 2003 at the Exhibition on Women In Space organized by ASITAF (the Italian Astrophilately Society) to commemorate the 50th Anniversary of Tereshkova. Mae Jemison was there to give a paper at the 8th IAA Symposium (International Academy of Astronautics) that was held in Turin in those days. After visiting the exhibition, Mae commented:

> "It is a reawakening of many memories and brings a smile to my face. I remember being young girl and hearing of Tereshkova's flight and how it proved my contention. I saw a picture of me during my flight and realized that so much of what we do may have an impact in places we don't expect. And it spurs me to continue the effort to expand involvement in space exploration."

The topic of the IAA Symposium on the Future of Space Exploration in Turin was "Towards the Stars" and Mae was there as the representative of the 100 Year Starship or 100YSS Foundation. "100YSS," she explains, "is an international, non-profit, non-governmental organization established to promote the research, technological development and societal capabilities required to enable human travel beyond our solar system to another star within the next 100 years."

100YSS was started with seed-funding through a competitive grant from DARPA (Defense Advanced Research Projects Agency) and support from NASA "won in February 2012 by the team I led", Mae adds. "100YSS' raison d'être is to foster the explosive innovation, radical leaps in knowledge, technical achievement and societal advances that will happen as a result of tackling such an audacious challenge." It's more of a thought experiment than a construction project:

> "100YSS is not focused on mounting its own interstellar mission at any particular time, but rather creating, inspiring and nurturing and maintaining an environment such that in the years to come, so a bold endeavor can be accomplished by someone."

The idea itself may spark some other pretty audacious proposals, such as the one by J. Craig Venter to send human genomes toward the stars and reconstruct them upon arrival: "Our overall vision is encompassed in the name of our proposal to DARPA—'An Inclusive Audacious Journey Transforms Life Here on Earth and Beyond'." Mae explains:

> "100YSS is inherently transdisciplinary. Successful human interstellar travel will only happen through engaging the full spectrum of human experience, including people across ethnicity, gender, disciplines and geography. And as importantly, the space derived knowledge, technological advances and innovations must be actively applied to enhancing life on Earth every step of the way. Working with international partners, we are building a global aspiration. By convening and supporting R&D, education, entertainment, and business consortia, 100YSS implements its commitment to inclusion. While the grand challenge is interstellar travel, again, the immediate and overarching goal is enabling explosive, beneficial innovation worldwide."

Why travel to the stars? Mae's answer is:

"This is a good question! Because it is very, very difficult and at the same time a comprehensible challenge, that is, I believe, an indelible part of the human spirit, with the potential to yield seismic benefits across the world. One might call it a "North Star" for high impact collaboration among nations, universities, governments, industry, non-profits and philanthropists. In so many ways just starting to address the myriad advances needed will create new industries and jobs in the years to come, accelerate creativity and give the world something to rally for that is bigger than any one of us individually or as a nation."

"When we explore space, we garner the greatest benefits here at home. People often take very much for granted (but would loath to surrender) the benefits space exploration has provided right here on Earth. I know the readers of Ad*Astra already know this, but I think it is important to state just a few examples of benefits we already have from space exploration. They include: global positioning satellites (GPS); remote sensing for water, minerals, and crop and land management; weather satellites, arms treaty verifications; high-temperature, light-weight materials; revolutionary medical procedures and equipment; pagers, beepers, and television and internet to remote areas of the world; geographic information systems (GIS) and algorithms used to handle huge, complex data sets; physiologic monitoring and miniaturization; atmospheric and ecological monitoring; and insight into our planet's geological history and future—the list goes on and on."

"Another way to look at space exploration is the remarkable boom across science and engineering fields—and the accompanying wave of social creativity, artistic expressions, and educational innovation—that followed the Apollo missions and each successive leap in space exploration. The challenge of traveling to another star system could generate transformative activities, knowledge, and technologies that would dramatically benefit every nation on Earth in the near term and years to come."

With NASA scaling back its manned space programs, the idea of a manned trip to the stars may sound audacious and appear fantastical:

"No more so than the fantasy of reaching the moon was in the days when H. G. Welles. published his 'First Man on the Moon', considerably less than 100 years before humans landed on the moon (1901 vs. 1969) at a time when we knew very little about propulsion, the universe, human health and sustainability while the rapidity of scientific advances were not nearly as great as today. Today the arc of our knowledge acquisition and technological advancement is much, much greater."

The truth is that the best ideas sound crazy at first. And then there comes a time when we can't imagine a world without them.

Which are the main issues to be addressed and the main capabilities to be developed?

"At this point, I do not consider establishing a roadmap—the task is to understand what capabilities will be required and our current capacities—rather creating a capabilities map. Some of what is needed is of course development of revolutionary technologies to generate, harness, control and store enormous quantities of energy reliably and safely; radical advancements in closed-loop life-support systems and sustainable habitats; understanding microbiomes; reimagining of financial investment and benefits; human behavior, health and governance; advances in robotics, automation, intelligent systems, and manufacturing techniques, just to name a few."

Really a broad vision, aiming high!

Born in Decatur, Alabama, on October 17, 1956, Mae Jemison became the first woman of color ever admitted into the astronaut training program. When she was three, her family moved to Chicago, Illinois, to take advantage of better educational opportunities there and it is that city that she calls her hometown.

Involved in extracurricular activities including dance and theatre productions, Mae began dancing at the age of 11 and wanted to become a professional dancer. Her mother encouraged her to follow on with her studies after college: "You can always dance if you're a doctor, but you can't doctor if you're a dancer."

After earning a double degree at Stanford University and graduating in Medicine at Cornell University in 1981, Mae joined the Peace Corps and worked as a staff physician in Sierra Leone and Liberia, where she also taught and did medical research. Following her return to the US in 1985, she made a career change. Inspired by Sally Ride, the first woman in space, and by Lieutenant Uhura, the first woman of color in space in the *Star Trek* television series (played by the African American actress Nichelle Nichols), she decided to follow a dream she had nurtured for a long time. In October of that year, she applied for admission to NASA's astronaut training program. The *Challenger* disaster of January 1986 delayed the selection process but, when she reapplied 1 year later, Mae was one of the 15 candidates chosen from a field of about 2000. Nichelle Nichols at that time was working with NASA as a recruiter.

When Mae was chosen on June 4, 1987, for Astronaut Group 12 (nicknamed GAFF: George Abbey Final Fifteen), she became the first African American female astronaut. On September 12, 1992, she flew into space with six other astronauts aboard the *Endeavour* on mission STS-47 as a mission specialist, in support of the Spacelab-J, a co-operative mission between the US and Japan. During her 8 days in space, she conducted experiments on weightlessness and motion sickness on the crew and herself. Altogether, she spent slightly over 190 h in space before returning to Earth on September 20.

In recognition of her accomplishments, she received several honorary doctorates.

Mae left the Astronaut Corps in March 1993. Although her departure from the agency was amicable, NASA was not thrilled to see her leave after investing in her, even in terms of marketing. She accepted a teaching fellowship at Dartmouth and also established the Jemison Group, a company that seeks to research, develop, and market advanced technologies, and the Dorothy Jemison Foundation for Excellence, named in honor of her mother. One of the goals of her foundation is promoting science and technology and providing an inspirational and educational message for young people to motivate them to fulfill their potential and especially encourage young girls toward careers in the sciences, by stimulating three C's of science—curiosity, creativity, and critical thinking.

Mae also appeared on an episode of *Star Trek*: *The Next Generation*, as Lieutenant Palmer in the episode "Second Chances," the first real astronaut ever to appear on *Star Trek*. Her book for children, *Find Where the Wind Goes*, is a testimony to the power of setting goals and the strength of character necessary to achieve them. And Mae adds: "I had to learn very early not to limit myself due to others' limited imaginations. I have learned these days never to limit anyone else due to my limited imagination."

References

Briggs, C.S. *Women Space Pioneers*, pp. 57–62. Lerner Publications, Minneapolis (2005).

Cavallaro, U. "Mae Jemison: Our Limit Are the Stars." *AD*ASTRA, ASITAF Quarterly Journal*, **18**(September), 9–11 (2013).

Gibson, K.B. *Women in Space: 23 Stories of First Flights, Scientific Missions and Gravity-Breaking Adventures*, pp. 112–118. Chicago Review Press, Inc., Chicago (2014).

Interview by the Author on July 5, 2013, and e-mails in the following days.

Kevles, T.H. *Almost Heaven: The Story of Women in Space*, pp. 129–133. The MIT Press, Cambridge, MA, and London, UK (2006).

Official biography of Mae Jemison, *jsc.nasa.gov* (March 1993).

22

Susan J. Helms: Three-Star General

Credit: NASA

U. Cavallaro, *Women Spacefarers*, Springer Praxis Books, DOI 10.1007/978-3-319-34048-7_22

Mission	*Launch*	*Return*
STS-54	January 13, 1993	January 19, 1993
STS-64	September 9, 1994	September 20, 1994
STS-78	June 20, 1996	July 7, 1996
STS-101	May 19, 2000	May 25, 2000
STS-102	March 8, 2001	
STS-105		August 22, 2001

Commemorative card of mission STS-101, signed by Susan Helms (from the collection of Umberto Cavallaro)

Susan J. Helms is a retired Lieutenant General in the United States Air Force (USAF). She has been the Commander of the Joint Functional Component Command for Space (JFCC SPACE), the operational space component of the US Strategic Command, located at Vandenberg Air Force Base (VAFB) in California, leading more than 20,500 personnel responsible for providing missile warning, space superiority, space situational awareness, satellite operations, space launch, and range operations.

Following in her father's footsteps, at age 18, Susan entered the USAF. Selected for the NASA Astronauts' Corps in 1990, she became the first US military woman in space, and is one of the six astronauts who for five times heard the countdown to lift-off from the inside of a spacecraft: after participating in four Space Shuttle missions, she finally served aboard the International Space Station (ISS) as a member of the long-duration Expedition-2 crew (in 2001).

The oldest of three sisters, Susan was born on February 26, 1958, in Charlotte, North Carolina, but her family moved early on to Portland, Oregon, which she considers her hometown. In Portland, she attended the elementary schools and learned to play the piano. When in NASA, she played the keyboard in the MAX-Q rock band, entirely made up of NASA astronauts ("Max Q" is the maximum aerodynamic pressure that the Shuttle experienced shortly after lift-off) and also carried a keyboard in space during her long-term expedition to the ISS.

Susan's father made his career in the Air Force and transmitted his passion for flying to her. During an interview, she said:

> "Definitely my interest in the Air Force has been there since probably the day I was born. And I basically decided at a young age that the whole thing of being in the military and particularly the Air Force appealed to me. There were a couple of strong points that I wanted. One of them was the opportunity to travel—little did I know how far that would go—and then also the opportunity for a stable career. I like the idea of moving every few years and seeing different places, and it appeared that my dad had had a very rewarding Air Force career. It looked like, as an engineer, I could also have the same."

Susan graduated from Parkrose Senior High School, Portland, Oregon, in 1976 and, that year, joined the Air Force Academy in Colorado Springs that, for the first time, was taking women. There were about 1500 cadets among the freshmen that year, including 157 women (only 98 of them graduated 4 years later). She experienced hostility from some of the male students, who resisted the sudden change or simply just didn't know how to handle it:

> "It was a challenge for everyone, from the highest generals to the freshmen. You have to realize that the previous three classes to us, all of them male students, expected to go to an all-male school. And then this changed on them right in the middle of schooling."

Most of the women decided to lie low. Susan received her Bachelor of Science degree in Aeronautical Engineering from the US Air Force Academy in 1980 and became a weapons engineer with the Air Force Armament Laboratory at Eglin Air Force Base, Florida, working on F-16 weapons systems. As the lead engineer, she worked on the dynamics of weapons separation, preventing the aerodynamics interference that can occur when weapons are released, causing them to return and hit the aircraft that jettisoned them.

In 1984, Susan was sent to Stanford University, where the event that changed the course of her life occurred: her meeting with Sally Ride, who visited the university during her second post-flight tour. There she learned that she wanted to become an astronaut, too:

> "As Air Force career officer, I basically flew in jets. When I had a taste of that, I got fairly addicted to it, and at some point along that career path, I decided, 'Well, how can I fly higher and faster than I am doing right now?' And that, combined with a couple of other brushes with NASA people, made me realize that this could be something I'd be very interested in doing, and I just put my application in, of course, never expecting to get selected. But when that happened, my Air Force career sort of took a turn to a NASA career, and I've had the chance to do just that: fly higher and faster. And that was what I was originally after."

After earning her Master of Science degree in Aeronautics/Astronautics from Stanford University in 1985, Susan served as an assistant professor of aeronautics at the US Air Force Academy. In 1987, Colonel Helms attended the Air Force Test Pilot school at Edwards Air Force Base, where she completed 1 year of training as flight-test engineer.

During the course, she met the guest speaker, Colonel Dick Covey, an astronaut who said to her: "I hope we see you in Houston some time." He was referring to the Johnson Space Center (JSC), home of astronaut training. "I took that as the final sign that I should probably apply for the astronaut program," Susan said. "So Covey was a big motivator in getting me to fill out the application and send it in." As a flight-test engineer, she was assigned as a USAF Exchange Officer to the Aerospace Engineering Test Establishment, Canadian Forces Base, Cold Lake, Alberta, Canada, where she worked as flight-test engineer and project officer on the CF-18 aircraft. She flew in 30 different types of US and Canadian military aircraft. In January 1990, she was leading the development of a CF-18 Flight Control System Simulation for the Canadian Forces when she knew that she had been selected for the astronaut program.

Susan became an astronaut in July 1991. She flew as member of four Shuttle missions—STS-54 (1993), STS-64 (1994), STS-78 (1996), and STS-101 (2000)—and served aboard the ISS as a member of the Expedition-2 crew (in 2001).

Susan became the first US military woman in space when, from January 13 to January 19, 1993, she participated in the *Endeavour* STS-54 mission. The primary objective of the mission was the deployment of a US$200 million NASA Tracking and Data Relay Satellite (TDRS-F) as part of the network used by NASA and other US government agencies for communications to and from the Space Shuttle, satellites, aircraft, and the ISS. This system had been designed to replace the existing worldwide network of NASA ground stations. Shuttle *Endeavour* also carried in its payload bay the DXS (Diffuse X-Ray Spectrometer) that collected high-quality X-ray spectra to investigate the origin of the Milky Way galaxy. In 1994, she took part in the *Discovery* STS-64 mission and served as the primary remote manipulator system (RMS) operator.

In July 1996, Susan was flight engineer aboard *Columbia* STS-78, where she also served as payload commander, on the longest Space Shuttle mission to that date: 17 days. Later that year, the STS-80 mission broke that record by 19 h. The mission included studies sponsored by 10 nations and five space agencies, and was the first mission to combine both microgravity studies and life-science investigation, thus serving as a model for future studies on board the ISS.

In 2000, Susan was mission specialist in the STS-101 mission, aimed at delivering to the ISS the cargo contained in the Spacehab logistic module and the Integrated Cargo Carrier pallet, and prepare the station for permanent occupation. During the mission, some critical hardware was repaired or replaced, including air filters, Zarya fire extinguishers, and smoke detectors, four suspect batteries on Zarya, the Radio Telemetry System memory unit, and other components of the communication systems. Susan's prime responsibilities during this mission were some critical repairs and the on-board computer network. She also served as the mission specialist for rendezvous with the ISS. After landing, it was discovered that, during re-entry, due to a damaged tile, superheated gas had entered into the left wing; had it gone deeper, the Shuttle could have been destroyed, as 2 years later would happen to *Columbia* STS-107.

Over the following years, Susan became the first woman to inhabit the ISS, as a member of the long-duration Expedition-2, the second crew to inhabit the International Space Station Alpha, which she reached aboard Shuttle STS-102. When she was asked to volunteer, she had little enthusiasm for the prospect, thinking that preparing for the mission would take almost a year, living in Russia in the winter, learning Russian, and leaving home for months in orbit; but she also knew that there would be a long wait for an alternative assignment, if any. "Just to get back into orbit, I said I would take the job," she explained later. In Houston, the memory of Shuttle–Mir was still alive. The mission carried into space for the first time "Leonardo," the pressurized logistic module known as MPLM, provided by ASI, the Italian Space Agency, and built in Turin, Italy, by Alenia Spazio. One of the goals of the mission was to unload five tons of experiments and equipment, and install a platform to be used to mount the Canadian-built robotic arm SSRMS (the Space Station Remote Manipulator System). The docking of the Shuttle was marked by a couple of hitches. Before docking, the Shuttle had to hover 400 ft (120 m) away from the Pressurized Mating Adapter-2 port as it awaited the array latch verification and proper lighting conditions for its final approach. A communications problem occurred just after docking, when signals could not be relayed for about 34 min.

Susan was welcomed onto the ISS by Commander Juri Usachev, the cosmonaut who had been in the Shannon Lucid's Mir crew, whom she knew well, as the two had trained together in Star City. Not surprisingly, Yuri immediately offered her the privacy of Destiny, the American Laboratory module, in which to keep her personal belongings, just as he had offered Lucid his own module on Mir. And this was particularly appreciated by Susan, since her entire life in that very moment was up in space: before leaving for space, she had packed up and stored all of her possessions and she did not have an Earth address anymore. "I was very disconnected from Earth," she said. "When we had to go fly for six months, I effectively just closed down my Earth life. I acted like it was a military deployment, I lived in space. It was my home."

On March 11, with colleague Susan Voss, Susan performed a world record extravehicular activity of 8 h and 56 min, the longest ever made up to that date, to prepare some necessary connections for the installation of the Leonardo module. It was again the veteran Susan Helms who performed, the following day, the delicate operation to attach the module to the space station, using the long robotic arm Canadarm-2. She stayed on board for about 5 months, exactly 163 days, which ended on August 22, 2001, with the return aboard STS-105. In contrast to the Shuttle's tight time schedule, similarly to "a quick business trip," the scheduling on the station was more relaxed and life was more similar to "a normal workday."

A veteran of five spaceflights, Susan has accrued 5064 h (211 days) in orbit, including a spacewalk of 8 h and 56 min. In July 2002, after a 12-year NASA career (and 22 years of military service), instead of retiring, Susan chose to return to the active-duty USAF and took a position at the Headquarters of the US Air Force Space Command. "The Air Force has always been so supportive of the things I wanted to do, and I guess I felt the time had come to come back and help with the military space program," she said.

In June of 2006, Susan was appointed a Brigadier General ("two-star General") and became Commander of the 45th Space Wing based in the Patrick Air Force Base, Florida, responsible for the processing and launch of US government and commercial satellites from Cape Canaveral Air Force Station, Florida. After earning the rank of Major General, in November 2006, she moved to the Offutt Air Force Base, Nebraska, where she served as the Director of Plans and Policy for the USSTRATCOM (United States Strategic Command). Appointed to the rank of Lieutenant General ("three-star General") in January 2011, she became Commander of the 14th Air Force (Air Forces Strategic) and Commander of the Joint Functional Component Command for Space (JFCC SPACE), the operational space component of the US Strategic Command, located at VAFB in California, leading more than 20,500 personnel providing tailored, responsive, local, and global space effects in support of national objectives.

In March 2013, Susan was nominated by President Barack Obama to become the vice commander of Space Command. However, the nomination was blocked by Senator Claire McCaskill, a lawmaker representative of the Democratic Party in the Armed Services Committee, due to an "unclear intervention" of General Helms, who, as the supreme authority of the Court-Martial, in February of 2012 overturned the sexual assault conviction of his subordinate, Captain Matthew Herrera, at VAFB. Military rape/sexual assault is a growing epidemic among the armed forces. It is estimated that there are 19,000 cases per year, of which only one in six is reported. Helms exercised her legal discretionary authority in the court-martial and decided not to approve the conviction due to the fact that—based on the evidence, her best judgment as a senior military officer, and her personal and professional integrity—she was not convinced "beyond-a-reasonable-doubt" that the burden of proof had been met. According to Senator McCaskill, such a decision made it more difficult for victims of sexual assault to seek justice. In the face of Susan's strong opposition, President Obama was forced to revoke the appointment. Susan submitted her resignation, which was accepted on April 1, 2014, and withdrew from public life.

References

Anthony, R. "Interview: Astronaut Susan Helms and DP J. Neihouse", *bigmoviezone.com* (April 2002).

Kevles, T.H. *Almost Heaven: The Story of Women in Space*, pp. 193–196. The MIT Press, Cambridge, MA, and London, UK (2006).

Knott, S. "Pioneering the Last Frontier", Air Force Space Command Public Affairs (March 28, 2003).
Official NASA biography of Susan Helms, *jsc.nasa.gov* (September 2012).
"Preflight Interview: Susan Helms", STS-101, *nasa.gov* (December 6, 2012).
Schogol, J. "With Nomination Blocked, 3-Star Applies for Retirement", *airforcetimes.com* (November 8, 2013).
Shayler, D.J.; Moule, I. *Women in Space – Following Valentina*, pp. 264, 270, 276, 292–293, 329–330. Springer/Praxis Publishing, Chichester, UK (2005).
Whitlock, C. "General's Promotion Blocked over Her Dismissal of Sex-Assault Verdict", *washingtonpost.com* (May 6, 2013).
Woodmansee, L.S. *Women Astronauts*, pp. 86–88. Apogee Books, Burlington, Ontario, Canada (2002).

23

Ellen Ochoa: "Reach for the Stars and Let Nothing Limit Your Potential!"

Credit: NASA

U. Cavallaro, *Women Spacefarers*, Springer Praxis Books, DOI 10.1007/978-3-319-34048-7_23

Mission	*Launch*	*Return*
STS-56	April 8, 1993	April 17, 1993
STS-66	November 3, 1994	November 14, 1994
STS-96	May 27, 1999	June 6, 1999
STS-110	April 8, 2002	April 19, 2002

Commemorative cover of mission STS-96, signed by Ellen Ochoa (from the collection of Umberto Cavallaro)

Inventor, physicist, and astronaut, Dr. Ellen Ochoa flew on four Space Shuttle missions and is presently serving as the Director of the NASA Johnson Space Center (JSC) in Houston, Texas, responsible for overseeing approximately 13,000 employees and contractors.

Born of a Mexican American father in Los Angeles, California, on May 10, 1958, Ellen has always considered her hometown to be La Mesa, near San Diego, where she grew up with her sister and three brothers. She says:

> "My father didn't speak Spanish with us at home. When I was growing up, my father believed, as many people did at the time, that there was a prejudice against people speaking their native language. It's really too bad, and I'm glad that things have changed in recent years."

One of the most influential people in her life was her mother, Rosanne, who was a firm believer in the value of education and held to the idea that a person can succeed at anything if he or she tries hard enough. Ellen says:

> "When I was a year old my mother started college. She had to raise five children primarily on her own and so she couldn't take more than one class each semester. She didn't graduate until 22 years later, but she did finish. Her primary focus was the enjoyment of learning. That's what I got from her example."

Ellen was an exceptionally good student, and in particular developed a love of math, graduating from high school at the top of her class. In addition, she was (and still is) very fond of music and earned recognition during her teen years as a flutist. At one point, she considered a professional career as a classical flutist, but decided it was more practical to pursue other careers. Her passion for the flute, however, never diminished. The photos of her taken while she plays flute floating on the Space Shuttle are well known.

Valedictorian of her 1975 graduating class at Grossmont High, Ellen was offered a partial scholarship to Stanford University in Palo Alto, near San Francisco, but chose to study at San Diego State University, which was more affordable. In 1980, she obtained her bachelor's degree in Physics, with top academic honors, once again the valedictorian of her class. She then went on to graduate school at Stanford University to study electrical engineering and was granted her master's degree in 1981 and her doctorate in 1985, all while performing as an award-winning soloist with the Stanford Symphony Orchestra.

At Stanford, Ellen met other graduate students who were interested in NASA's astronaut training program, and realized that she might qualify for it as well.

> "I never considered being an astronaut as an option because when I was growing up there were no female astronauts. It wasn't until the first six female astronauts were selected in 1978 that women could even think of it as a possible career path."

She decided to apply when she completed her doctorate. In her doctoral dissertation, Ellen studied an innovative optical inspection system to detect defects in objects with

Fig. 23.1 Ellen playing Vivaldi on the flute while serving as a mission specialist on her first mission, STS-56, in 1993. Credit: NASA

repetitive features. She, along with her advisors at Stanford, received a patent on the process.

Her doctorate complete, Ellen began her career as a research optical engineer at the Sandia National Laboratory in Livermore, California, developing optical systems. She earned two more patents: one for an optical object recognition method and another for a method for image noise removal. She also published several scientific articles on the subject. While there, she learned she had become one of 100 finalists under consideration for the NASA astronaut training program. She had the opportunity to interview in 1987 but wasn't selected. She decided she wanted to work for NASA in some capacity and, in 1988, she was hired at the NASA Ames Research Center in California and soon became chief of the Intelligent Systems Technology Branch, where she managed a staff of 35 engineers working on advanced development of space-based computational systems. She said later:

> "I would advise everyone to set their aspirations high, and then shoot for them. I don't think it matters if you reach that one lofty goal. In reaching high you will encounter other opportunities that can lead to interesting career paths and an exciting life."

In 1989, Ellen was again interviewed for the astronaut program. In January 1990, she learned that she had been selected for NASA's 13th Astronaut Candidate Class, along with 22 other candidates. The group included five women.

In April 1993, Ellen flew on the STS-56 mission on Space Shuttle *Discovery* as a mission specialist for Spacelab ATLAS-2 (Atmospheric Laboratory for Applications and Science-2), thus becoming the first Hispanic American woman in space: "I was in training for 3 years before my first mission, which isn't that long of a wait. Some astronauts have waited 10, or more years before they could finally go into space!" The primary purpose of the STS-56 mission was to study the Sun's energy during an 11-year solar cycle and to learn how changes in the irradiance of the Sun affect Earth's environment and climate. Ellen used a robotic arm (RMS) to deploy and retrieve the Spartan satellite that carried instruments to investigate the relation of the solar wind to solar flares and determine the amounts of solar radiation in low Earth orbit: "I have worked the robot arm on all four of my space missions, and I really love it. It's challenging to do, but lots of fun!"

In November 1994, Ellen served as payload commander for the *Atlantis* STS-66 mission and continued the research with Spacelab ATLAS-3, another data-collecting mission on solar energy, studying the problem of the ozone hole and ozone depletion. Among her many responsibilities was using the Shuttle's robotic arm to retrieve the Cryogenic Infrared spectrometer and telescope instruments mounted on the CRISTA-SPAS satellite, which had been deployed 8 days earlier by fellow crewmember Jean-François Clervoy with the European Space Agency (ESA). Following her flight, she received an award for her technical achievements from by the Hispanic Engineer National Achievement Awards Corporation.

For the next few years, Ellen worked on the development of the International Space Station (ISS). She recalls:

> "When I came back from my second flight NASA had joined forces with Russia to redesign the space station that we had been planning with other countries for a few years. So it became my job to lead astronaut office support for that program. We were negotiating with the Russians on how we were going to operate, so I took my first trips to Russia—to Star City, where the Russian Mission Control and Cosmonaut Training Center are. I met a lot of folks there, and we talked about what language we were going to speak in training and on the orbit and how we were going to work together as an international team. That was a fascinating part of being an astronaut that I never imagined when I first thought about it as a career."

In 1999, after spending many years contributing to the development of the ISS, Ellen was named a mission specialist and flight engineer for the *Discovery* STS-96 mission: the first Shuttle flight to dock to the ISS, to prepare the station for a permanent resident crew in early 2000. At that time, she was the mother of a 1-year old son: "My son turned a year old the day we flew down to Cape Kennedy for launch. Luckily, he was only a baby, so he didn't realize I wasn't there for his birthday." The 10-day mission, using the Spacehab in a Dual-Module configuration, delivered four tons of equipment and supplies to the space station. Ellen was loadmaster, tasked with coordinating the operations. She again handled the robotic arm, alternating with the Canadian Space Agency's astronaut Julie Payette:

> "I worked with the help of cameras and monitors because we were docked in a way that prevented me from seeing the robot arm. This made things more difficult, but then again, everything I've done on actual missions in space has always been easier than when I first tried it during training."

Following STS-96, Ellen served as CapCom (capsule communicator) in NASA's Mission Control Center in Houston, communicating with astronauts who were working in space. In 2002, she made her fourth mission to space on STS-110, again on *Atlantis*. By that time, the space station had grown and crews had lived on board for a couple of years. The primary goal of the mission was to add the first piece of the station's nearly 300-foot-long backbone, "Segment S-0," or the center-integrated truss assembly of the Integrated Truss Structure. The station's 11 integrated trusses consist of a linearly arranged sequence of connected trusses on which various unpressurized components are mounted, such as logistics carriers, radiators, solar arrays, and other equipment such as GPS antennas. Ellen, with Dan Bursch and Carl Walz, members of the space station's Expedition-4 crew, operated the Canadarm, the station's first robotic arm, to install the S-0 and supported the colleagues who worked outside the ISS during three of the four spacewalks necessary to complete the task. "During that flight," she recalls, "my second son turned two, so I got to say happy birthday to him from orbit."

After this last mission, Ellen was appointed Deputy Director of Flight Crew Operations at the JSC—the organization that manages the Astronaut Office and aircraft operations. She was in Mission Control when, on February 1, 2003, the Space Shuttle *Columbia* STS-107 was lost during re-entry, killing seven people. In 2006, she was named Director of Flight Operations. The following year, she was appointed Deputy Director of JSC and, in November 2012, became Director, in charge of overseeing about 13,000 employees.

Ellen—frequently asked to speak to students and teachers about her career and the success she has enjoyed as NASA's first Hispanic woman astronaut—regards this part of her job as an unexpected bonus: "I do as much speaking as I am allowed to do," she says. For years, she undertook two speaking engagements per month, the maximum number NASA allowed its astronauts. She has delivered hundreds of such talks, relishing the many chances she has to inspire young people to study mathematics and science. "I never thought about this aspect of the job when I was applying, but it's extremely rewarding," she noted in the Stanford University School of Engineering Annual Report, 1997–1998. "I'm not trying to make every kid an astronaut, but I want kids to think about a career and the preparation they'll need." She especially enjoys speaking to children with similar backgrounds to hers:

> "I think that it's important for children to have a role model to see what they can grow up to be. It's important they know that if they work hard, they can be and accomplish whatever they want. I am proud to be an example of that."

Ellen now has four schools named after her—one in Pasco, Washington; two in the Los Angeles, California area; and one in Grand Prairie, Texas—and is viewed by many young Latinos as a role model. Married with two children, she still plays the flute, as she did in high school.

References

"Ellen Ochoa", *encyclopedia.com.*

Gibson, K.B. *Women in Space: 23 Stories of First Flights, Scientific Missions and Gravity-Breaking Adventures*, pp. 119–124. Chicago Review Press, Inc., Chicago (2014).

Gueldenpfenning, S. *Women in Space Who Changed the World*, pp. 58–66. The Rosen Publishing Group, New York (2012).

Hasday, J.L. *Ellen Ochoa*, Chelsea House Pub., New York (2007), 106 pp.

Interview released to *scholastic.com* during the Hispanic Heritage Month in 1999.

Kevles, T.H. *Almost Heaven: The Story of Women in Space*, p. 188. The MIT Press, Cambridge, MA, and London, UK (2006).

Nevarez, G. "Ellen Ochoa, First Latina in Space, Makes History Once More", *huffington-post.com* (January 28, 2013).

"Ochoa Honored as Hispanic 2008 Engineer of the Year", *nasa.gov* (October 10, 2008).

"Ochoa Named Johnson Space Center Director: Coats to Retire", Press Release NASA J12-020, *nasa.gov* (November 16, 2012).

Official NASA biography of Ellen Ochoa, *jsc.nasa.gov* (March 2014).

Personal contacts with the Author by e-mail in April 2016.

"Preflight Interview: Ellen Ochoa", STS-96, *nasa.gov* (April 7, 2002).

Stanford University School of Engineering Annual Report, 1997–98 (February 23, 2002).

Stanford Women in Space. *Sandstone & Tile*, **38**(1), 3–12 (2014).

The Albert V. Baez Award for Outstanding Technical Achievements in Service to Humanity. *Hispanic Engineer & IT 1995*, **11**(3), 26–27 (1995).

Traynor, S. "Intervista a Ellen Ochoa", *amazing-kids.org*.

Woodmansee, L.S. *Women Astronauts*, pp. 89–91. Apogee Books, Burlington, Ontario, Canada (2002).

24

Janice Voss: Visits the International Space Station as a Cygnus Craft

Credit: NASA

U. Cavallaro, *Women Spacefarers*, Springer Praxis Books, DOI 10.1007/978-3-319-34048-7_24

Mission	*Launch*	*Return*
STS-57	June 21, 1993	July 1, 1993
STS-63	February 3, 1995	February 11, 1995
STS-83	April 4, 1997	April 8, 1997
STS-94	July 1, 1997	July 17, 1997
STS-99	February 11, 2000	February 22, 2000

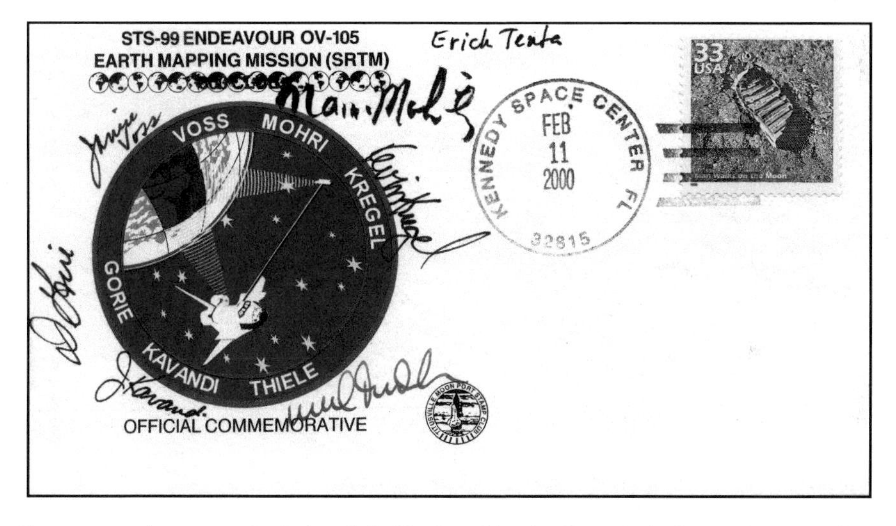

Commemorative cover of mission STS-99, signed by the Crew, including Janice Voss (from the collection of Umberto Cavallaro)

"Janice devoted her life to space and accomplished many wonderful things at NASA and Orbital Sciences, including five Shuttle missions. And today, Janice's legacy in space continues. Welcome aboard the ISS, Janice." With these words, she was welcomed on July 16, 2014, after the docking of CYGNUS CRS-2 to the International Space Station (ISS), on the 45th anniversary of the historical launch of Apollo 11. The Cygnus pressurized automated cargo spacecraft was christened "SS *Janice Voss*" in honor of Janice, who flew five Shuttle missions during her prolific astronaut carrier, and worked for Orbital Sciences before joining NASA.

During her career, Janice Voss had flown to Mir but never flew to the ISS. Frank Culbertson, Orbital's vice president and a former NASA astronaut, said:

> "Janice was a friend of many of us, both in the Orbital and NASA communities. We wanted to honor her and her family by naming this spacecraft for her. We think it is a fitting tribute to a really fine engineer and outstanding astronaut."

Janice Elaine Voss was born in South Bend, Indiana, on October 8, 1956, but considered Rockford, Illinois, to be her hometown. She began to be interested in space an early age. Her mother, Louise Voss, explains in an interview:

> "All started at the age of six, when Janice picked up a book at the local library: it was 'A Wrinkle in Time' by Madeleine L'Engle, a fantasy where one of the main characters is a scientist who happens to be a woman."

Janice became interested in this kind of reading and was an avid reader of both fiction and non-fiction by Isaac Asimov, Robert Heinlein, and Arthur Clarke. Some of her passions for life also included volleyball, dancing, and flying.

Janice began working for NASA in 1973 as a co-opted intern at the Johnson Space Center (JSC), in the frame of an experimental "job-school" program, when she was just 16 and a freshman at Purdue University. She worked in the Directorate for Engineering and Development, and contributed to the development of simulations for the Shuttle. In 1975, after obtaining her Bachelor of Science degree in Engineering Science from Purdue University, she returned for 1 year to JSC as a technical instructor of navigation systems for the crews.

In 1977, Janice obtained a Master of Science degree in Electrical Engineering from the Massachusetts Institute of Technology (MIT) and joined Orbital Sciences Corporation—the same company that launches to the ISS Cygnus spacecraft whose pressurized module is produced in Italy by the Turin-based Thales Alenia Space—where she contributed to the mission integration and flight operations support for an upper stage called the Transfer Orbit Stage (TOS). In the autumn of 1992, the TOS launched from a Titan III the unlucky Mars Observer, which later malfunctioned despite a successful launch. Again, in September 1993, the TOS launched from the Space Shuttle STS-51 the Advanced Communications Technology Satellite (ACTS). The primary and backup pyrotechnic devices inside the Super*Zip were erroneously fired simultaneously, resulting in minor damage to both the TOS and the Shuttle's payload bay. However, the ACTS satellite deployed successfully and functioned normally on orbit and the Shuttle landed safely despite the anomaly.

In 1987, Janice earned her doctorate in Aeronautics/Astronautics at the MIT. She was selected as a NASA astronaut for Group 13 in 1990, together with two other female astronauts: Nancy Currie and Ellen Ochoa. Initially assigned to the Astronaut Office Mission Development Department, she worked at the Spacelab and Spacehab modules, and then specialized in developing the robotic arms.

Janice was one of six women to have logged five spaceflights. She flew five missions in 7 years, spending a total of 49 days in space, and traveled 18.8 million miles in 779 Earth orbits. In each of her missions, she flew with another woman.

In 1993, with colleague Nancy Currie, Janice was the mission specialist of the Shuttle *Endeavour* STS-57 mission which retrieved EURECA—the European science lab, also built in Italy by Alenia Space—that had been freely floating in space for almost 1 year. This mission also was the maiden voyage of the Spacehab module, a 9600-pound pressurized laboratory—built in Italy for Spacehab Inc.—mounted in the orbiter's payload bay. Spacehab was the first commercial laboratory launched into space, its primary purpose being to offer industrial and academic researchers access to space. During the mission, Voss supervised 22 experiments that focused on the growth of animal organisms in space. According to the original plans, the mission was supposed to last 8 days but, due to adverse weather conditions, its landing was delayed by nearly 2 days.

In 1995, Janice participated in the STS-63 mission, a historic NASA mission that marked the first Shuttle rendezvous with Mir, which flew around the Russian space station without actually docking, to test communications and perform the dress rehearsal of the docking that would have been attempted with the subsequent mission. For the first time, the Shuttle was piloted by a woman, Eileen Collins. In that mission, Janice operated *Discovery*'s robot arm to grasp an astronomy satellite being deployed.

In 1997, Janice flew in space twice within a few months, together with Susan Still, piloting the Shuttle for the first time. In fact, in April, she took part in the STS-83 mission, which began with a delay of 24 h because of the need to improve the thermal insulation of a cooling pipe in the Shuttle cargo bay. On the third day, a malfunction was discovered in one of the three fuel cell electric generators and it was decided to turn it off immediately to avoid an explosion, and to consequently switch off the Spacelab equipment and abort the mission. In 30 years of Space Shuttle history, this was the only mission that had to be interrupted for technical reasons and the only time an entire crew was launched twice to achieve the same mission. The mission was repeated in July as mission STS-94, with the same crew and the same "payload." In both missions, Janice was payload commander in charge of 33 experiments. The crew set more than 140 small fires in insulated chambers to test the behavior of fire in weightlessness and to gain a better understanding of how fire and heat work in space and also to address safety concerns after fire had flared aboard the Mir station 5 months earlier. Janice also coordinated experiments on how plants adapt to extraterrestrial flight, using a greenhouse containing about 50 spinach, clover, sage, and periwinkle plants.

In February 2000, with her colleague Janet L. Kavandi on her second flight, Janice participated in the STS-99 mission—once again aboard *Endeavour*—that realized what to this day remains the most accurate digital topographic map of our planet. Again, she served as payload commander and worked on the Shuttle Radar Topography Mission, which mapped Earth's land surface at unprecedented resolution levels.

After this mission, Janice set her spacesuit aside, although the ISS was in her heart. She predicted:

> "I think the world will see 2001 as a major turning point in history, the time when our space odyssey took off that is when we began having people in space continuously for an entire year, with our shuttle flights and the International Space Station."

From October 2004 to November 2007, Janice moved from Houston to Moffett Field in California, at the NASA Ames Research Center, where she was the Scientific Director of the interstellar satellite Kepler, the NASA Space Observatory launched in March 2009 to hunt for exoplanets. As of October 2016, Kepler had identified 2331 exoplanets, plus another 4696 candidate planets awaiting confirmation.

In 2007, Janice returned to the JSC in Houston to lead the payload effort for NASA's Space Station Division of the Astronaut Office, with a focus on the ISS. Her great attention to the scientific education of young people led her to become a champion of the space program and to spend time on campus and around the country talking about what it takes to have the "right stuff," whether as an astronaut or as one of the many more who support the program from Earth. In an interview, she said:

> "In an era when no one who knows who the astronauts are as individuals—unless I'm wearing my flight suit, most people don't recognize me—everyone still wants to hear our story. And I have been able to use that spark of interest to help children and high school students understand the importance of studying science and math."

In 2009, Janice donated all of her personal documentation—including professional papers and videos of her spaceflights and interviews, but also records of her childhood and school report cards—to the archives of the library of the Purdue University: "Purdue," she commented, "has always made its astronauts feel like they are a special part of its family." She died of breast cancer, aged only 55, on Monday February 6 in Scottsdale, Arizona, where she was receiving treatment.

References

Cavallaro, U. "Janice Voss Visits the ISS as a Cygnus Craft." *AD*ASTRA, ASITAF Quarterly Journal*, **22** (October 2014), pp. 7–9.

Hevesi, D. "Janice Voss, Shuttle Astronaut and Scientist, Dies at 55", *nytimes.com* (February 10, 2012).

Kremer, K. "Cygnus Commercial Resupply Ship 'Janice Voss' Berths to Space Station on 45th Apollo 11 Anniversary", *universetoday.com* (July 16, 2014).

Norberg, J. "Purdue's 'Space Odyssey' Loses Astronaut Janice Voss", *purdue.edu* (February 7, 2012).

Official biography of Janice Voss, *jsc.nasa.gov* (March 2012).

Pearlman, R. "Janice Voss, Veteran of 5 Space Shuttle Flights, Dies at 55", *today.com* (August 2, 2012).

Pearlman, R. "SS Janice Voss Takes Flight: Station-Bound Spaceship a 'Fitting Tribute' to Astronaut", *collectspace.com* (July 13, 2014).
Ward, D. "An Early Interest in Science Fiction Put Purdue Grad on the Road to Space Travel", *roundaboutmadison.com* (April 2001).
Woodmansee, L.S. *Women Astronauts*, pp. 92–93. Apogee Books, Burlington, Ontario, Canada (2002).

25

Nancy Currie-Gregg: A Passion for Flying and Concern for Safety

Credit: NASA

U. Cavallaro, *Women Spacefarers*, Springer Praxis Books, DOI 10.1007/978-3-319-34048-7_25

Mission	*Launch*	*Return*
STS-57	June 21, 1993	July 1, 1993
STS-70	July 13, 1995	July 22, 1995
STS-88	December 4, 1998	December 15, 1998
STS-109	March 1, 2002	March 12, 2002

Nancy Currie-Gregg had served in the US Army as a helicopter instructor pilot and officer, before becoming an astronaut in July 1991. A veteran of four Space Shuttle missions, she has logged 999 h (41.5 days) in space. She was a mission specialist and flight engineer on STS-57 in 1993; STS-70 in 1995; STS-88, the first International Space Station (ISS) assembly mission in 1998; and STS-109, the fourth Hubble Space Telescope servicing mission in 2002.

Nancy Jane Sherlock Currie-Gregg was born in Wilmington, Delaware, on December 29, 1958, the youngest of four siblings. Her father was in the Army Air Corps and, when she was a little girl, her family moved to Troy, Ohio, which she considers to be her hometown:

> "I knew I wanted to fly since from a very early age: I mean, I just dreamed about flying probably from the time I could walk. My father was a bombardier on a B-29 in WWII, stationed on Tinian Island, which is where the Enola Gay launched off. I think that it was vicariously that I developed this love of flying because he always used to talk to my two brothers about it. I was the one that kind of listened. I was the one that wore his flight jacket to school. I was the one that took all his pictures of his airplanes in and went to air shows with him."

A career in the space program, or even in the uniformed military services, was however unthinkable during her childhood. And it really wasn't a concrete goal until much later in Nancy's life. She says:

> "I guess looking back on it I find it amazing because I was born in 1958 and in the late 50's and 60's a goal of becoming a military pilot or certainly an astronaut was

> not something that could not ever be a reality for a woman at that time. But no one ever told me that."

Only at the end of her time in high school did the military begin to accept female aviators, and only whilst she was at Ohio State University, studying biological sciences, did NASA begin to accept female astronauts:

> "So I was very, very fortunate that I was born at just the right time because in the mid 70's, when I was in high school, they opened up military flying to women. Of course in the late 70's, NASA hired the first women astronauts and so I was literally in the right place at the right time all the way through my career."

When she was in high school, Nancy used to read stories about the "medevac" pilots in Viet Nam and how they had abstracted the wounded from these landing zones in Viet Nam and she decided she wanted to do that: "I had a love for medicine. I had a love for flying. I thought what a great combination. So that was my original plan."

Upon receipt of her degree in 1980, Nancy entered the Army with a dream of becoming an aviator. Women at that time were gradually being introduced into combat positions and she went directly to flight school:

> "I went through air defense training at Fort Bliss, Texas, and then right into flight school. I had never flown anything in my entire life; And the first day they put me in a helicopter—and flying a helicopter is not all that easy or intuitive—and just as soon as I got in the aircraft, I said, 'This is for me! This is what I wanted to do'."

Nancy became a Master Army Aviator and was eventually assigned as an instructor pilot at the US Army Aviation Center. Other positions followed: section leader, platoon leader, and brigade flight-standardization officer. By the end of her military career in 2005, after serving in the Army for 23 years, she had logged over 3900 h in the air aboard various rotary-wing and fixed-wing aircraft: "It was the most fun, the most enjoyable job I've ever had in my entire life! I just loved training flight students and sharing with them the same enjoyment of flying and the discipline of flying that I had."

Meanwhile, Nancy had followed with her studies and in 1980 she received a bachelor's degree, with honors, in Biological Science from the University of Columbus, Ohio, and a Master of Science degree in Safety Engineering from the University of Southern California in 1985. She explains:

> "When I was in flight school, we actually had an accident that killed my instructor pilots and two of the guys that I flew with every day. It was just kind of a strange coincidence that I wasn't in the aircraft. And it was at that time that I decided to kind of devote a portion of my career and my academic life to safety and safety engineering. So, that's why I got a master's in safety."

Later, during her NASA life, Nancy went on to earn a doctorate in Industrial Engineering with an emphasis in automated systems and human factors engineering from the University of Houston in 1997. She adds:

> "Once I came down to Houston I then received a doctorate in industrial engineering. Again with an emphasis on human factors and safety engineering. Something I've

always held very close to my heart because I did see what catastrophic things can happen, due to human error in the cockpit, or human error combined with a malfunction in the aircraft."

In 1986, Nancy unsuccessfully applied for the first time for admission into that year's astronaut candidate class:

"My first time I applied was actually the year the Challenger happened, the application process was cancelled at that point. They resumed it again in '87. In '87 I was not selected but they asked me to come and work at NASA. I really thought it was a standard rejection call. You know, congratulation here, we didn't select you this time but we'd like you to come and work here. They were very serious. I came down and worked here for two and a half years as a flight simulation engineer on our shuttle training aircraft and then I was selected in 1990 to be an astronaut."

In September 1987, a few months after the birth of her daughter, Stephanie, in January that year, Nancy moved to the Johnson Space Center (JSC) as a flight simulation engineer on the Shuttle Training Aircraft. Selected as an astronaut in 1990, she completed the Astronaut Candidate Training Program in 1991. She worked as a spacecraft communicator (CapCom, providing a communications interface between ground controllers and flight crews), lead flight crew representative for crew safety and habitability equipment, and was chief of the Robotics Branch in the Astronaut Office, highly contributing to the analysis and development of the man–machine interface of the robotic arm. She co-authored 11 scientific papers on human factors engineering. She has also served in a variety of senior management positions at the JSC, including Chief of JSC's Habitability and Human Factors Office; Senior Technical Advisor in the JSC's Automation, Robotics and Simulation Division; and Deputy Director of Engineering and Chief Engineer in the NASA Engineering and Safety Center.

Nancy participated in four Space Shuttle missions. In 1993, she, then known as Nancy Sherlock, was first assigned to the *Endeavour* STS-57 mission, as flight operator and mission specialist. She was responsible for operating the Shuttle's robotic arm to retrieve the EURECA platform (which, built by Alenia Spazio in Italy, had floated in space for almost a year) and place it back into the *Endeavour* bay, while her colleague David Low conducted a spacewalk to manually position the EURECA communications antennas for latching. For the first time, the commercial laboratory Spacehab (built in Italy by Alenia Spazio) was carried in space, with 22 flight experiments on materials and life sciences research. She spent every single moment of her little spare time looking out of the window. She said later:

"I can't imagine taking a book into space other than I did fly a bible on all of my flights. I can't imagine spending time doing something other than watching the sunrise on orbit. Every hour and a half you see a sunrise and it is the most incredible experience you have ever seen. As the sun comes up on the horizon you just see every conceivable color painted on the horizon. You see these incredible reds and oranges and yellows and greens, even pinks and purples. Come over the south Pacific and see a lightning storm at night and it is almost like someone set up this giant electrical charge and you just see these flashes just dancing all across the ocean."

In 1995, Nancy participated in the *Discovery* STS-70 mission, which deployed the final NASA TDRS (Tracking and Data Relay Satellite), a massive spacecraft that completed the constellation of NASA's orbiting communication satellite system. The crew also conducted a myriad of biomedical and remote sensing experiments. She said during an interview:

> "We joke in the office that we're human guinea pigs for a week or two. The problem is that we need the International Space Station. In ten days or fourteen days in a shuttle you really can't get that long term research that really is required in something in like biomedicine."

At that time, the ISS did not exist. They started to build it up in 1998, Nancy's next mission. In 1998, she was in fact flight engineer and mission specialist in the STS-88 mission—the first Space Shuttle mission to carry hardware to space for the assembly of the ISS. The mission carried into space "Unity," the first American ISS module, which was connected to the Russian module "Zarya" that had been launched into space a few weeks earlier. Nancy had once again the heavy responsibility of operating the robotic arm. Flying to the ISS was the fulfillment of Nancy's dreams, but there was a lot of pressure, as she outlined in her preflight interview: "If I missed it, you know, we may not have a space station. I mean, this is our critical piece!" Everything went well and her dream came true. She said later:

> "Definitely, without a doubt the most memorable experience of my professional life has been to be a part of the crew, to enter the international space station for the first time and to make that call to the ground, 'Houston, this is the International Space Station'."

In 2002, Nancy took part in the *Columbia* STS-109 mission that was the fourth mission to service the Hubble Space Telescope. She was once again flight engineer: in her career, she had served as flight engineer on four Shuttle missions—more than any other female astronaut. Her primary role in this mission was to operate the Shuttle's 50-foot robot arm to capture and redeploy the Hubble telescope following the completion of numerous upgrades and repairs, and to support a series of five consecutive spacewalks performed by four crewmembers to help move equipment and astronauts who replaced both solar arrays and the primary power control unit, and installed the new Advanced Camera for Surveys (that replaced the Faint Object Camera) and a scientific instrument cooling system to upgrade the Hubble's capabilities. "I can't think of another mission," she said during the pre-flight interview, "that will be a better one to be on as the arm operator because, essentially, every single day except launch and landing, we're using the arm on this flight."

This mission was the last successful flight of *Columbia*, which, in early 2003, was destroyed during its re-entry into the atmosphere, killing the entire crew of seven. Following this tragedy, Nancy was selected to lead the Space Shuttle Program's Safety and Mission Assurance Office, assisting with NASA's Return to Flight efforts.

Nancy retired from Army in May 2005 after achieving the rank of Colonel and became then Deputy Director of the JSC Engineering Directorate. After her husband, Dave Currie, passed away in July 2011 from kidney cancer, in December 2015, Nancy married the award-winning broadcast journalist and corporate manager, Tim Gregg. She currently serves as a Principal Engineer in the NESC (NASA Engineering and Safety Center) at the JSC, Houston. She concurrently holds an appointment as an Adjunct Associate Professor in the Department of Industrial Engineering at North Carolina State University. "I want," she said in an interview. "to inspire young people to come this job because it is the most incredible job I could ever imagine. I want to give them the same opportunities that I've had."

References

Army Officer Makes Fourth Shuttle Flight. *The Eagle: United States Army Space & Missiles Defense Command*, **9**(2), 11–12 (2012).

Evans, B. "A Legacy of Women in Space: Twenty Years since STS-57", *americaspace.com* (June 22, 2013).

Georgsson, A. "Women and Space" (interview with Donna Fender and Nancy Currie), *digital.lib.uh.edu* (1999).

Official NASA biography of Nancy Currie-Gregg, *jsc.nasa.gov* (August 2014).

Personal contacts with the Author by e-mail in April 2016.

"Preflight Interview: Nancy J. Currie", STS-88, *spaceflight.nasa.gov* (April 7, 2002).

Woodmansee, L.S. *Women Astronauts*, pp. 90–92. Apogee Books, Burlington, Ontario, Canada (2002).

26

Chiaki Mukai: The First Japanese Astronaut

Credit: NASA

U. Cavallaro, *Women Spacefarers*, Springer Praxis Books, DOI 10.1007/978-3-319-34048-7_26

Mission	*Launch*	*Return*
STS-65	July 8, 1994	July 23, 1994
STS-95	October 29, 1998	November 7, 1998

Commemorative cover of mission STS-65, signed by Chiaki Mukai (from the collection of Umberto Cavallaro)

Chiaki Mukai was one of the first three Japanese astronauts selected in 1985, the first group of the Astronaut Corps of the Japanese Space Agency NASDA, which was later merged into the Japan Aerospace Exploration Agency (JAXA). Besides being the first Asian astronaut to fly into space, she also was the first Japanese citizen to make two trips: a remarkable achievement in a country where few women can hope to reach executive positions, much less outer space. The choice of Chiaki Mukai was hailed in Japan as a tangible sign of respect that finally the woman had conquered, in a traditionally male-dominated society, although initially—she remembers—were asked questions like "How do you feel to be a woman astronaut?" to which she once replied: "Did you ever ask a man how he feels to be a man astronaut?"

In the Foreword of the book *Women in Space—Following Valentina*, Chiaki Mukai writes:

> "I have often been asked if it was my childhood dream to become an astronaut. The answer is 'no'. The thought never occurred to me until I was thirty-two years old. When Sputnik orbited the Earth in 1957, I was only five years old, and as aware of the significance of the event as most other Japanese were: that is to say, not at all. The 'space firsts' that marked the next decades inspired me to read the biographies of the history-makers; but that was the extent of my interest in space exploration, which seemed to be another world entirely. My childhood dream was much more immediate and personal. I wanted to be a doctor, and to help those, like my younger brother, suffering from diseases. He had aseptic necroses—a rare disease which made his leg bones brittle. Our family watched him struggle to walk, and the teasing by other children made our hearts heavy with sadness. My parents eventually took him to a big university hospital in Tokyo, and as his condition improved, so did my determination to become a doctor. When I was ten, a composition I wrote in school, entitled 'What will I be in the future?', promised as much. I left my parents' home at fourteen, and moved to Tokyo to prepare for medical school."

Chiaki Mukai was born on May 6, 1952, in Tatebayashi, a small village located near Tokyo, in the prefecture of Gunma, which is now a Tokyo suburb, where her father taught in a junior high school and her mother ran her own shop. She was the oldest of four siblings.

In 1977, Chiaki became a doctor, specializing in cardiovascular surgery, and started to teach at the Medical School of the Keiō University, Tokyo—one of the main Japanese universities and the oldest in the country. There she met Dr. Makio Mukai, whom she married in 1982.

In the Foreword of the abovementioned book Chiaki says:

> "One morning in December 1983 as I was relaxing in my office after a night on duty in the intensive care unit, a newspaper article caught me by surprise. The Japanese space agency was looking for candidates to fly onboard the Space Shuttle in 1988. I literally shouted, 'Gee! Can someone from Japan actually fly in space?' I thought (stereotypically) that space travellers had to be either American or Russian. I did not know that a German had flown on Spacelab-1 just days earlier, much less that a Czech, a Pole, another German, a Bulgarian, a Hungarian, a Vietnamese, a Cuban, a

Mongolian, a Romanian and a Frenchman had already flown to Soviet space stations. I did not even know that we had had a Japanese space agency—the National Space Development Agency—since 1969. The article held another surprise for me: the candidates were to have scientific backgrounds and conduct experiments in space. But were not astronauts always pilots and aviators? With a shock, I realised that science and technology had progressed to the point where ordinary people living and working on Earth were actually able to do the same kind of work in space. We were now entering the era of space utilization."

"I became more intrigued with the possibility of seeing our beautiful blue planet from outer space with my own eyes. Would such a magnificent sight deepen my way of thinking and expand my concept of life itself? At the same time, I was fascinated by the possibility of using the spaceflight environment—especially weightlessness—for research purposes. Here was an opportunity to contribute my medical expertise to the space programme."

In 1985, Chiaki was selected as one of three Japanese payload specialist candidates for the First Material Processing Test (FMPT) which flew in the Spacelab-J mission, aboard STS-47. But she did not fly on that mission—where she was backup payload specialist and served as a Spacelab communicator for crew science operations—and started her long space odyssey. She explains:

"My path into space had highs and lows. The first high was being one of three candidates, selected from 533 applicants, for a Spacelab-J mission in 1988. But the first low, four months later, was the loss of the Space Shuttle Challenger. In the wake of the tragedy and the ensuing uncertainty about the future of the space programme, I spent days in consideration and soul-searching over whether to abandon my second dream and return to my first: the medical field."

The Chernobyl accident in the Soviet Union a few months later led Chiaki to dramatically reflect on the vulnerabilities of technology. Eventually, she took the difficult decision to remain in the space program. In 1988, she achieved a doctorate in Physiology at the Keiō University in Tokyo:

"Other lows followed, as our mission was delayed repeatedly into 1992, and then when I was selected as the back-up and not as a member of the prime crew. Of course I was disappointed for a while, but the lows became highs as the training put me in an advantageous position to understand the mission as a whole, and to witness how many people it took, all working together, to make the mission successful. This preparation served me well when I finally achieved my second dream and flew into space in 1994."

After over 10 years of training and delays, Chiaki finally embarked on the Shuttle STS-65 mission, and spent 2 weeks, establishing, at the time, the longest record of endurance of a woman in space. She was in space in July 1994 when the 25th anniversary of the first Moon landing of Apollo 11 was celebrated. The research of the IML-2 Spacelab, the second International Microgravity Laboratory, was also more ambitious than in IML-1, in which her Canadian colleague, Roberta Bondar, was involved. During the mission, 82 experiments were conducted dealing with life science (human physiology, space biology, radiation biology, and bioprocess) and microgravity science (material and fluid science, research on microgravity environment, and countermeasures). Chiaki was responsible for the experiments on the cardiovascular system, the vegetative nervous system, and bone and muscle metabolism. She fondly remembers the sensations felt in microgravity, the lightness, the strange sensations of the eyes, ears, and touch, which are the senses that tell us how we are positioned in space, and especially she remembers the feeling of heaviness she felt in her arms when she returned to Earth:

> "I knew the gravity that exists on earth. But I was not aware of it. After being in space, your body adapts to microgravity. When you get back, you feel your body is so heavy. Your body is pulled to the center of the earth. When I came back from space, I developed a special sense for about a day. I felt the weight of the business card and the paper."

In 1998, Chiaki flew again—becoming the first Japanese astronaut to fly in space twice—on board the STS-95 mission, in which the veteran US astronaut and Senator John Glenn, the pioneer who in 1962 was the first American to fly in Earth orbit of Mercury, returned to space aged 77. Relying upon her medical expertise, Chiaki worked with the Senator to study spaceflight and its relationship to the aging process. Glenn said that Chiaki had "more energy than anyone I know of." The appreciation was mutual: "I have learned a lot from his positive attitude," said Chiaki, referring to John Glenn. "He's so energetic. I was fortunate to work with such an American hero."

On the seventh day on *Discovery*, during a conversation with Japan's Prime Minister Keizo Obuchi and Yutaka Takeyama, the head of the Science and Technology Agency, Chiaki read a poem that she had written about being in space: "Weightlessness—repeating somersaults in space—as many times as I like." NASDA asked the Japanese public to think of two lines with seven syllables to finish her poem, making it a "tanka"[1] with an overall rhythm of 5–7–5–7–7. More than 40,000 people, also from abroad, took part in the competition and submitted a total of 144,781 verses. The youngest submitter was 5 years old and the oldest was 101.

During the mission, Chiaki conducted various experiments on modification in the human organism and aging in microgravity. She also studied sleep in space and repeated the experiment already performed during the Neurolab mission on Shuttle STS-90 to evaluate the effectiveness of melatonin in astronauts and explore ways to combat the insomnia complained of in space. She logged a total of 566 h (over 23 days) in space.

Chiaki was assigned to a third mission to the International Space Station (ISS), scheduled for launch in 2004, and was preparing for the mission an experiment that she had proposed on the effects of microgravity and radiation, while coordinating, as deputy mission scientist for STS-107, the scientific operations of that mission. The mission—after several delays—flew in 2003 and ended tragically with the destruction of the Shuttle

[1] *Tanka* are five-line poems of 31 syllables split into a pattern of 5–7–5–7–7. This form has been in use in Japan since at least the seventh century and is still an important and very popular style for serious and amateur poets today.

Columbia over the skies of Texas during re-entry. After that, she decided to abandon her "second dream" and left NASA.

From 2004 to 2007, Chiaki worked as a visiting professor at the International Space University (ISU) in Strasbourg, France. A few years later (in 2015), she would be awarded with France's highest honor—the Legion of Honor—for her contribution to strengthening cooperation between France and Japan in the field of space exploration.

In 2007, Chiaki returned to Japan and, until March 2011, she directed the Space Biomedical Research of JAXA that meanwhile had taken the place of NASDA. In April 2011, she became Advisor of the JAXA's Managing Director and, in July, was appointed the first Director of the newly organized JAXA Center for Applied Space Medicine. She produced over 60 scientific publications. In April 2015, she assumed the position of Vice President of the Tokyo University of Science, her responsibilities including the promotion of internationalization and women's affairs. From the beginning of 2016 she is chairing a subcommittee of the U.N. Committee on the Peaceful Uses of Outer Space, based in Vienna, Austria.

References

Donwerth-Chikamatsu, A. "Chiaki Mukai, Japan's First Woman Astronaut", *anniedonwerth-chikamatsu.com* (November 7, 2015).

Gibson, K.B. *Women in Space: 23 Stories of First Flights, Scientific Missions and Gravity-Breaking Adventures*, pp. 176–180. Chicago Review Press, Inc., Chicago (2014).

Gueldenpfenning, S. *Women in Space Who Changed the World*, pp. 83–90. The Rosen Publishing Group, New York (2012).

"ISU Faculty, Dr. Chiaki Mukai Honored by France", *isunet.edu* (February 4, 2015).

Kevles, T.H. *Almost Heaven: The Story of Women in Space*, pp. 126–128. The MIT Press, Cambridge, MA, and London, UK (2006).

Mukai, C. "Foreword". In: D.J. Shayler and I. Moule (eds), *Women in Space – Following Valentina*, pp. XIII–XV. Springer/Praxis Publishing, Chichester, UK (2005).

Mukai, C. "Kibo Promises Development in Space Medicine", *jaxa.jp* (2003).

Official JAXA biography of Chiaki Mukai, *jaxa.jp* (January 8, 2013).

Official NASA biography of Chiaki Mukai, *nasa.gov* (October 2003).

Personal communication with the Author through e-mail in April/May 2016.

"Preflight Interview: Chiaki Mukai", STS-95, *spaceflight.nasa.gov* (January 21, 2003).

Shimbun, A.S. "Best Closing Lines for Mukai's Poem Awarded", *web-japan.org* (March 1999).

Thornton, E. "Our Home Planet Is So Beautiful: It's Fragile … and Has Dignity", *businessweek.com* (June 14, 1999).

27

Elena Kondakova: The First Woman to Take Part in a Long-Duration Space Mission

Credit: NASA

U. Cavallaro, *Women Spacefarers*, Springer Praxis Books, DOI 10.1007/978-3-319-34048-7_27

Mission	*Launch*	*Return*
Soyuz TM-20	October 3, 1994	March 22, 1995
STS-84	May 15, 1997	May 24, 1997

Elena Kondakova was the third Russian woman to fly in space, 10 years after the second, Svetlana Savitskaya, although, unlike her, Elena did not have an influential father. Her only link to the space world—she insists, however, that it had no influence on her selection as a cosmonaut—was her marriage in 1985 to Russian cosmonaut Valeri V. Ryumin, 18 years her senior. Ryumin, a three-time space traveler, after being on the Salyut space station for more than 300 days, and twice named a Soviet Union Hero, had been appointed Director of the Russian Shuttle–Mir program in 1992. Not only did he not support her, Elena says, but he was for a long time her nightmare. She explained:

> "My husband vehemently opposed this idea. He had already been in space three times. He finally agreed to let me apply because he was convinced that I would not pass the medical tests. When I did pass them, he told: 'I never doubted that I had a healthy wife.' But he was not pleased with the success."

At the time, the rules set by Ryumin, who, as deputy chief designer at NPO Energia, had signed an order forbidding the assignment of women cosmonauts to long-duration missions, were still in force.

Like Svetlana Savitskaya had done 10 years before, Elena also set a record with her first flight: she was in fact the first woman to perform a long-duration mission in space and, with her 169 days on Mir, she established the record of endurance for a woman in space. Observers didn't miss that the assignment of Elena suddenly sprung up immediately after the announcement that Norm Thagard would be the first American astronaut to participate in a long-duration flight in two decades, and would beat the records set by his colleagues on Skylab. According to American observers, this was just the latest in a long history of cynical propaganda used by Russia to score political points over the US, pretending to apply disproportionate levels of political importance to their female cosmonauts.

Elena Vladimirovna Kondakova was born in Mitischi, Moscow Region, on March 30, 1957, the year of Sputnik. She was raised near Kaliningrad, where OKB-1 was based: the historic Special Design Bureau 1 of Korolev that, after the Fall of the Wall, became the private NPO Energia and today is known as RSC-Energia—the state organization that traditionally leads Russian aerospace production and is involved in the building of all spacecraft and all the components of the former Soviet, and now Russian, space stations. Both her parents worked there: her father headed one of Energia's labs and her mother worked there as an accountant. So Elena did not learn from science-fiction movies. She says:

> "I lived in a real city where real people worked, real cosmonauts returned after their flights. I saw around me a real thing. We were very proud that it was in our city that spacecraft were built and cosmonauts trained."

Elena's parents sent her and her older brother to the Moscow Bauman High Technical College where she graduated in Engineering in 1980. "They knew," she says, "that I was mathematically and technically inclined." A lover of the quiet life, in both the sports she practiced (cycling and river fishing) and the entertainments she enjoyed (theater and reading), unlike her female predecessors Tereshkova and Savitskaia, she did not learn to jump with a parachute or pilot a plane. Upon graduation, in 1980, she started to work in RSC-Energia, completing science projects, experiments, and research work. She was soon noticed for her skills and recommended for her candidacy as a cosmonaut. But she hadn't much support from her family: "My father worked in this area and dealt with actual tests. He knew how really dangerous it was." And he did not want his only daughter to be in danger.

Elena decided, however, to apply to enter the space program and, in 1989, she was selected as a cosmonaut candidate by RSC-Energia main Design Bureau and was sent to Gagarin Cosmonaut Training Center to start the space training. She was assigned to the 17th Mir mission "Euromir-94" and was flight engineer when she blasted off on October 4, 1994, on board the Soyuz TM-17 spacecraft to reach the orbital complex Mir. Ryumin protested often and loudly, and provided many interviews maintaining that he would have preferred a wife who worked only a few hours a day, took care of their child, and always had dinner waiting for him. In one interview, he said:

> "It's my opinion that a wife should stay at home, not at work and not in spaceflight. I think the majority of men will support me because the majority of us would prefer that everything in our homes is taken care of, and everything is quiet."

In the end, he accepted the situation and, since life in Russia is a life without babysitters, Energia reassigned Ryumin to his home to take care of their daughter until the return of her mum. On her return, Elena said:

> "In principle, life is the same as here on Earth: we have an eight hour working day and we have time for personal rest, we have time when we can watch movies and what is going on Earth and life is easier onboard because you don't have to do laundry there, you don't have to cook there. So I think that, for a woman, being in space is kind of a vacation from home work."

Elena was on Mir when, in February 1995, STS-63, the Space Shuttle *Discovery*—with Eileen Collins as a pilot—performed its first rendezvous test with Mir, without docking. She says:

> "I was very happy because, when we were on orbit in February, STS-63 approached the station with Jim Wetherbee, but they only approached us to a distance of ten meters. We wanted so badly for them to dock to our station and visit us and we were so certain that we could just see them but we didn't have an opportunity to work together."

Jim Wheterbee, the Commander of Shuttle STS-63, recalls that, during the approach of the Shuttle towards Mir, he saw Elena waving and holding up to the window a little doll, a little cosmonaut doll, and "at 30 ft you could see their eyes and their smiles and they were waving and we were waving." Mike Foale, who eventually flew with Elena during mission STS-84, reported: "Elena had described how unexpectedly beautiful it was when the shuttle came up and how disappointed they were when flew away without actually docking."

The expedition had to perform lot of maintenance on board Mir. As Elena explained in 1996:

> "When the station was launched initially, we had planned to fly the station for only three years. Of course, like any mechanical system, device, or machine, it has its own resource which should be extended at some point. Periodically, we have to replace devices and elements and conduct unplanned activities, but it proves one more time that our technology has great achievement; we had planned to fly it for three years and it has been in flight for 11 and we plan to extend this flight even more."

In May 1997, Elena made her second spaceflight aboard the US Shuttle *Atlantis* and returned to Mir for 9 days within the Shuttle–Mir program. She was the seventh Russian cosmonaut to fly on the Shuttle as mission specialist during the *Atlantis* STS-84 mission, with Eileen Collins as pilot for the second time. That expedition eventually earned her a medal from NASA. "I picked a winning lottery ticket," she said. She was the first Russian cosmonaut to fly on both Soyuz and the Shuttle. In an interview with *Voice of Russia*, Elena said that she felt safer on the Russian spacecraft:

> "We all remember the disaster of US shuttle Challenger in 1986. Unlike Russian spacecraft, American ones lacked reliable rescue system. Our astronauts also did go through very risky situations but life-saving equipment at Soyuz capsules was very good. It helped Vladimir Titov and Gennady Strekalov survive an accident during a spaceflight in 1983."

On March 23, 2001, Elena traveled with several other cosmonauts and private citizens to the South Pacific, where they watched the Mir space station re-enter Earth's atmosphere. It was a sad thing for her and many others to watch the end to the place they had called home for a time.

Elena logged over 178 days in space altogether. Since 1999, she has served as a deputy in the Duma, the lower house of the Russian parliament.

References

Borenstein, S. "Husband's Remarks Don't Fly with Cosmonaut", *orlandosentinel.com* (May 3, 1997).

Evans, B. "A Cog in a Political Machine: The Career of Svetlana Savitskaya", *americanspace.com* (February 10, 2012).

Gibson, K.B. *Women in Space: 23 Stories of First Flights, Scientific Missions and Gravity-Breaking Adventures*, pp. 62–65. Chicago Review Press, Inc., Chicago (2014).

Interview with Elena Kondakova, NASA SP-4225, *history.nasa.gov* (September 6, 1996).

Intervista, *voiceofrussia.com* (April 10, 2011).

Kevles, T.H. *Almost Heaven: The Story of Women in Space*, pp. 145–151. The MIT Press, Cambridge, MA, and London, UK (2006).

Official NASA biography of Elena Kondakova, *nasa.gov* (July 1997).

28

Eileen Collins: The First Woman "Shuttlenaut" in the Driver's Seat

Credit: NASA

U. Cavallaro, *Women Spacefarers*, Springer Praxis Books, DOI 10.1007/978-3-319-34048-7_28

Mission	*Launch*	*Return*
STS-63	February 3, 1995	February 11, 1995
STS-84	May 15, 1997	May 24, 1997
STS-93	July 23, 1999	July 27, 1999
STS-114	July 26, 2005	August 9, 2005

Commemorative cover of mission STS-114, signed by Eileen Collins (from the collection of Umberto Cavallaro)

Eileen Collins broke the last gender barrier in the NASA Astronaut Corps when, in 1990, she was accepted as the first woman astronaut-pilot and started in NASA, her career punctuated by firsts. She made history when, in February 1995, she became the first woman to pilot a spacecraft, the Space Shuttle *Discovery* STS-63, the first of four Shuttle missions, during which she accumulated a number of records. Four years later, in July 1999, while the world was celebrating the 30th anniversary of the first Moon landing, she would become the first woman to command a Space Shuttle. It had taken 20 years for NASA to embrace wholeheartedly the idea of admitting women into their Astronauts Corps, and it took another 20 years to entrust to a woman with the command of a spacecraft.

Eileen Marie Collins was the second of four children, born of Irish parents on November 19, 1956, in Elmira, New York: an appropriate birthplace for a would-be pilot, since the city is known as the "soaring capital" of the US. Some of her earliest and fondest childhood memories were visiting the Harris Hill with her parents to watch planes take off or standing around the local airport. She was inspired by the Mercury astronauts and dreamed of flying but, at that time, there were no such opportunities for a young girl. She says:

> "When I was very young and first started reading about astronauts, there were no women astronauts, but I still wanted to do it. And I didn't think there'd be any reason that a woman couldn't do it. Maybe I was thinking that women just hadn't asked."

Eileen soon realized that her goal was unusual for a woman: at the time, most women stayed home and brought up children. There were plenty of women nurses and teachers, but few engineers or military officers. So she decided not to advertise her plan. She says:

> "I never told anybody I wanted to be an astronaut or pilot. I consciously never talked about it because I knew people would say, 'You can't do that.' And I just didn't want to fight it: it wasn't worth it. Even when I started my flying lessons—and this would have been when I was between my junior and senior year in college—I didn't tell my friends. I don't think I even told my parents."

After graduating from Elmira Free Academy High School in 1974, Eileen went on to a community college, as she was needed at home to help out with the family, and, in 1976, she obtained an associate degree in Math and Science from the Corning Community College. Meanwhile, she took a part-time job, saving the money until she could pay for flying lessons on a little Cessna 150 at the Elmira Corning Regional Airport, but did not have time to take her private pilot's license. As soon as she had learned that the armed forces had started opening doors to women in 1976, she applied to enter the Air Force. While, in 1978, after she had obtained her degree in Mathematics and Economics from the University of Syracuse, she was about to take the last flight examination, the US Air Force called her, 15 days ahead of schedule. She enlisted and was assigned to Oklahoma's Vance Air Force Base, where she attended the Undergraduate Pilot Training, which that year was admitting women for the first time. Right at the beginning of the course, the base was visited by new astronauts of the 1978 class, which, for the first time, had accepted women astronauts. "I decided," she says, "that I wanted to be part of our nation's space program. It's the greatest adventure on this planet, or off the planet, for that matter. I wanted to fly the Space Shuttle."

After graduating from the pilot's training, Eileen took the final check ride and finally got her private Federal Aviation Administration (FAA) pilot license. The following year, she became an instructor pilot for the T-38, the high-performance military jet, also used at that time by NASA for astronauts' training and private astronauts' transport. For 3 years, she was the only woman who was an instructor in the Squadron. She says:

> "The years I was there I was very careful that I did a good job so I could really leave a good impression for the women to follow. I didn't want the women to have a hard time of it. I wanted to make it as easy for them as I could."

From 1983 to 1985, Eileen moved to the Travis Air Force Base in California, where she became commander and instructor pilot of Lockheed C-141 Starlifter, a military strategic transport aircraft that served only the US military and was never exported. In 1986, she left all the activities for 1 year and concentrated on studies at the Air Force Institute of Technology and earned a Master of Science degree in Operations Research from Stanford University. She was eventually assigned to the US Air Force Academy in Colorado as an assistant professor in mathematics and as a T-41 instructor pilot.

In 1987, Eileen married Pat Youngs, a civilian pilot in Delta Airlines, whom she had met when they were both flight instructors in California. In 1989—"by going to school in the evening for two years," she said in an interview—she earned a Master of Arts degree in Space Systems Management from Webster University, and applied to become an astronaut. Meanwhile, she was accepted in the prestigious school for military test pilots of Edwards Air Force Base, with 800 colleagues (she was one of the first four women ever admitted). She was highly motivated to finish the course, since, during the pilot training, she had learned that she had been selected for the 13th group of NASA astronauts: the *Hairballs* group, so nicknamed from the black cat that was used on an early patch design, then rejected by NASA, that they carried for luck during the selection, with reference to their "lucky 13." In June 1990, she was the second woman to graduate as a test pilot at Edwards.

Eileen had applied both as a mission specialist and as a pilot, and, in the phone call from Houston, John Young told her "You are going to be a pilot. You are going to be our first woman pilot." The requirements for test pilots were more stringent that those for mission specialists. "It was like a dream come true," she says. She moved to Houston in July and became the first woman pilot ever selected—a prestigious job: for the general public, the pilot continued to be regarded as the *true* astronaut, the one with the "the right stuff" according to the successful expression which became common saying after the famous book by Tom Wolfe. With the arrival of Collins, those at Houston began to joke about the double meaning of "right stuff."

Eileen discourages people from looking at what other astronauts are into and choosing that. The exact opposite worked for her: when she joined the corps, there were no astronauts in her field: operations research: "I said I think I can fill a void, and I think they bought it." It's paid off, too, she said, since much of her background ties in directly with the operation of the Shuttle.

After the astronaut training, Eileen's first job was in Space Shuttle engineering for about a year and she then moved for 15 months to the Kennedy Space Center as "Cape Crusader," supporting the people at the Cape, getting the orbiters ready to fly and the crews that had to be launched or had returned from their mission. "It was just a super job because I got to work in the actual Space Shuttle itself," she says. She also served as

CapCom (capsule communicator) for a few months. She gradually assumed various responsibility positions within the Astronaut Office and soon realized that her career path was quite different from that of a mission specialist. As a pilot, she would become a commander—a job that would put her in a position to choose the crew and in orbit, where there is a strict command chain, she would be at the top. And she also was potentially Chief of the Astronauts Office. And everyone knew, and behaved accordingly.

In February 1995, Eileen actually became the first woman ever (apart from Tereshkova in her solo flight) to pilot a spacecraft, the Shuttle *Discovery* STS-63, which was the first flight of the new joint Russian–American space program. For the occasion, she invited to the launch, as guests of honor, the legendary pioneering "Mercury-13" women to whom she felt particularly close, as they had dreamed of flying and were qualified when she still was in kindergarten. But their country didn't allow them to fly and they had been beaten by a Russian who, much less qualified than them, had the honor of making history as "the first woman in space." Seven took up her invitation.

Mission STS-63 was a "Near-Mir" mission—the first test rendezvous with the Russian space station in which the Shuttle approached to 10 meters without actually docking:

> "We were the first Americans to see the Russian Space Station. We didn't dock, but we did a close approach rendezvous to 30 feet. Tested out the shuttles, navigation, communication, and flying systems to make sure that the subsequent flight, which would be the first docking would be ready to go."

Janice Voss also flew in the same mission as mission specialist.

After 1 year of maternity leave following the birth of her daughter, Bridget, in May 1997, Eileen was for the second time Shuttle pilot in *Atlantis* STS-84, the sixth Shuttle–Mir mission:

> "We were the shuttle crew that docked with Mir in between the two major accidents that they had: the fire they had in March '97, and collision with a Progress vehicle they had in June '97. We got to be really friends and co-workers with that Russian crew in space."

Eileen flew with Elena Kondakova. One of the mission's goals was to bring Linenger back to Earth after his dramatic stay on Mir: during the fire, he had feared never getting back home. After that flight, Eileen became the Chief of the Space Hardware Branch in the Astronaut Office, which included the International Space Station (ISS) and the Space Shuttle.

In July 1999, with mission *Columbia* STS-93, Eileen became the first woman to command a Shuttle. "My daughter," she said in an interview, "thinks that all moms fly the space shuttle." At that time, Bridget was only 3 years old. There was a tremendous amount of publicity around this mission because of the first woman commander, both before and after the mission. The mission was scheduled for launch on July 20, the day of the 30th anniversary of the first Moon landing, and a large crowd had exceptionally gathered to see the night start of the first female commander and occupied all the hotels within 50 miles.

But the countdown was stopped at T–7 s, only to discover soon after that there were in fact no technical problems of a hydrogen leak in an engine of the Shuttle, as a faulty sensor had indicated. Once the engines are stopped, however, it takes 48 h to restart them. And, at half past one at night, all the people had to go back to their hotels. The next launch was postponed again due to adverse weather conditions and an electrical storm. It lifted off

3 days later, at night, using the latest "launch window" (otherwise the launch would have had to have been delayed for weeks).

The goal of this mission was to take up the largest and most powerful X-ray telescope that had ever been launched into space: the US$1.5 billion-dollar Chandra X-ray Observatory, initially known as AXAF (Advanced X-ray Astrophysics Facility), designed to enable X-ray exploration from high-energy regions of deep space, such as exploding stars, quasars, and black holes. The astrophysicist Cady Coleman was the mission specialist responsible for deploying Chandra. It wasn't, however, a nominal launch. During the lift-off, a voltage drop shut down the redundant main-engine controllers on two of the three engines. Eileen maintained a cool head and, with quick thinking, managed to guide *Columbia* to orbit. The orbit attained was, however, seven miles short of that originally planned due to premature main-engine cut-off an instant before the scheduled cut-off. The Shuttle landed at Kennedy Space Center in one of the rare night landings in the program's history.

After that mission, Eileen had her second child. As for many working mothers, the balancing act required was, for her, the hardest part of the job. The last mission she flew kept her away from home for 5 weeks because the launch was delayed: "But it was manageable. I would say there's a lot of people in professional jobs around this city, around the country, that spend more time away from their families than I ever did as an astronaut."

In 2005, Eileen ended her remarkable military career and retired from the US Air Force with the rank of Colonel, after logging 6751 flight hours in 30 different models of aircraft. She commanded her last mission in July 2005: the STS-114 mission that marked the "Return to Flight" of the Space Shuttle after the loss of *Columbia* on February 1, 2003. The launch was delayed by over 2 years due to the *Columbia* accident, but the opportunity to visit the ISS is why she decided to keep flying at a point in her career when many astronauts were retiring from spaceflight: "Our team had prepared this mission for four years. It's hard to wait," she said, recalling that she was about to give up, but the desire to visit the ISS made her want to keep flying. "I had never gone to the Station, and I really wanted to go to there. I really wanted to be part of the Station mission." On the eve of the launch, the debate on the Shuttle's safety was rekindled in the media. One of the most difficult challenges that Eileen had to face was to reassure her daughter, who now was seven and was following the debate with some apprehension. She had to do the same thing with her crew, earning the nickname "Mum." The future of the space program depended upon the success of this "Return to Flight" mission; as a result of taking every precaution, there were many delays.

Collins showed her incredible skill when she performed for the first time a new security procedure: she was the first pilot to make the "Rendezvous Pitch Maneuver" in which she took the Shuttle close to the ISS and slowly flipped the Orbiter a full 360° on itself, to make the heat shield visible while the ISS crew took hundreds of high-resolution photographs of the heat-resistant tiles, eventually sent back to Houston for analysis to identify any possible critical damage. This maneuver, since then, became habitual to decide whether the Orbiter was safe for re-entry.

After the mission, in May 2006, Eileen left NASA. "It has been wonderful," she said, "but the number one thing for me now is to spend time with my family." Her daughter Bridget was then 10 and her little boy Luke was 5 years old. She said that she also hoped her retirement would give newer astronauts an opportunity to fly before the Shuttle fleet itself retired in 2010:

"It's important that these young people get a chance to fly. It's very important to the country to have more people that have flown in space because we take that spaceflight experience with us, which is a valuable thing to have when you go on to design future spacecraft and educate young people. I know there are qualified women out there who would love to do this job, and I encourage them to look at this job and to realize that I have had an extremely rewarding career with a lot of flexibility. I'm married. I've had two children while I was in the astronaut office. In the 16 years I was here, I've flown four missions and had two children, and I've been able to do that without too much heartache."

A veteran of four spaceflights, Eileen has logged over 872 h (over 36 days) in space. But she still has some regret, since, during her spaceflights, there has been little time to just enjoy being in space: "Someday I would like to go into space as a tourist, and have the time to have fun." Hoping for developments in the field of civilian spaceflight, she adds: "I would like to see more people traveling to space someday. I would like to see space tourism blossom. It's such an incredible experience."

References

Briggs, C.S. *Women Space Pioneers*, pp. 79–86. Lerner Publications, Minneapolis (2005).

Dean, B. "Eileen Collins: An Astronaut's Endless Endeavor", *nasa.gov* (January 8, 2007).

"Eileen Collins – NASA's First Female Shuttle Commander to Lead Next Shuttle Mission", *nasa.gov* (October 4, 2003).

Gibson, K.B. *Women in Space: 23 Stories of First Flights, Scientific Missions and Gravity-Breaking Adventures*, pp. 125–133. Chicago Review Press, Inc., Chicago (2014).

Gueldenpfenning, S. *Women in Space Who Changed the World*, pp. 67–73. The Rosen Publishing Group, New York (2012).

Kevles, T.H. *Almost Heaven: The Story of Women in Space*, pp. 171–176. The MIT Press, Cambridge, MA, and London, UK (2006).

Malik, T. "NASA's First Female Shuttle Commander Retires from Spaceflight", *space.com* (May 1, 2006).

Mattson, W.O. and Davis, V. "Curation Paper10 – Interview with Eileen Collins", *nmspacemuseum.org* (Winter 2014).

Official NASA biography of Eileen Collins, *nasa.gov* (May 2006).

Contacts with Author by e-mail in June 2016.

Wilson, J. "Eileen Collins – NASA's First Female Shuttle Commander to Lead Next Shuttle Mission", *nasa.gov* (October 4, 2003).

Woodmansee, L.S. *Women Astronauts*, pp. 96–97. Apogee Books, Burlington, Ontario, Canada (2002).

29

Wendy Lawrence: The First Woman Astronaut of the US Navy

Credit: NASA

U. Cavallaro, *Women Spacefarers*, Springer Praxis Books, DOI 10.1007/978-3-319-34048-7_29

Mission	*Launch*	*Return*
STS-67	March 2, 1995	March 18, 1995
STS-86	September 25, 1997	October 6, 1997
STS-91	June 2, 1998	June 12, 1998
STS-114	July 26, 2005	August 9, 2005

Wendy Lawrence was the third American astronaut to arrive at Start City to train for a Shuttle–Mir mission and was the first woman of the US Navy to fly into space.

Wendy Barrien Lawrence was born in Jacksonville, Florida, on July 2, 1959. After watching the historic walk on the Moon, the 10-year-old Wendy was instantly bitten by "the space bug" and took a giant leap of her own—sitting in front of a black-and-white television, she decided that she wanted to be an astronaut: "I was actually probably more fortunate than most other kids in that my dad was involved in the selection process for the original group of astronauts, so I had some inside information." Her father had in fact been a finalist for the 1959 Mercury astronaut selection.

After graduating from high school, following her father, who was pilot in the US Navy and received a Congressional Medal of Honor during the Vietnam War, Wendy entered the US Academy at Annapolis in 1977 and was awarded a Bachelor of Science degree in Oceanographic Engineering in 1981. In 1982, she became a US Navy pilot:

> "I grew up in a Navy family: both my father and my mother's father were Navy pilots. My mother's father had been in the Navy, had gone to the Naval Academy and gone into naval aviation. My dad went to the Naval Academy, went into naval aviation. So I grew up surrounded by planes, and certainly developed a fascination for them. I also knew that a lot of the first astronauts had gone to the Naval Academy and been naval aviators, so I thought, kill two birds with one stone; I'll go to the

> Naval Academy and I'll become a naval aviator, always keeping in the back of my mind that where I wanted the path to lead was to NASA."

Wendy was one of two women who participated in combat actions on the Indian Ocean, assigned to the HC-6, the 6th Helicopter Combat Support Squadron. In 1988, she received her master's degree in Oceanographic Engineering at the Woods Hole Oceanographic Institution (WHOI) and was assigned to Anti-submarine HSL-30, the 30th Helicopter Squadron Light, as officer in charge of the ALPHA detachment. In October 1990, she passed at the Naval Academy as an instructor and novice women's crew coach. She logged over 1500 flight hours in six different types of helicopters and over 800 shipboard landings. She was selected by NASA as an astronaut of the 14th group in March 1992. Her technical assignments within the Astronaut Office included flight software verification in the SAIL (Shuttle Avionics Integration Laboratory), Astronaut Office Assistant Training Officer, and Astronaut Office representative for space station training and crew support.

Wendy participated in four Shuttle missions. Her first mission was in 1995 when she flew as mission specialist aboard *Endeavour* STS-67 that, for the second time, delivered into space the ASTRO-2, the second dedicated Spacelab mission to conduct astronomical observations in the ultraviolet spectral regions, using three unique instruments to observe objects ranging from some inside the Solar System to individual stars, nebulae, supernova remnants, galaxies, and active extragalactic objects. Her on-board duties included stowage and crew equipment and filming the crew.

After this mission, Wendy was assigned as John Blaha's backup for the NASA-4 long-duration Mir mission. Four days before leaving for Russia to start her training, she accidently discovered a memo sent from the Russian space agency—that nobody in the office had seen before—which said that the minimum height for Soyuz had been changed from 160 cm to 164 cm, to use the Russian Orlon spacesuit, which is the suit that Russians use for their spacewalks. As she was exactly 160 cm tall, she wasn't qualified anymore. She already had the ticket to Moscow and her chief, Bob Cabana, proposed that she should go to the Star City anyway as Russian crusader or Director of Shuttle–Mir Operations. She agreed and helped Shannon Lucid and John Blaha to prepare their mission. Meanwhile, Valery Ryumin (Program Manager for the Russian Phase-1) wrote a letter granting her a waiver and stating that she could start training for a flight to Mir, provided that she would never be considered for a spacewalk in the Russian spacesuit. So she could start her training for the sixth increment, scheduled for September 1996, on board *Atlantis* STS-86. The biggest challenge of that period was the language, since all the training was in Russian:

> "Our instructors were very committed to their task, they knew their area of expertise very well, and they spent a lot of hours making sure that they're able to answer the most in-depth question from the crewmembers that they're training, so they were truly professionals. The frustrating part was trying to do it in another language, when you felt like you didn't have the level of proficiency that you needed; we weren't 100 percent sure of what the person was saying, being able to catch maybe 80 percent of the conversation and then the frustration of not being able to exactly ask the question that you wanted to ask, because your vocabulary didn't enable you to do that."

The whole nature of the program suddenly changed when, in June, the Progress resupply vehicle collided with the Station's Spektr module and damaged a radiator and one of four solar arrays on Spektr. For a while, it looked as though none of the America astronauts would get a chance to go up to Mir anymore:

> "It looked like it would probably be the end of the program, but the Russians persevered. They became focused on repairing Spektr with space walks. They were very determined that that was the course of action that they wanted to take, and based on that, both sides felt that all three crew members on board should be able to do an EVA in a Russian suit, and it would be an opportunity for U.S. astronauts to get some more space walk experience, particularly in the Russian suit. So in July '97 Bob Cabana replaced me with my backup. He felt pretty adamant about making sure that my participation in the Phase One Program at that point would be rewarded: I would fly on STS-86 and fly on STS-91. I knew that I was getting two space flights out of it, that wasn't a bad deal for me. Dave Wolf got a space walk out of it, and I got to fly twice and got to fly on 91 as the flight engineer, which is a job that I really enjoy doing, so I think we were both satisfied with how everything worked out."

Wendy was impressed by the Russian space station: "It looks so different than Shuttle. Shuttle is clean and systems look good. And Mir, it looks old, it looks tired, and there's all this stuff." People working at the Control Center in Moscow really had no concept of what Mir looked like on the inside. They had lost track. Valery Ryumin himself, who had helped in designing Mir, had said: "After about three years of operation, we completely lost track of what is on board Mir."

Wendy was responsible for transferring five tons of cargo, stow, scientific experiments, and spare parts to repair the damaged Mir module. At the end of the mission—after 6 days of intense work—Shuttle *Atlantis* undocked and performed a 46-min fly-around visual inspection of Mir to check the repaired module:

> "The fly-around on STS-86, I think will always be very memorable. It's like, we have these two 100-ton spacecrafts that are literally doing this dance around one another, while both of them are traveling around the Earth at 17,500 miles per hour. And then Mir, the modules are white, and with the sun shining on them, they absolutely glisten, and then that, set against the very, very blackness of space, is a strikingly beautiful sight. I think it's a memory I'll have all of my life. But then the fact that we were doing this in space and that two different countries that had formerly been enemies were now closely cooperating, so that we could maneuver these vehicles around one another."

In 1998, Wendy flew aboard *Discovery* STS-91 and was flight engineer during ascent entry and then in charge of all of the 2.7 tons of logistics transferred between the two vehicles: a very demanding task. She says:

> "It always goes very quickly on orbit, unfortunately. That's the philosophy of a short-duration mission, is to maximize every moment while you're up there. With a long-duration mission, you know that you have time and you don't have to work as

furiously to get everything done. During our ten days, we were very, very busy, I think at times to the detriment of being able just to enjoy each other's company, to go over and spend some time in the Mir base block and get a chance to talk with your Russian counterparts. Certainly it's difficult to grab a couple of minutes just to look out the window. You have to fight to protect that time, also."

Surprisingly, Mir looked better than it had 5 years before:

"The cosmonaut crews over the years had kept it in good shape. In fact, Mir looked better than it did on STS-86. So we have to give them compliments because it takes a lot of effort to keep that thing up and running."

This was the ninth Shuttle–Mir mission, which ended Phase One, the first phase of the collaboration program between the US and Russia:

"Closing the hatch was hard for me, because it was the end of a program that I had spent many, many years participating in, but I think we went out on a good note. it was a great way to close out what has been a very successful program."

Wendy's fourth and last spaceflight was the *Discovery* STS-114 mission in 2005—the mission to the International Space Station (ISS) that, under the command of Eileen Collins, marked the American "Return to Flight" after the devastating tragedy of *Columbia* in 2003. The *Columbia* tragedy was for her a hard experience. She said:

"It's hard to lose one friend … but to lose seven of them at once is just absolutely devastating. I think I was very well aware of the risk, but since the Shuttle is an experimental vehicle and we are continuing to learn things about flying in space and how difficult that is, I'm not sure, on a daily basis, I was aware of all the risks that are out there. I can honestly say that I never really thought that entry posed more of a risk than ascent; now I'm very much aware of that. And as far as commitment to my work, I think the key message that I've taken away from the accident is that we can't take anything for granted."

One important goal of this first mission after the accident was to show that NASA had implemented the Columbia Accident Investigation Board (CAIB) recommendations to the best of its ability and that astronauts had the inspection capability, that they could determine the health of the Shuttle's TPS (Thermal Protection System), and, if necessary, they had techniques that they could use to repair tiles. They used lasers to detect very small damage to the leading edge and sent down images to the ground. And, in fact, the following day, during a spacewalk on the exterior of a spacecraft in flight, Stephen Robinson for the first time demonstrated repair techniques. During this mission, Wendy operated the ISS robot arm to transfer five tons (over 11,000 pounds) of equipment and supplies on the station, and subsequently remove and stow on the Multi-Purpose Logistics Module (MPLM) Raffaello 3.2 tons (over 7000 pounds) of experiments and unneeded equipment and trash materials for return to Earth.

Altogether, Wendy logged 1225 h (over 51 days) in space. After 14 years, she left NASA in June 2006 to pursue interests in private-sector spaceflight and, in November that year, she joined Andrews Space, Inc. (integrator of aerospace systems and developer of advanced space technologies based in Houston, Texas) as a Senior Advisor for Human Spaceflight and Crew Safety. She also participated in the initial development of the Rocketplane Kistler K-1 crew and cargo module.

References

"2004 Preflight Interview: Wendy Lawrence", *nasa.gov/vision/space/preparingtravel* (January 20, 2004).

Kevles, T.H. *Almost Heaven: The Story of Women in Space*, pp. 167–168. The MIT Press, Cambridge, MA, and London, UK (2006).

Nevills, A. "STS-114 Mission Specialist Wendy Lawrence: 'Hanging 10' for Shuttle Countdown", *nasa.gov/vision/space/preparingtravel* (March 7, 2005).

Official NASA biography of Wendy Lawrence, *jsc.nasa.gov* (August 2006).

"Preflight Interview: Wendy Lawrence", STS-114, *nasa.gov/vision/space/preparingtravel* (February 23, 2005).

Shayler, D.J.; Moule, I. *Women in Space – Following Valentina*, pp. 272–273, 280, 284, 300, 328–329, 341. Springer/Praxis Publishing, Chichester, UK (2005).

Woodmansee, L.S. *Women Astronauts*, pp. 98–99. Apogee Books, Burlington, Ontario, Canada (2002).

Wright, R. "Interview with Wendy Lawrence", *spaceflight.nasa.gov/oral-histories* (July 21, 1998).

30

Mary E. Weber: From Skydiving to Stellar Strategies

Credit: NASA

U. Cavallaro, *Women Spacefarers*, Springer Praxis Books, DOI 10.1007/978-3-319-34048-7_30

Mission	*Launch*	*Return*
STS-70	July 13, 1995	July 22, 1995
STS-101	May 19, 2000	May 29, 2000

A lover of math and science since she was in college in 1983, Mary E. Weber also became an avid skydiver and participated in US national championships, logging thousands of skydives in a few years. "It's a sport that you can do for many years," she says, "and there's always room for improvement and always another challenge around the corner. That's what I liked about both fields, science and aviation."

Mary Ellen Weber was born on August 24, 1962, in Cleveland, Ohio. After graduating from Bedford High School in Bedford Heights, Ohio, in 1980, she joined the faculty of Chemical Engineering of the Purdue University, led there by the reputation of the university which—she discovered later—Neil Armstrong, Eugene Cernan, "Gus" Grissom, and Roger Chaffee had helped to create: "I became interested in Purdue," she says, "because I love math and science and figuring how thing work, and Purdue is the top engineering school in the Country and I wanted to learn from the best." She took part in an internship program with Ohio Edison, Delco Electronics, and 3M. After majoring in Chemical Engineering with honors from Purdue in the spring of 1984, she went on to earn a Ph.D. in Physical Chemistry from the University of California at Berkeley in 1988, exploring the physics of gas-phase chemical reactions involving silicon. While at the university, she discovered the space program and decided to send a request to NASA for an application form for the astronaut program:

> "When I was growing up, I never even considered being an astronaut. That was not something women did. It didn't even enter my mind as a possibility. It wasn't until I was in graduate school in chemistry and I was very much into science and research. I'd also gotten involved in a lot of aspects of aviation, and space just seemed like the perfect adventure and the perfect mix of all the things that I loved: science and aviation."

Mary waited, however, until she had completed her Ph.D. and accomplished other things, including her pilot's license, before sending in her application in 1991:

> "I'm the kind of person that just likes to try new things and likes to experiment and likes to be bold. I like to try different things, and that's true in science, that's true in my hobbies. And I think all of those experiences led me to try the biggest, the grandest adventure that I could imagine."

Mary then joined Texas Instruments (TI) to research new processes for making computer chips. TI assigned her to a consortium of semiconductor companies, SEMATECH, and subsequently to Applied Materials, to create a revolutionary reactor for manufacturing next-generation chips. She received one patent and published nine papers in scientific journals. In 1992, she was selected for the 14th NASA astronaut group.

During the 1990s, Mary participated in several US National Skydiving Championships and was awarded silver medals in 1991, 1995, and 1997. In 2002, she participated in the largest completed freefall formation world record, with approximately 300 people. She recalls:

> "It was actually in Russia. It's tough to get 300 people to go out to dinner at the same place, so it's quite an endeavor. You have to find a site with planes that are large enough, with facilities. It's a logistical nightmare. Everybody has to be in the formation in their predetermined slot, and we had some stragglers who didn't get into their slot before time ran out."

She then had to suspend the skydiving activities—which she later resumed—because of the NASA rules which forbid astronauts from practicing hazardous activities.

Mary participated in two Shuttle missions, STS-70 and STS-101, logging more than 450 h (almost 19 days) in space. After only 2 years in the Astronaut Corps—she was among the youngest to fly in space—she was selected for her first Space Shuttle mission, STS-70 *Discovery*, which was launched from Kennedy Space Center on July 13, 1995, only 6 days after the landing of Shuttle *Atlantis*, marking the fastest turnaround between flights in the history of the program. This mission was known as the "All Ohio" flight, since four of the five crewmembers were from Ohio: Mary herself, the Commander "Tom" Henricks, and the mission specialists Donald A. Thomas and Nancy J. Currie. Mary was responsible for controlling the complex docking module and overseeing final "capture" of the International Space Station (ISS). She explains:

> "We go into an orbit that's slightly different than the actual space station's orbit. And we have to use Newton's laws of gravity and orbital mechanics and speed up and slow down appropriately so that our two orbits eventually are going to be the same orbit, in order to match the station's. It's a complex effort, as we have two vehicles that are going over 17,000 miles (over 27,000 km) per hour."

Mary was also responsible for the checkout and launch of the 7th TDRS, the last communication satellite of the TDRSS (Tracking and Data Relay Satellite System), the network of American communications satellites designed to replace the existing network of ground stations that had supported all of NASA's manned flight missions since Mercury, Gemini, and Apollo. She also performed pioneering biotechnology experiments never before possible, growing in microgravity human colon cancer tissues. In her very intensive activity, Mary also operated the Shuttle's robotic arm to maneuver the fellow spacewalkers around the Shuttle and station to retrieve and install equipment. Her view was obstructed, since the station was docked within inches of the crew compartment windows—a unique situation at the time. She therefore developed new techniques to use only camera views and animation during these very delicate operations. She was also in charge of the Italy-built Spacehab logistic module installed in the back of the Shuttle's bay, which provided storage and living space, and she oversaw the internal transfer of thousands of pounds of equipment to the station.

In 2000, Mary flew on her second spaceflight, STS-101: an early construction mission for the ISS. One of the goals of the mission was to deliver to the space station over two tons of equipment contained in the Spacehab module and install the components for life-support systems and complete electrical wiring, by operating both inside and outside the station. During the mission, some critical hardware was repaired or replaced, including air filters, Zarya fire extinguishers, and smoke detectors, four suspect batteries on Zarya, a Radio Telemetry System memory unit, and other components of the communication systems. Mary also maneuvered the long robotic arm to give support to her colleague Jim Voss during extravehicular activities, and to move the cargo from Spacehab to the station. In the same mission, another woman astronaut also flew: Susan Helms.

The scope and complexity involved in building the station, which ultimately would exceed the size of a football field, has been unmatched throughout the history of space exploration. Before her mission, Mary said:

> "This is an unbelievable endeavor that we, NASA, are trying to do with the other international partners. Technologically, it is more complicated than anything we've ever done before, and culturally, bringing together the world, having the world act as a single unit to do something to advance our whole society, is something that's never been done before. There are going to be things that you can't anticipate, no matter how meticulously you plan. Pretty soon, we'll realize the dream, and we are going to have an orbiting laboratory in space."

The significance and uniqueness of this STS-101 mission led the American TV channel A&E to produce a behind-the-scenes documentary, *Mission Possible*, chronicling the crew's 18 months of preparations. After returning from the mission, Mary received a master's degree of Business Administration from Southern Methodist University.

During her 10-year career with NASA, Mary held different positions. She served as Kennedy Space Center *Cape Crusader*, participating in critical launch, landing, and test operations, and helped in developing standards and methods for crew science training. She was then Chairman of the procurement board for the Biotechnology Program contractor and she also served as a key liaison in government relations and on an oversight team that revamped the US$2 billion space station programs across the country. She was involved in commercializing technologies, looking for ways to move the NASA space experiments to the marketplace. As part of a team reporting to NASA's chief executive, she worked directly with a venture capital firm to successfully identify and develop a business venture leveraging space technologies to assess prospective businesses for their market potential, feasibility, and various risks. In addition, Mary was the Legislative Affairs liaison at NASA Headquarters in Washington, DC, interfacing with Congress and traveling with NASA's chief executive.

After leaving NASA in December 2002, Mary served for 9 years as Vice President of the Southwestern Medical Center of the University of Texas, an internationally renowned research center, hospital complex, and medical school in Dallas, Texas. In 2012, she founded Stellar Strategies, LLC, a company focused on providing consulting in strategic communications which offers leadership development for leaders and teams in high-risk high-stakes operations to minimize risk and handle emergencies in such critical environments as oil industries, chemicals, or transportation, and prepares for making the right decisions in time-critical situations by adopting strategies and proven techniques from spaceflight, skydiving, and aviation. She also serves on a number of boards, including among others the NASA Advisory Council Committee on Technology, Innovation and Engineering—which advises NASA on it's portfolio of future technologies—and the Board of Trustees of Texas Health Presbyterian Hospital in Dallas, which experienced the first Ebola patient in the US.

Aviation, in many aspects, is still a passion for Mary, who still is an active competitive skydiver with over 5000 skydives, 18 medals to date at the US national championships, and a world record. She continues to be—after more than 25 years—a member of the widely known Deguello skydiving team, which has been a formidable contender at the US National Skydiving Championships throughout this time, medaling every year and overshadowing all other amateur teams from around the US: the only non-professional 16-person team to successfully beat professional teams including the Army Golden Knights. She is also an instrument-rated pilot, with 800 flight hours, including 600 on jets.

Mary is married to Jerome Elkind, who, after applying twice to become an astronaut and being finalist in both Group 15 and Group 16, founded in 2009 Stellar Generation, LLC, a company focused on producing ultraclean affordable fuel and renewable energy.

References

"About Mary Ellen Weber", *stellarkeynotes.com.*

Dismukes, K. "STS-101 Preflight Interview: Mary Ellen Weber", *spaceflight.nasa.gov* (April 7, 2002).

Norberg J. *Wings of Their Dreams: Purdue in Flight*, pp. 341–342. Purdue University Press, West Lafayette, Indiana (2003).

Official NASA biography of Mary E. Weber, *nasa.gov* (October 2004).

Woodmansee, L.S. *Women Astronauts*, pp. 100–101. Apogee Books, Burlington, Ontario, Canada (2002).

31

Cady Coleman: The Colonel Playing the Flute in Space

Credit: NASA

U. Cavallaro, *Women Spacefarers*, Springer Praxis Books, DOI 10.1007/978-3-319-34048-7_31

Mission	*Launch*	*Return*
STS-73	October 20, 1995	November 5, 1995
STS-93	July 23, 1999	July 27, 1999
Soyuz TMA-20	December 15, 2010	May 23, 2011

Cady Coleman's father was in the military. She says:

> "I grew up in a family with a dad who did undersea exploration. He was part of a program when we first designed habitats where men could live under the sea, and so I actually thought that exploration was normal."

Catherine "Cady" Grace Coleman, the second of four siblings, was born on December 14, 1960, in Charleston, South Carolina, and lived in a number of places around the country: San Francisco, Virginia Beach, and Washington, DC. She moved to Fairfax, Virginia, when she was 12. After graduating from Wilbert Tucker Woodson High School, Fairfax, in 1978, she was, until mid-1979, an exchange student at Røyken Upper Secondary School in Norway with the AFS Intercultural Programs. Back in the US, she joined the Massachusetts Institute of Technology and, in 1983, she received a Bachelor of Science degree in Chemistry.

While many boys grew up wanting to be astronauts, Cady said that she looked with interest and curiosity at the space program, but she had never thought of becoming an astronaut herself—that was not something within the reach of a girl, when she was young—until she happened to meet Sally Ride while she was working at her dissertation at the Massachusetts Institute of Technology in 1983. She recalls:

> "After Sally Ride made her first flight, she came to MIT and she talked to the women students, and I just looked at her and I thought, I want that job. She was a scientist and at the same time she was also somebody that was helping to explore the universe, and she got to fly jets, scuba dive, all these things that I loved, and I was actually just so inspired to meet her and it made a big difference to me. I'd seen a lot of astronauts on TV, in pictures; none of them looked like me. It was a bunch of guys that seemed a lot older to me and they didn't have much hair, and it just didn't really make me think, that could be me, and then I meet Sally Ride and I think, maybe that could be me."

This was the turning point in her life, when she knew what she wanted to do: "I was interested in being in the military, I was interested in science, I loved chemistry, and at the same time I always wished for just a little bit more adventure than just the laboratory." The career of the astronaut would offer her the opportunity to fly and explore and she decided "I want that job!"

That same year (1983), while she had almost finished her multidisciplinary dissertation in Polymer Science and Engineering, Cady joined the US Air Force ROTC, the Reserve

Officers' School of the Air Force, as a second lieutenant, where she met Pamela Melroy and the two became great friends. She took active service in the US Air Force in 1988 after earning her doctorate and was assigned as a chemical researcher at the Materials Directorate of the Wright Laboratory at Wright-Patterson Air Force Base in Ohio, where she worked in polymer chemistry and optical applications in computer science and data storage. In this position, she also contributed to the analysis of the surface of the LDEF (Long-Duration Exposure Facility)—the experimental platform that NASA had placed in Earth orbit in April 1984, deployed by Space Shuttle *Challenger* during the STS-41C mission, and left freely floating in space for nearly 6 years to study the effects of long-term exposure to space of materials, components, and systems, in view of the design of the space station. The platform was then recovered in January 1990 by the Space Shuttle *Columbia* during mission STS-32, in which two women astronauts also flew—Marsha Ivins and Bonnie Dunbar, who had actually recovered it using the robotic arm.

While she was at the Wright-Patterson, Cady volunteered to be a test subject in the centrifuge program at the Crew Systems Directorate of the nearby Armstrong Aeromedical Laboratory, which was doing medical trials for NASA. "In several occasions," she recalls, "the five women who were part of the twenty people volunteer panel outperformed the men." She set several endurance and tolerance records during her participation in physiological and new-equipment studies. She decided to go back to school and, in 1991, received her doctorate in Chemical Engineering from the University of Massachusetts and applied to enter the NASA Astronaut Corps. She was selected for the 14th group of astronauts in 1992.

Initially assigned to the SAIL (Shuttle Avionics Integration Laboratory) for flight software verification, Cady subsequently served in the Astronaut Office Payloads and Habitability Branch, working with experiment designers to ensure that payloads can be operated successfully in the microgravity environment of low Earth orbit. As the lead astronaut for long-term habitability issues, she led the effort to label the Russian segments of the International Space Station (ISS) in English and also tracked issues such as acoustics and living accommodation aboard the station. She served for a number of years as a CapCom (capsule communicator) in Mission Control for both the Space Shuttle and the ISS. She represented the Astronaut Office in the Tile Repair Team for NASA's "Return to Flight" after the *Columbia* accident and she also served as the Chief of Robotics for the Astronaut Office, tasked with overseeing astronaut robotics training and the integration of crew interfaces into new robotics systems.

Fig. 31.1 The patch of the STS-73 mission features the shape of the Cupola because the research performed in the United States Microgravity Laboratory (USML) was paving the way for space station science. At that time the International Space Station still didn't exist: it was built up starting from 1998, 3 years later. Credit: NASA

Cady is a veteran of two Space Shuttle missions (STS-73 and STS-93) and one long-duration mission on the ISS (Expedition 26/27). In all three missions, she was responsible for operations with the robotic arm.

Mission STS-73, the second USML (United States Microgravity Laboratory) flight, was one of longest Shuttle missions. The launch was postponed six times, two times due to adverse weather conditions and other four due to technical problems. During the 16-day scientific mission, experiments on material science, biotechnology, combustion science, and fluid physics were conducted in the Spacelab. (At that time, the ISS still didn't exist: it was built up starting from 1998.) In an interview, Cady says (Fig. 31.1):

"If you look at our patch you'll see actually the shape of the Cupola, because our job was to go and pave the way for space station science, to allow scientists to really maximize the resource of microgravity up there."

Cady experienced the hectic schedule of the short-duration Shuttle missions. She says:

"I personally interacted with 30 experiments, many of which were in fluid physics and crystal growth. This is where my training as a scientist really came in handy, as it is a challenge to work quickly but well. It is like having 30 customers in 16 days. There isn't time for mistakes. You need to go up there and do your best."

In order to provide the best stability for the experiments, the *Columbia* was flown in a "gravity gradient attitude," with its tail pointed towards Earth. Except for the pilot, the crew consisted of only scientists. Only the commander Kenenth Bowersox and the payload commander Kathryn Thornton had flown before.

During her second mission STS-93, Cady, as the lead mission specialist, was responsible for deploying the Chandra-X-ray Observatory, the NASA orbital telescope designed to enable scientists to study exotic phenomena such as exploding stars, quasars, and black holes, and to improve our knowledge and understanding of the universe. This was the first mission ever commanded by a woman, Eileen Collins, who—like Cady—had been a test pilot of military jet aircrafts.

In October 2004, Cady participated in the seventh NEEMO expedition (NASA Extreme Environment Mission Operations), the NASA program aimed at investigating the survival in a hostile, alien place for humans to live, during space-exploration simulation missions performed in the Aquarius underwater laboratory, the world's only undersea research station. Operated by Florida International University (FIU), Aquarius is located 5.6 km (3.5 miles) off Key Largo in the Florida Keys National Marine Sanctuary and is deployed next to deep coral reefs 62 ft (19 m) below the surface. During this expedition, Cady lived and worked underwater for 11 days. She retired from the US Air Force in 2009, with the rank of colonel.

In July 2010, Cady was flight engineer on the Russian Soyuz TMA-20 spacecraft. As part of the long-duration expedition 26/27—in which the Italian astronaut Paolo Nespoli also participated—she spent 159 days aboard the ISS, contributing to many scientific research experiments, including the investigation of the deterioration of the human skeletal structure:

"We are lab rats, and I don't mind being a lab rat. It is part of what we go up there to find out. How does space really affect us? One of the really exciting areas I think is osteoporosis—what happens to bones and bone health as we get older. Up in space we don't necessarily get older any faster but we do by floating around in, in microgravity, we don't actually put stress on our bones and we lose bone mass at a very high rate, much higher than your average 70 year old woman with osteoporosis, and that means in just a few months, we can understand what happens to bones when they start to, quote-unquote, dissolve, and how do they rebuild themselves, and how can we prevent that from happening. Here on the Earth we can look at those questions but they take a few years to look at, and we're looking a population that often has a number of different kinds of health problems in addition to osteoporosis, whereas we as astronauts often have a fairly clean medical history.

"Another example is the investigation on our circulatory system. When we go up to space where our hearts do not have to actually pump blood from our feet all the way up to our head against the force of gravity, and it means that we can observe our hearts under some different circumstances that allow some of those really subtle factors to be better understood, and so on our mission we're doing a very comprehensive experiment called the Integrated Cardiovascular Experiment, and it's a whole bunch of different investigations all over the body that have to do with circulation, with breathing, with blood chemistry associated with those, with the physical effects on your organs. We're going to do ultrasound on each other to look at our hearts when we first get up there, after a month, after 2 months, after 3 months, because this different behavior of the heart up in microgravity means that it's going to actually start being a different size, perhaps, have a different strength in muscle volume, and we are doing a number of tests to measure all those different factors."

Her long-term expedition started on December 15, 2010—the day after her 50th birthday—when she lifted off from the Baikonur Cosmodrome aboard the Soyuz TMA-20 and returned to Earth on May 23, 2011, aboard the Soyuz TMA-19, landing in the steppes of Kazakhstan. She was in space when important anniversaries, such as the 50th anniversary of the enterprises of Yuri Gagarin and Alan Shepard, the first American in space, and the 30th anniversary of the first Shuttle flight, were commemorated.

An accomplished flutist, Cady carried some flutes with her on the ISS and, on April 12, 2011, to salute 50 years of human spaceflight, she performed the first space–Earth musical duet, playing—with the musician Ian Anderson, the founder of the Jethro Tull rock band—an arrangement of the famous Bach's *Bourée* which Anderson and Jethro Tull performed during their 1969 US tour as Neil Armstrong and Buzz Aldrin stepped on the Moon. Anderson played his part from Perm, Russia, while Cady played her part from 220 miles above Earth. The two parts where played separately and then joined together (Fig. 31.2).

Cady also gave advice and tips from space to actress Sandra Bullock while she was shooting *Gravity*, the science-fiction movie directed by Alfonso Cuarón, that won seven Oscars at the 86th Academy Awards and compliments from NASA. According to Coleman,

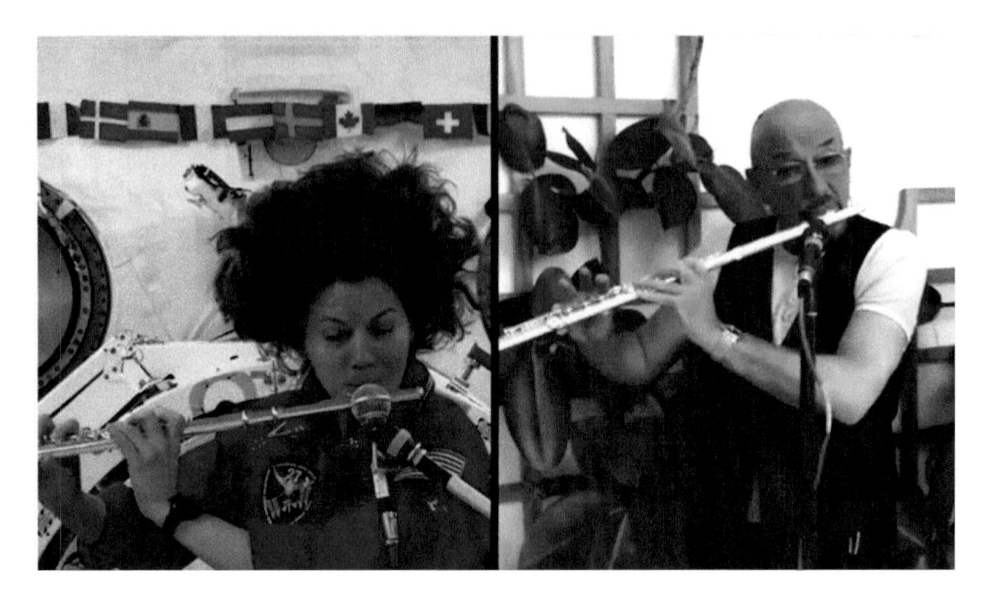

Fig. 31.2 Flute duet—astronaut Cady Coleman playing on the International Space Station and Ian Anderson playing on ground. Credit: NASA

although life and movie are two different things, having put the spotlight on the astronaut's work helps to make us think:

> "Our planet sits in a neighborhood within the universe, and we are all space explorers. I think space movies, in general, bring that message home to us. Whether we live with our feet on the planet or whether we live on the space station, we are all space travelers, and we are a people of space exploration."

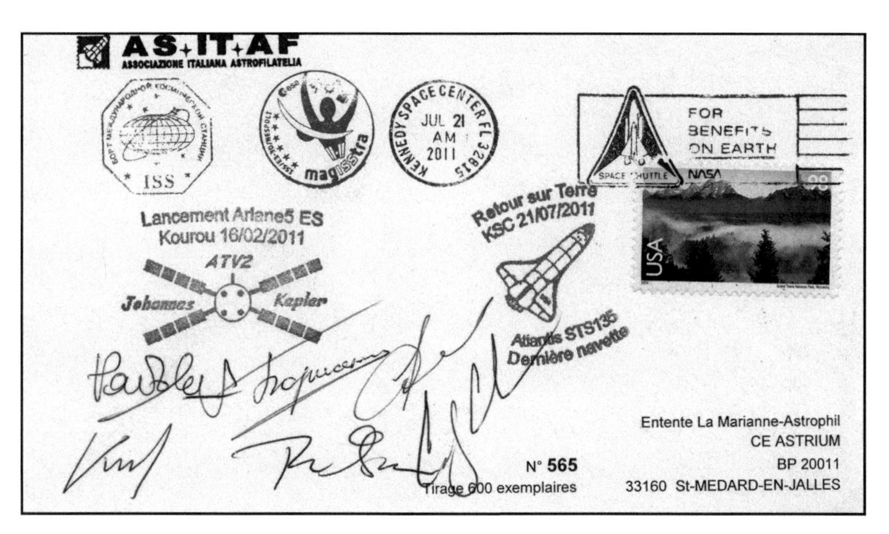

Cover flown in space, carried onboard the ISS from the ATV-2 and signed by the crew of Expedition 26, including Cady Coleman (from the collection of Umberto Cavallaro)

Expedition 26/27 hosted a record number of visiting vehicles to the ISS: a total of two Space Shuttle missions (STS-133 and STS-134), three Russian Progress supply ships, the second ATV (Automated Transfer Vehicle) "Johannes Kepler" from the European Space Agency and the second Japanese HTV supply vessel "Kounatori" ("Oriental stork" in Japanese). Notably, Coleman was the lead robotic arm operator for the capture of Kounatori, performing the second-ever free-flyer robotic capture aboard the ISS.

Cady is married to the glass artist Josh Simpson.

References

Dismukes, K. "How to Become an Astronaut 101", *spaceflight.nasa.gov/outreach* (April 7, 2002).

Frevele, J. "The First Ever Space–Earth Flute Duet", *themarysue.com* (April 16, 2011).

Gibson, K.B. *Women in Space: 23 Stories of First Flights, Scientific Missions and Gravity-Breaking Adventures*, pp. 134–138. Chicago Review Press, Inc., Chicago (2014).

Kevles, T.H. *Almost Heaven: The Story of Women in Space*, pp. 182–186. The MIT Press, Cambridge, MA, and London, UK (2006).

Moskowitz, C. "Back on the Ground: Q&A with Astronaut Cady Coleman", *space.com* (June 21, 2011).

Moskowitz, C. "Gravity's Astronaut Describes the Trials of Space", *scientificamerican.com* (October 1, 2013).

Official NASA biography of Cady Coleman, *jsc.nasa.gov* (January 2012).

Pearlman, R. "Astronaut Gave 'Gravity' Advice to Sandra Bullock from Space", *collectspace.com* (September 13, 2013).

"Preflight Interview: Catherine Coleman [before Expedition 26]", STS-93, *nasa.gov* (October 28, 2010).

Shayler, D.J.; Moule, I. *Women in Space—Following Valentina*, pp. 275, 277, 292. Springer/Praxis Publishing, Chichester, UK (2005).

Woodmansee, L.S. *Women Astronauts*, pp. 101–102. Apogee Books, Burlington, Ontario, Canada (2002).

32

Claudie Haigneré: The First European Space Agency Female Astronaut

Credit: ESA

U. Cavallaro, *Women Spacefarers*, Springer Praxis Books, DOI 10.1007/978-3-319-34048-7_32

Launch		*Landing*	
Soyuz TM-24	August 17, 1996	Soyuz TM-23	September 2, 1996
Soyuz TM-33	October 21, 2001	Soyuz TM-32	October 31, 2001

Commemorative cover of mission Soyuz TM24, signed by Claudie André-Deshays Haigneré Haigneré, André-Deshays Claudie (from the collection of Umberto Cavallaro)

A "Spationaute" of CNES, the French Space Agency, Claudie Haigneré participated in two missions that led her both to the Russian space station Mir and to the International Space Station (ISS). Claudie entered the European Space Agency (ESA) Astronaut Corps in 1999, thus becoming the first female astronaut in the European Space Agency, and also was the first woman ever to qualify as Soyuz Commander. She was the fourth and last woman to visit Mir (after Helen Sharman, Helena Kondakova, and Shannon Lucid), and the first European woman to visit the ISS. To the interviewer who asked her in 2001 whether the presence of women in space would make a difference, she answered:

> "The flights of short duration have shown that from a physiological point of view, there are no significant differences between men and women. As for long missions, relatively few women have been in space so far, and we have not enough data. From a psychological point of view, instead, several studies have focused on human behavior in isolation and on life in confined spaces. All of them agree that a mixed crew works better—in organizing their work, in making decisions, in managing

conflicts, in maintaining relationships with the staff of the Control Centre. Men and women are different but complementary. You have to let them live and work together in their own way, without trying to make them do things in the same way. When we'll explore the planets, it will be a big step forward for the entire human race. And the human race is made up of men and women."

Claudie André-Deshays Haigneré was born in Le Creusot (France) on May 13, 1957—"A few months before Sputnik!", she is keen to highlight. She graduated at 15:

> "I have always enjoyed going to school. I have a sister who is 2 years older than me, but we were in the same class, because I skipped 1 year at nursery, learning to read over her shoulder and then I skipped the second grade. In high school my sister was a 'normal' girl who liked to go to parties, while I was leaving with my bike and my dictionary on the rack to make translations of Latin with my friends."

After participating in competitive athletics throughout her school career, Claudie wanted to enroll at INSEP, the Higher Institute of Physical Education, but she was not accepted because of her age. Having to wait 2 years, she chose a course of study related to physiology and anatomy, which could have come in handy later, once she had resumed her sport-related studies. So she enrolled at the Faculty of Medicine of the University of Paris. Coming of age, however, she decided to complete the course and, at 20, she had already completed her university studies in Medicine. She then obtained certificates (Certificats d'Etudes Spécialisées) in Biology and Sports Medicine (1981), Aviation and Space Medicine (1982), and Rheumatology (1984). In 1986, she gained a diploma (Diplôme d'Etudes Approfondies) in Biomechanics and Physiology of Movement and, in 1992, completed her Ph.D. thesis in Neuroscience, accumulating so many degrees that she earned the nickname "Bac +19" among her friends.

From 1984 to 1992, after graduating in Rheumatology, Claudie served for 8 years in the Rheumatology and Rehabilitation Department at Cochin Hospital in Paris, where she became a respected researcher, involved in research and the application of diagnostic and therapeutic techniques in rheumatology and sports traumatology. Although she always maintained a great interest in space, as in so many other areas, the idea of becoming an astronaut came suddenly, while she was serving in the hospital and met by chance a tender of CNES (Centre National d'Etudes Spatiales, the French Space Agency) that was searching for scientists, without any gender indication, for pursuing its programs of microgravity research in space. It was love at first sight, and immediately she sent in her dossier, along with 1000 other candidates, 100 of whom were women. In September 1985, she was selected as one of the seven

French astronaut candidates in the second Group of CNES. She was involved in the development and preparation of scientific experiments for CNES in the field of human physiology, in particular with the "Physalie" and "Viminal" experiments flown on the Franco-Soviet "Aragatz" mission to the Mir station in 1988, with Jean-Loup Chrétien. Her research topics were human adaptation of motor and cognitive systems in weightlessness.

Claudie continued her medical career and, from 1985 to 1990, she also worked in the Neurosensory Physiology Laboratory at the Centre National de la Recherche Scientifique in Paris. In 1989, she was appointed Scientific Coordinator responsible for the research programs of CNES Life Sciences Division in Paris, and took part in the preparation of the experiments on the adaptation of human beings to microgravity, sent to Mir in 1992, during the Franco-Russian "Antares" mission. In 1993, she was assigned to the backup crew of the Mir mission "Altaïr" (the Soyus TM-17 mission to Mir) and moved to Star City for training. "The first challenge," she said, "was to learn Russian in 1 month. I had to learn it well, not only for technical training, but also to culturally integrate in Star City." She did not fly in that mission, however, but, during this mission, she was responsible for monitoring the biomedical experiments from the Mission Control Centre in Kaliningrad, near Moscow. She then helped to coordinate experiments for the Euro Mir 94 mission of ESA astronaut Ulf Merbold and prepare the scientific experiments for the 1996 Franco-Russian "Cassiopée" mission, to which she was assigned in July 1994.

Claudie reached Mir on August 17, 1996, aboard Soyuz TM-24 and became the first French woman in space. She was welcomed aboard Mir with the traditional ritual of bread and salt with which the Russians used to welcome new guests. The party then continued with fresh cucumbers, tomatoes, lemon, and cheese offered by Claudie. Several times, she recalled the unforgettable feeling for the peculiar scents they emit into space. On board was Shannon Lucid, who had participated in the MIR-21 expedition and was now waiting for the Shuttle STS-79 to come to retrieve her. It was the first time that two women were aboard Mir at the same time and that Mir had hosted two non-Russian guest cosmonauts at the same time. Shannon had prepared the bed for her and a place for her belongings: "She was for us like a mum," Claudie recalls. Shannon also helped her to unravel the intricacies of the station and, in the early days, was to help on more than one occasion: "A very feminine attention that I wouldn't expect. It was hard to leave Shannon."

Supported by more than 200 French scientists and engineers, Claudie conducted a very intensive research program in four fields: life sciences (physiology and biological development), fluid physics, new technologies, and student experiments. Because of her professional background in rheumatology, biomechanics, and physiology, she could significantly contribute to the successful results obtained. Although some problems were experienced (some of the experiments hardware required in-flight repairs), an unexpected amount of data was gathered. To reach the mission goals, Claudie worked up to 20 h a day, and it was hard to find any time for rest. She would prefer a much longer mission as, she commented, "It had taken a week to learn to live onboard the station and use the facilities and a longer mission would allow to utilize them to the full potential." She returned to Earth 16 days later on board Soyuz TM-23, accompanied by the "two Yuris," who had been members of the Expedition MIR-21 together with Shannon Lucid.

After her return, Claudie lived in Russia for several years with various positions, including working in Moscow as the French representative of the Franco-Russian company Starsem, training as backup to astronaut Jean-Pierre Haigneré, her future husband, whom she married later in Star City in 1999, and supporting his "Perseus" mission as crew interface coordinator at the Mission Control Centre in Korolyev. In the meantime, she was also certified as Soyuz Return Commander, thus becoming the first woman commander of the Soyuz (and the only one so far).

On November 1, 1999, Claudie entered the ESA's Astronaut Corps, based in the European Astronaut Centre (EAC) in Cologne, Germany, and was involved in the development of the Microgravity Facilities of the European Columbus laboratory. "As astronauts," she explains, "our career is certainly marked by flights into space, but between the flights we carry an extensive professional career both as engineers and as scientists, experts in the development of ESA's technical projects, as well as teachers." In other words, even astronauts spend most of their time here on Earth engaged in different tasks.

In December 2000, Claudie was assigned to the "Androméde" mission and also trained with Nadezhda Kuzhelnaya (backup of flight engineer Konstantin Kozeyev). She lifted off on October 21, 2001, aboard the Soyuz TM-33, becoming the second ESA astronaut (after the Italian astronaut Umberto Guidoni) and the first European woman to visit the ISS. Post-flight reports were full of praise for her work, both from the scientific world and from French authorities. The French Minister of Research (with responsibilities for space) expressed his esteem and admiration for her skill, dedication, and courage.

After her last space mission, Claudie returned to France and embraced a political career. She was appointed Minister for Research and New Technologies in the second center-right government of Jean-Pierre Raffarin (2002–2004) and later became Minister for European Affairs in the third Raffarin government (2004–2005) and secretary of the Programme for Franco-German Cooperation. In November 2005, she was appointed Advisor to the ESA Director General for space policy, to support the new strategies that the agency would have taken in front of the geopolitical changes in the international arena.

From April 2009 to February 2015, Claudie was President of Universcience, the public institution that was founded by grouping together two centers of excellence: the Cité des Sciences et de l'Industrie in Paris and the Palais de la Découverte, aimed at becoming the most important French Scientific Center. "I want to make this institution," she said, "a point of reference at National, European and global level for the dissemination of scientific and technical culture. And of course a special place will be devoted to space." The goal of Universcience, connected with the Ministry for Culture and the Ministry for Research, is to attract young people to scientific studies and careers and to strengthen the dialogue between science and society where scientific and technological progress is triggering a rapid change in every area. Claudie says:

> "The common thread of all my life, in all missions and functions I was entrusted in has always been to love science, to help everyone understand that science and technology developments are the heart of our daily lives and provide the tools for our future, to support young people in their desperate search, and to direct new talent to these exciting careers."

Not to mention that the research also involves self-discovery and understanding the world around us. When asked whether she had any regret, Claudie answers:

"Actually, if I could go back I would apply more in studying philosophy. Today my role leads me to reflect on the meaning of things, on the transmission of knowledge and sometimes I regret having neglected these areas when I was at school. I was quite attracted to science where I could cope well and I was successful, and—honestly—philosophy at the time did not attract me much. Even I didn't understand it deeply. But then I tried to recover by reading everything. I guess I'm one of the best customer of Amazon.fr!"

From February 2015, Claudie returned to ESA as an advisor to the Director General. She has been decorated in France and in Russia, with many awards by the French National Order of Merit, Légion d'Honneur, and the Russian Order of Friendship of Peoples. Many observers recognize her merit of having contributed—with her work in the Russian space program—to the good relations that have been established between the space programs of the two countries.

References

Cavallaro, U. interview, Milano, Italy, October 28, 2014.

Cavallaro, U. "Claudie Haigneré, the First ESA Female Astronaut". *AD*ASTRA, ASITAF Quarterly Journal*, 23 (December 2014), pp. 10–12.

Chesnel, S. "Claudie Haigneré: 'A 20 ans, j'étais en 5e année de médecine'", *letudiant.fr* (June 15, 2010).

ESA, "An Interview with Claudie Haigneré", *esa.int* (October 16, 2001).

Gibson, K.B. *Women in Space: 23 Stories of First Flights, Scientific Missions and Gravity-Breaking Adventures*, pp. 181–186. Chicago Review Press, Inc., Chicago (2014).

Kevles, T.H. *Almost Heaven: The Story of Women in Space*, pp. 163–168. The MIT Press, Cambridge, MA, and London, UK (2006).

Official biography of Claudie Haigneré, *esa.int*.

Shayler, D.J.; Moule, I. *Women in Space—Following Valentina*, pp. 326–327, 332–334. Springer/Praxis Publishing, Chichester, UK (2005).

Woodmansee, L.S. *Women Astronauts*, pp. 103–105. Apogee Books, Burlington, Ontario, Canada (2002).

33

Susan Still Kilrain: The Second Woman to Ever Pilot the Shuttle

Credit: NASA

U. Cavallaro, *Women Spacefarers*, Springer Praxis Books, DOI 10.1007/978-3-319-34048-7_33

Mission	*Launch*	*Return*
STS-83	April 4, 1997	April 8,1997
STS-94	July 1, 1997	July 17, 1997

Susan Still Kilrain is a veteran of two Space Shuttle flights. She is one of only three women to have piloted the Shuttle.

Susan Leigh Still Kilrain was born on October 24, 1961, in Augusta, Georgia (USA). As the only sister among brothers, she learned to stand up for herself. As a young girl, she did not think about becoming an astronaut or to enlist into the military. "All the women in my life," Susan said in an interview, "were nurses, hairdressers, or secretaries." During her twenties and thirties, she enjoyed martial arts, triathlons, and playing the piano, but also flying. It all started with flying lessons when she was a teenager: "I've always been interested in flying. I learned how to fly when I was 16 and I did my first solo flight after only 4 h of instruction. To me, this was just sort of the natural progression." She got her private pilot's license when she was 17 and her instrument rating at 19.

After graduating from Walnut Hill High School, Natick, Massachusetts, in 1979, Sue joined the Embry-Riddle Aeronautical University, Daytona Beach, Florida, and earned a bachelor's degree in Aeronautical Engineering in 1982. While pursuing her master's degree, she worked as a wind tunnel project officer for Lockheed Corporation in Marietta, Georgia, but she wasn't completely satisfied with her job. In January 1985, her boss arranged for her to speak with his friend Dick Scobee, who died one year later as Commander of the *Challenger* STS-51-L. Scobee advised her to join the military as a pilot if she wanted to increase her chances of being accepted into the astronaut program. She took that advice and joined the US Navy in 1985. She quickly learned that the Navy wasn't a female-friendly organization. They were quite reluctant to accept women into what had only recently been a male bastion. She decided to simply tune out what she didn't want to hear, to ignore their reluctance, and to keep an eye on her goal. She soon realized that, while the good aviators didn't care, the most negative reactions to her presence came mostly from poor aviators, who felt threatened by her skill. She said:

> "I realized that I was a woman in a man's world, so I was going to be an outsider. My whole philosophy was not to make waves. My goal was to be an astronaut. I wanted to fit in without accepting unacceptable behavior."

Despite the overall cool reception from her fellow pilots, she qualified as Naval Aviator in 1987, and learned how to take off and land on an aircraft carrier. "I am lucky to have been born in this time. Just in the nick of time doors opened before me," she said. She was then selected to be a flight instructor in the TA-4J Skyhawk, and later flew EA-6A Electric Intruders for Tactical Electronic Warfare Squadron 33 (VAQ-33) in Key West, Florida. After completing US Naval Test Pilot School at NAS Patuxent River, Maryland, in 1993, she reported to Fighter Squadron 101 (VF-101) in Virginia Beach, Virginia, for F-14 Tomcat training. She logged over 3000 flight hours in more than 30 different aircraft.

At this point, Sue was ready to apply to become a NASA astronaut. In 1994, she was selected for the 15th Astronaut Group and, after completing the "Ascan training," she was assigned to the Vehicle Systems and Operations Branch of the Astronaut Office, and served as CapCom (capsule communicator) at the Mission Control Center. In 1997, she flew in space as a Shuttle pilot, twice within a few months. She was the second woman ever to pilot a Shuttle, after Eileen Collins. In April, she took part in the STS-83 mission which delivered into space MSL-1 (Microgravity Science Laboratory), a collection of 33 microgravity experiments housed inside the Spacelab—the European laboratory built in Italy. The goal of the mission was to test some hardware, facilities, and procedures being developed in view of the long-term research program to be performed on the future International Space Station (ISS). The mission, which was supposed to last for 16 days, was cut short due to concerns about one of three fuel cell power generation units that provided electricity and water to *Columbia*. Shortly after on-orbit operations began, the differential voltage in one substack of fuel cell no. 2 began trending upward. There were three fuel cells on each orbiter, each containing three substacks. Although one fuel cell produced enough electricity to conduct on-orbit and landing operations, Shuttle flight rules required all three to be functioning well to ensure crew safety and provide sufficient backup capability during re-entry and landing. After unsuccessfully trying to correct the anomaly, on the third day, the Mission Management Team ordered to power down the Spacelab and to terminate the mission, marking only the third time in the Shuttle program history that a mission had ended early (the other two being STS-2 in 1981 and STS-44 in 1991). Despite the shortened flight, very valuable information and experience were gathered and two fire-related experiments were conducted: NASA was in fact concerned with accidental fires in space, as had happened a few months earlier aboard Mir.

With delays in the ISS construction leaving ample room in the Shuttle schedule, in the days following *Columbia*'s return, the Mission Management Team made the unprecedented decision to leave the equipment installed in *Columbia* and to re-fly this mission with the same crew and the same "payload." The re-flight was first designated STS-83R and then renamed STS-94 (the next available unused Shuttle mission number at the time). The crew patch was updated with the re-flight, changing the outer border from red to blue and changing the flight number from 83 to 94. The new mission, piloted again by Susan L. Still, lifted off 3 months later in July 1997, and lasted for 16 days, as originally planned. The crew responsibility remained the same for both missions, but their experience during the short STS-83 mission helped them to improve productivity of the new flight that, a few weeks later, resulted in a wealth of data. In addition to carrying out her piloting

Commemorative cover of mission STS-94, signed by Susan Still Kilrain (from the collection of Umberto Cavallaro)

responsibilities, Susan was the primary mission photographer for Earth observations. She realized that, on a Spacelab science mission, while the mission specialists had every moment scheduled, the pilot was often under-tasked.

Following the two missions, Susan married and became pregnant, taking herself out of assignment to a third mission. While pregnant with her second child, Susan elected to move to Washington, DC, to be close to her husband and served as the Legislative Specialist for the Shuttle for the Office of Legislative Affairs at NASA Headquarters, Washington, DC. Susan retired from the Astronaut Office in December 2002 and from the Navy in June 2005: she became a stay-at-home mother of four children.

Susan logged 471 h in space altogether. Her regret was the retirement of the Space Shuttle that "made space travel a common occurrence," she said. "It's the only reusable space vehicle, and it may end up being the only one ever." Another regret was the uncertainty of the US space program:

> "Other countries are certainly going to, and China is ramping up its own program. We're not in a space race like we were in the'60s, but we'll have no say in the matter if we have no program at all."

Susan is now a successful and inspiring speaker.

References

Briggs, C.S. *Women Space Pioneers*, pp. 86–91. Lerner Publications, Minneapolis (2005).

Interview with Susan Kilrain, *secure.collegefortn.org*.

Kevles, T.H. *Almost Heaven: The Story of Women in Space*, pp. 176–178. The MIT Press, Cambridge, MA, and London, UK (2006).

Official biography of Susan Still Kilrain, *jsc.nasa.gov* (June 2005).
Pavey, R. "End of Shuttle Era Disappoints Augusta Astronaut", *chronicle.augusta.com* (May 14, 2010).
Personal contacts by e-mail in May 2016.
Shayler, D.J.; Moule, I. *Women in Space—Following Valentina*, pp. 277–278. Springer/Praxis Publishing, Chichester, UK (2005).
Woodmansee, L.S. *Women Astronauts*, pp. 105–106. Apogee Books, Burlington, Ontario, Canada (2002).

34

Kalpana Chawla: The Regret of Not Officially Visiting Her Native Country

Credit: NASA

U. Cavallaro, *Women Spacefarers*, Springer Praxis Books, DOI 10.1007/978-3-319-34048-7_34

Mission	*Launch*	*Return*
STS-87	November 19, 1997	December 5, 1997
STS-107	January 16, 2003	–

Indian-born and naturalized US citizen, Kalpana Chawla was one of the victims who perished during the disaster of Shuttle *Columbia* STS-107, which disintegrated over Texas on February 1, 2003, upon re-entry into Earth's atmosphere.

Kalpana Chawla was born in Karnal, Punjab, on March 17, 1962, about 120 km (70 miles) north-west of Delhi, in the State of Haryana, North India, in an archetypal family with strong drive. In a country that prized the birth of sons more, she was the youngest of four siblings, after two sisters, Sunita and Dipa, and a brother, Sanjay, with whom she shared an obsession with planes and the dream of becoming a pilot. Most biographies report that Kalpana was born on July 1, 1961. As the authoritative biography by her husband Jean-Pierre Harrison reveals, that date entered her official records because it was used to enroll her in school 1 year in advance. Her family were refugees from Pakistan, who settled in Karnal after partition in 1947. Her father, one of the few survivors in his family, managed to reach India safely and, starting from scratch as a self-taught technologist, ventured into a variety of trades and finally set up his own tire factory and succeeded in becoming a leading industrialist of Karnal.

From her father, Kalpana acquired his attitude of never giving up and not knowing failure. She was originally called "Montu" by the family. At the age of three—as her husband explains in her biography—she chose for herself the name "Kalpana," meaning "idea" or "imagination" in Sanskrit. Open-minded in her thoughts and unique in her attitude, she learned karate (and became eventually a black belt) and enjoyed hiking, backpacking, and reading. Her mother remembers her as extremely intelligent, hard-working, and sometimes a rebellious teenager: "… as she grew into a young woman she cut her hair short and never put on any make-up. She refused to cook, never ironed her clothes and began to wear trousers or jeans." She did not want to take from Mother Earth for her survival more than was absolutely essential. Sharing an anecdote during a visit of Sunita Williams (the second Indian-American female astronaut) in India, Kalpana's father said:

> "Kalpana was once late from work and I asked her what took her so long. After coaxing, Kalpana revealed that she had gone to fix her broken shoes. I asked her why did she do so as she could have bought a new pair. To this, she replied that by doing so she had saved an animal's life and given employment to a person."

From a very young age, Kalpana felt attracted to the starry sky and spent her evenings, she remembers, lying on the roof and admiring the firmament. And she was drawn by flight. After a thrilling joyride at a club, flying became her first love. She became as excited as a child whenever she saw planes or talked about them. She says:

> "I was very lucky that we lived in a very small town which had a flying club. Me and my brother, sometimes we would be there on bikes looking up. Every once in a while, we'd ask my dad if we could get a ride in one of these planes. And, he did take us to the flying club and get us a ride in the Pushpak and a glider that the flying club had. Also growing up, we knew of J. R. D. Tata, who had done some of the first mail flights in India. And also the airplane that he flew for the mail flights now hangs in one of the aerodromes out there that I had a chance to see. Seeing this airplane and just knowing what this person had done during those years was very intriguing. Definitely captivated my imagination. And, even when I was in high school if people asked me what I wanted to do, I knew I wanted to be a flight engineer."

Dreaming at becoming a "flight engineer," Kalpana undertook her pre-university and pre-engineering studies at Dayal Singh College, Karnal. When her father, whose business often kept him out of Karnal, became aware of her plans, he tried his best to dissuade her. Conservative by temperament, he felt that a girl had no career prospects in engineering and advised her to become a doctor or a school teacher. Firmly supported by her mother, however, she joined the Punjab Engineering College in Chandigarh. She was the first girl ever to enroll in the Aeronautical Engineering course, and one of the first seven girls to undertake any engineering course, and became the first woman in the college to earn a Bachelor of Science degree in Aeronautical Engineering in 1982.

After her degree in Cardigarh, Kalpana applied to several American universities and was accepted by the University of Texas at Arlington for a master's degree in Aeronautical Engineering. "In the back of my mind," she said later, "I knew that the US would have more airplanes than we would at home." Initially, her father did not agree to allowing his daughter to live alone in a strange country and, since he happened to be away for business for more than a month, in the male-dominated household, no one else could take a decision. So Kalpana reluctantly accepted a teaching job at the Punjab Engineering College. When her father, now a rich and powerful entrepreneur, returned home and realized how determined his daughter was to go to the US, and knew that the last date for admission to Arlington was in 3 days' time, he organized Kalpana's passport, visa, and tickets for the following day. As the British Airways flight he had booked was suddenly cancelled, her father then called friends in the US and arranged for Kalpana to be admitted behind schedule. Reportedly, the university even organized a pickup for Kalpana, accompanied by her brother, from the airport. In the first hour of her arrival in the University of Texas at Arlington, in September 1982, Kalpana met Jean-Pierre Harrison, who eventually became her flight instructor and whom she married in December 1983. Because of this marriage, there was tension with her family for a while. They only grew closer again when she joined the space program.

At last, Kalpana could fulfill her childhood dream of learning to fly planes, and took Commercial Pilot's licenses for single- and multi-engine land and seaplanes, and gliders, devoting a lot of time and money to maintain all those licenses: "I have a very cheap lifestyle in everything else," she would explain years later and, joking, added: "I hope NASA doesn't find out about the car I drive."

After completing her master's degree in Aeronautical Engineering at the University of Texas at Arlington in 1984, the couple moved to Boulder, Colorado, where Kalpana earned her doctorate in Aerospace Engineering in 1988. In the same year, she began working at the NASA Ames Research Center, where she did computational fluid dynamics (CFD) research on Vertical/Short Takeoff and Landing concepts and applied for US citizenship—which she acquired in April 1991. After the completion of the Ames project, in 1993, she joined Overset Methods, Inc., Los Altos, California, as Vice President and Research Scientist. There, she noticed a bulletin announcing a new class of astronauts and decided to apply, figuring: "you're not going to win a lottery, until you buy a ticket!" She tried twice before she won the lottery and, at the beginning of 1995, was selected to join the Group 15 astronauts and moved to Houston.

After spending 10 years in the most beautiful and progressive areas of the US, Kalpana was not willing to return to Houston—with its oppressive heat and humid weather for 6 months a year, it "remembered Los Angeles with the climate of Calcutta"—but the enthusiasm for the new job was stronger. She was initially assigned as crew representative to work technical issues for the Astronaut Office EVA/Robotics and Computer Branches. Then she served in the development of Robotic Situational Awareness Displays and testing Space Shuttle control software in the Shuttle Avionics Integration Laboratory (SAIL). She completed the training for spacewalks but could not be certified because of her small size. She took part in two Shuttle missions, both on board the *Columbia*: STS-87 and STS-107.

Kalpana became the first Indian-born woman in space when she lifted-off with Space Shuttle STS-87. She served as mission specialist and flight engineer with the task of supporting pilots during the ascent phase of the Shuttle—a task that, given her passion for flying, made her particularly euphoric: "I felt like Alice in Wonderland." This was a completely new and exhilarating experience: "The only thing I feel is my thoughts. There is nothing else touching me, telling me I have limbs."

While in space, Kalpana had the privilege of speaking with the Indian Prime Minister I.K. Gujral, who is said to have commented later that talking to Kalpana in orbit was for him the high point of 1997. As mission specialist, she was the prime robotic arm operator responsible for deploying the reusable free-flying astronomical observatory Spartan (Shuttle Point Autonomous Research Tool for Astronomy) aiming at collecting data from the Sun. She wasn't successful, however, and the satellite went out of control and had to be recovered through a contingency spacewalk. After 5 months of investigation, NASA found that the mistake was the result of many small errors, including some problems in the software of the interfaces. The crew had planned to visit India in its post-flight tour, but suddenly relations between the US and India became strained because of India's underground nuclear tests on May 11 and 13, 1998. And this was one of Kalpana's deepest regrets, as she could not visit her native country in any official capacity because, as an astronaut, she was a representative of the US government.

In January 2003, Kalpana participated in the ill-fated *Columbia* STS-107 mission that ended tragically 16 min before landing at the Kennedy Space Center, while returning after its successful 16-day science mission. This mission had been originally scheduled for launch on May 11, 2000, and then had been repeatedly delayed due to scheduling conflicts and technical problems such as the discovery of cracks in the Shuttle's engine flow liners in July 2002. And the launch date was ultimately pushed to January 16, 2003. As the training was delayed and became extended, she would often remind her crew: "Man, you are training to fly in space. What more could you want?"

The launch went off perfectly. Kalpana's family was in Houston to witness the launch. This mission was one of the rare missions between 1998 and 2006 not committed to construction of the International Space Station (ISS). It also marked the maiden flight of the SPACEHAB Research Double Module or RDM (built By Alenia Spazio in Turin, Italy) in which the crew—divided into two shifts: the blue shift and the red shift—worked 24 h a day on a variety of experiments: 59 in total. All the experiment went well: a great scientific mission during which *Columbia*'s crew may have detected, among others, a new atmospheric phenomenon, dubbed a "TIGER" (Transient Ionospheric Glow Emission in Red). The crew was in high spirits the day they were to return to Earth. One of the last images of the STS-107 crew in orbit, recovered from wreckage inside an undeveloped film canister, shows Kalpana happily floating with the crew in the weightlessness and saying "You are just your intelligence" (Fig. 34.1).

Kalpana's colleague, Janet Kavandi, who at that time was deputy Director of Flight Crew Operations at Johnson Space Center, remembers her as the "sweetheart" of the group: "Everyone she met was her best friend," said Janet, during the Memorial dedicated to Kalpana at the Nedderman Hall in the University of Texas Arlington College of Engineering, where Kalpana had earned her master's degree in Aerospace Engineering. A bad feeling seemed to surface during the last in-flight call that Kalpana had with her family while in orbit. Her sister Sunita recalls:

> "She asked mom 'Mamu, shall I show you something?' We could just see her doing something, but for a few seconds we didn't know what she was actually doing. Then out came a picture of mom and dad from 35 years ago, when she was a kid. Then she took out a picture of her and JP [her husband Jean-Pierre Harrison] and showed it to him. Then she took out a picture of us three sisters, and she said, 'You are all with me'."

This reminded her of a card she had sent a few times before: "One day a spaceship is going to kidnap me."

Among many memorials to Kalpana, a peak in a mountain chain on Mars, an asteroid (51826 Kalpanachawla), an Indian series of meteorological satellites, students' and colleges' awards, a NASA supercomputer, and a song from the musical group Deep Purple's Steve Morse named after her. In his novel *Before Dishonor*, novelist Peter Allen David named a shuttle "the Chawla."

Kalpana was posthumously awarded the Congressional Space Medal of Honor, the NASA Space Flight Medal, and the NASA Distinguished Service Medal. In her brother

Fig. 34.1 One of the last images of the STS-107 crew in orbit, showing Kalpana happily floating in the weightlessness and saying "You are just your intelligence." Credit: Getty Images

Sanjay Chawla's words, "to me, my sister is not dead. She is immortal. Isn't that what a star is? She is a permanent star in the sky. She will always be up there where she belongs."

References

Cavallaro, U. "Ten Years since Loss of Space Shuttle Columbia: Lost But Not Forgotten." *AD*ASTRA, ASITAF Quarterly Journal*, 16 (March 2013), pp. 1–3.

Chaudhary, A. (ed.) "A Success Story of Kalpana Chawla", Biyani's Group of Colleges, *www.sanjaybiyani.com* (2011).

Gibson, K.B. *Women in Space: 23 Stories of First Flights, Scientific Missions and Gravity-Breaking Adventures*, pp. 191–196. Chicago Review Press, Inc., Chicago (2014).
Gueldenpfenning, S. *Women in Space Who Changed the World*, pp. 91–100. The Rosen Publishing Group, New York (2012).
Gulati, M. "In Conversation with Banarsi Lal Chawla", *www.thezine.biz* (September 19, 2015).
Harrison, P. *The Edge of Time: The Authoritative Biography of Kalpana Chawla*. Harrison Publishing, Los Gatos, CA, USA (2011), 236 pp.
Kevles, T.H. *Almost Heaven: The Story of Women in Space*, p. 220. The MIT Press, Cambridge, MA, and London, UK (2006).
Official biography of Kalpana Chawla, *jsc.nasa.gov* (May 2004).
"Preflight Interview: Kalpana Chawla", STS-107, *spaceflight.nasa.gov* (November 12, 2002).
Shayler, D.J.; Moule, I. *Women in Space—Following Valentina*, pp. 280–282, 298–299. Springer/Praxis Publishing, Chichester, UK (2005).
"Space Agencies Must Fulfil Kalpana Chawla's Dream: Sunita Williams", *ndtv.com* (February 27, 2016).
Woodmansee, L.S. *Women Astronauts*, pp. 107–108. Apogee Books, Burlington, Ontario, Canada (2002).

35

Kathryn Hire: The First American Woman Assigned to a Combat Aircrew

Credit: NASA

U. Cavallaro, *Women Spacefarers*, Springer Praxis Books, DOI 10.1007/978-3-319-34048-7_35

Mission	*Launch*	*Return*
STS-90	April 17, 1998	May 3, 1998
STS-130	February 8, 2010	February 21, 2010

Kathryn "Kay" Patricia Hire was born on August 26, 1959, in Mobile, Alabama, the youngest of the three daughters of a land surveyor and a draftsman. She was raised on the Gulf Coast: "I was very fortunate to have access to Mobile Bay and the Gulf Coast and we were especially interested in anything had to do with the water: water skiing, swimming, surfing, sailing." She used to look at craters on the Moon using her father's surveying equipment and started to develop a love for aviation and space: "I was just totally fascinated with space," she says.

After attending St. Pius X Grade School, in Mobile, Alabama, Kay graduated in 1977 from Murphy High School. She credits her Mobile teachers for nurturing her curiosity. She remembers:

> "I had such a great foundation at St. Pius the Tenth Catholic School, not only academically but just as a whole person. The way the teachers instilled curiosity and absolutely encouraged us to explore that helped me ask a lot of questions about math, science and certainly be curious about space and space travel. This is where it all started for me. At Murphy High School we had such great teachers and again, they inspired us to work harder. And, then there was a little bit of competition among the students who wanted to be the best."

Kay heard rumors that military academies were about to take women. As she wanted to pursuit her scholarship without weighing heavily on the family budget, she applied for the Naval Reserve Officer Training Corps (ROTC) program and went to Annapolis for an interview, where she was accepted and entered the US Naval Academy that "truly broadened her horizons," she recalls. "Growing up in Mobile," she adds, "I was very drawn to the water, spent a lot of time as a kid on the water. So the Navy certainly interested me. So it really wasn't a tough decision." In 1981, she graduated with Bachelor of Science degree in Engineering Resources Management from the US Naval Academy, and her attention was caught in April 1981 by the launch of the first Space Shuttle. As she really wanted to fly but was not keen-eyed enough to become a pilot (20/20 vision was required), she went through flight training and became one of the first women naval flight officers: "The Navy," she explains, "had had women pilots since 1973, but they had just opened to women the Naval Flight Officer position, the 'back-seaters', the year before I graduated."

In 1982, Kay earned her Naval Flight Officer Wings and was assigned as a navigator/communicator to the Oceanographic Development Squadron Eight (VXN-8), and participated in worldwide oceanographic research missions to collect data during North Atlantic, European, and Caribbean operations aboard specially configured Orion P-3C Update III maritime patrol aircraft. Over 3 years, she flew to 25 different countries: "I joined the Navy to see the world," she comments, "so that was just a tremendous experience." After this experience, from 1986 to 1989, she instructed other naval flight officers in navigation over land and over water "and long range over water navigation." During those 3 years, she taught over 600 students in the US Navy "Naval Flight Officer Training," progressing from Navigation Instructor through Course Manager to Curriculum Manager and was awarded the rank of Air Force "Master of Flying Instructor," while concurrently assigned as a navigator flight instructor in the USAF T-43A aircraft.

After serving in the US Navy for seven-and-a-half years, Kay left full-time service and, continuing in the aviation field as a Navy Reserve Officer, in May 1989, she began work at the Kennedy Space Center (KSC) as a Space Shuttle engineer for Lockheed Space Operations Company, and also joined the Florida Institute of Technology, where, in 1991, she earned a Master of Science degree on Space Systems Technology. In the same year, she was certified in Lockheed as a Space Shuttle Test Project Engineer (TPE). During her 6 years at KSC, she was first Orbiter Mechanical Systems Engineer, then TPE, and eventually Supervisor of Launch Pad Access Swing Arms, serving as "Cape Crusader" and Astronaut Support, processing Space Shuttles from landing through ground preparations and launch countdowns for over 40 missions, and working closely with flight crews. She says:

> "It was a very exciting time. It's very much like a sports team that has been training, preparing for years and you have all your trainers. Like Olympic athletes. They're very talented, but they also have a team of people behind them, like coaches and all the support people that help them achieve the best that they possibly can in their own sport."

In that position, Kay also headed the checkout of the Extravehicular Mobility Units (spacesuits) and Russian Orbiter Docking System. She decided that she wanted to submit her application to become an astronaut. She says:

> "It was really just a wonderful job for me. It was very exciting to be working directly on the space shuttle flight hardware and preparing so many different missions. So during that time I went ahead and made my application."

This was before finishing her master's degree and she received notification that she was on a good path but still needed to get a little more experience and finish her education.

Meanwhile, Kay had continued her Navy Reserve duty with various units based in Florida, Louisiana, and Texas, and was routinely deployed for flight operations throughout the North Atlantic, Europe, and the Caribbean. She became the first American woman assigned to a combat aircrew when, on May 13, 1993, she reported to Patrol Squadron 62 (VP-62). "I always felt," she says, "the combat restrictions would go away, end eventually they did. When that happened I was right there ready to take advantage of the new opportunities."

When Kay applied again for the NASA Astronaut Corps in 1994, she was called in for an interview. She was selected by NASA for the 15th Astronaut Group in December 1994. She was at the time a supervisor at the KSC. The news arrived in the afternoon during the Space Center's annual year-end picnic. She had to hold a press conference and all the colleagues started to clump around to congratulate her.

> "These people that I had known and worked with for many years, came up to me and what they had were paper plates, and they thrust their paper plate at me and said: 'Can I get your autograph?' I thought they were kidding. I burst out laughing, and they were serious. They were really upset. I got a pen and started signing paper plates. I felt so weird in signing paper plates for people that I'd known and worked right besides for several years."

Kay left the Cape and reported to the Johnson Space Center (JSC) in Houston in March 1995, still a commander in the Navy Reserve. She initially served in the Shuttle Avionics Integration Laboratory (SAIL) and eventually became its head. She then worked as shuttle payloads, flight crew equipment, and spacecraft communicator. "Serving as Capsule Communicators or CAPCOM in Mission Control," Kay says, "gives a lot of insight into how the ground portion of our team is functioning while we are on orbit."

Kay flew in space two times: in 1998 and in 2010. In 1998, she was mission specialist in mission STS-90 known as "Neurolab," the last mission of the Spacelab Long Module (LM) payload prior to the start of the construction and habitation of the International Space Station (ISS):

> "This was the only NASA mission that was dedicated solely to just one area of science and that was neurology. There were twenty-six primary neurological experiments conducted. We took along with us animals that were test subjects and we ourselves were test subjects and it was quite interesting. The interesting overall results from the Spacelab, was that the human central nervous system is very complicated. We already knew that, but it's also very adaptive. It's amazing how quickly the human body adapts to the zero gravity. Also the animals that we took with us adapted very quickly to this new environment and then when we came back to Earth, all of us adapted very quickly again back to the 1 g environment here on the surface of the Earth."

After this mission, Kay was recalled to active naval duty from 2001 to 2003 to support Operation *Enduring Freedom* and Operation *Iraqi Freedom* during second Gulf War as a member of US Naval Central Command staff at MacDill Air Force Base, Florida. On December 1, 2002, she was promoted to her current rank of Captain in the US Navy. After those operations, she resumed her part-time reserve status.

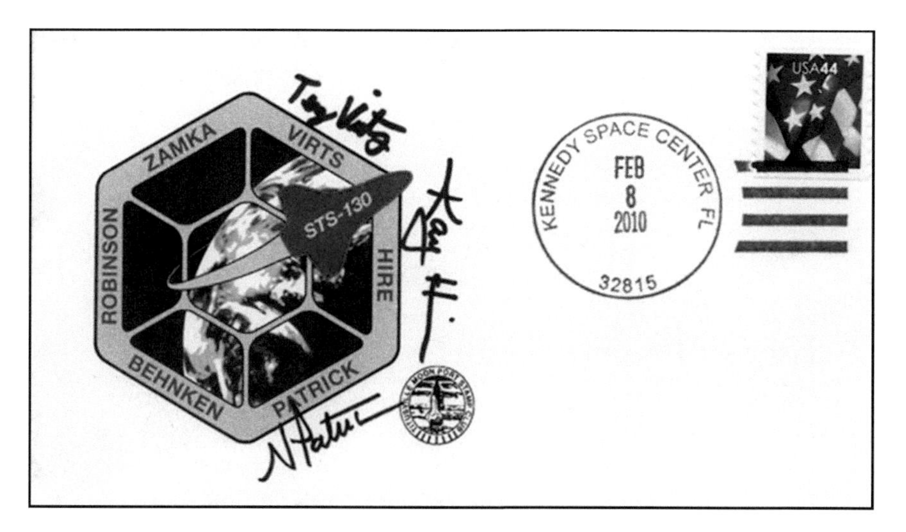

Commemorative cover of mission STS-130, signed by Kay Hire and other Crew Members (from the collection of Umberto Cavallaro)

In 2010, Kay flew with mission STS-130 as arm operator, in charge of supporting the installation on the ISS of the Node-3 Tranquility and of Cupola, which were built and integrated in Italy by Thales Alenia Space. In the NASA pre-flight interview, she said:

> "When people look at STS-130 and they see the six crew members, we are just a small representative of the entire team. To make this mission happen and to make it successful, we have thousands of people working all across the country, at NASA centers and even at subcontractors and vendors scattered all across the country to provide individual pieces and parts of not only the space shuttle but also the payload that we carry and every bit of controlling the mission once we are on orbit. It takes all of these different people, also our trainers and instructors that prepare us for the mission, and not to mention our international partners as well. Our payload is coming from Italy and we've had just tremendous support from the folks in Italy that have provided this fantastic payload for us to bring to the International Space Station to enhance the capability there so it's not just the six crew members that you see. We are just representatives of this entire effort that is just fantastic."

Using the robotic arm, Kay also transferred—as the loadmaster—four-and-a-half tons of materials and supplies for the station. She has logged over 711 h in space.

References

Byrd, R. "The Interview: Navy Capt. Kay Hire", *fox10tv.com*.
"Interview with Astronaut Kay Hire", *kids.usa.gov* (February 1, 2016).
Kevles, T.H. *Almost Heaven: The Story of Women in Space*, pp. 183–184. The MIT Press, Cambridge, MA, and London, UK (2006).

Official NASA biography of Kathryn Hire, *jsc.nasa.gov* (March 2010).
"Preflight Interview: Kathryn P. Hire, Mission Specialist", STS-130, *nasa.gov* (January 27, 2010).
Shayler, D.J.; Moule, I. *Women in Space—Following Valentina*, pp. 283–284. Springer/Praxis Publishing, Chichester, UK (2005).
Woodmansee, L.S. *Women Astronauts*, pp. 108–109. Apogee Books, Burlington, Ontario, Canada (2002).

36

Janet Kavandi: The Rewards of Perseverance and Tenacity

Credit: NASA

U. Cavallaro, *Women Spacefarers*, Springer Praxis Books, DOI 10.1007/978-3-319-34048-7_36

Mission	*Launch*	*Return*
STS–91	June 2, 1998	June 12, 1998
STS–99	February 11, 2000	February 22, 2000
STS–104	July 12, 2001	July 24, 2001

Janet Kavandi is a veteran of three Space Shuttle missions. After serving as NASA's Deputy Chief of the Astronaut Office, Director of Flight Crew Operations, and Deputy Director of Human Health and Performance at the Johnson Space Center (JSC), Dr. Kavandi is now Director at the NASA's Glenn Research Center. The center is named after Senator John Glenn, the first US astronaut to orbit Earth in 1962 in the Mercury Friendship 7 capsule.

Janet Lynn Kavandi was born in Springfield, Missouri, on July 17, 1959. Her favorite studies in school were math and science, and especially astronomy. In her early years, she did a lot of reading on her own about space. She said:

> "Space has been a subject that's intrigued me since I was a child. I've always been interested in space and astronomy. My father and I used to sit outside at night and look at the stars and talk about the first people that were going into space. We wondered aloud what it would be like to be up there looking down at the Earth. In school, whenever I had a choice of subjects to write about, I would always pick something to do with space or astronomy, black holes, or quasars, or something like that. All the space launches I watched as a child, especially the moon landing, fascinated me. So growing up it was something that I thought would be fascinating to be able to do. Of course at that time, there were no girls in space."

Janet graduated as valedictorian from Carthage Senior High School in 1977. She received her Bachelor of Science degree in Chemistry from Missouri Southern State College in Joplin in 1980 and a Master of Science degree in Chemistry from the University of Missouri in Rolla in 1982. Following graduation, she accepted a position at Eagle-Picher Industries in Joplin, Missouri, as an engineer in new battery development. In 1984, she joined the Power Systems Technology Department of the Boeing Aerospace Company in Seattle, Washington, where she served for 10 years. While working for Boeing, she saw the IMAX movie "*The Dream Is Alive*," released in June 1985, about NASA's Space Shuttle program. The film was narrated by Walter Cronkite and directed by Graeme Ferguson. She recalled:

> "I became seriously interested in becoming an astronaut after the first females were selected to the Space Shuttle program in the late 70s, and I was even more inspired to become an astronaut when I saw this film. So I decided to try my hand at it and send in my application."

After sending in her astronaut application, Janet began working toward her doctorate in Analytical Chemistry at the University of Washington, Seattle, studying the development of a pressure-indicating coating that uses oxygen quenching of porphyrin photoluminescence to provide continuous surface pressure maps of aerodynamic test models in wind tunnels. Her work on pressure-indicating paints resulted in two patents, and she presented several papers at technical conferences and in scientific journals. She earned her doctorate in Analytical Chemistry in 1990. She recalled:

> "I sent in my first astronaut application shortly after the Challenger accident, I was inspired by the bravery and dedication of those fallen astronauts and wanted to be part of an organization where people were willing to risk their lives in pursuit of a goal bigger than their individual aspirations. I returned to graduate school to obtain my Ph.D. and renewed my application every year until I was asked to interview in 1994. Fortunately, I was accepted on my first interview and became a member of the 15th class of astronauts in 1995."

Janet started her Astronaut Candidate (ASCAN) training in March 1995. As a mission specialist and veteran of three Space Shuttle missions, she visited both the Mir space station and the International Space Station (ISS), logging more than 33 days in space. In 1998, she and her colleague, Wendy Lawrence, worked together as mission specialists on STS-91, the ninth and final Shuttle–Mir docking mission that concluded the joint US–Russian Phase-One program. She recalled:

> "One of my fondest memories is my first view of the Earth from space. I floated up from the middeck to the flight deck to take a photo of the large orange external fuel tank as it separated from the Space Shuttle. I was in awe of the beauty of the planet and how blue it appeared against the black background of space. Later, during a night pass, I remember watching how lightening appeared to "pop" below us as we flew over. I also remember how quickly we adjusted to microgravity and how comfortable it became to interact with one another in any orientation—like having a meal upside down, for instance."

One of the main goals of the mission was to transfer more than 500 kg (1100 lb) of water and almost 2130 kg (4700 lb) of cargo experiments and supplies to Mir. STS-91 also carried a prototype of the Alpha Magnetic Spectrometer (AMS) designed to measure antimatter in cosmic rays and search for evidence of dark matter in the universe. (A full research system named AMS–02 would eventually fly to the ISS during the STS-134 mission in May 2011.) Other experiments conducted by the Shuttle crew during the mission included a checkout of the orbiter's robot arm to evaluate new electronics and software and

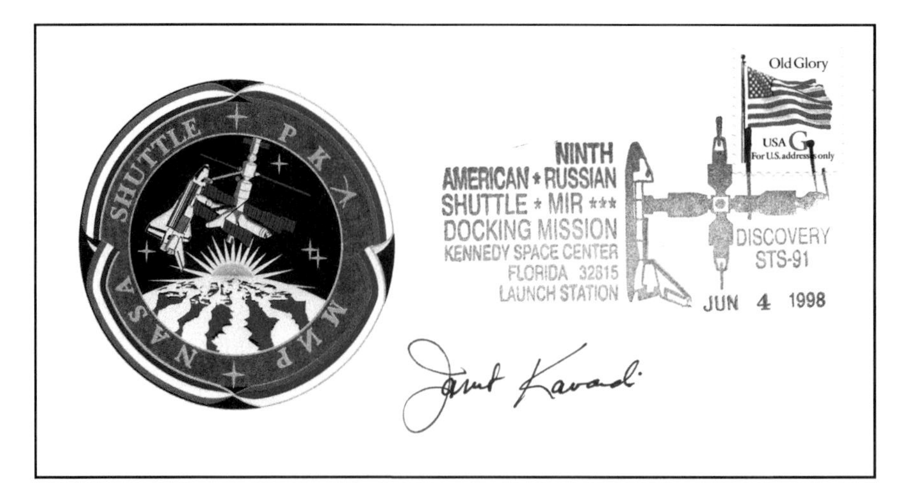

Commemorative cover of mission STS-91, signed by Janet Kavandi (from the collection of Umberto Cavallaro)

the orbiter space vision system for use during assembly missions for the ISS. Following the mission, Janet worked as a capsule communicator (CapCom) in NASA's Mission Control Center.

On her second mission, Janet served aboard STS-99 (February 11–22, 2000), the Shuttle Radar Topography Mission (SRTM), which mapped more than 47 million miles of Earth's land surface to provide data for a highly accurate three-dimensional topographical map. This was the last *Endeavour* mission entirely devoted to scientific experiments. Starting with the following Shuttle mission, STS–104, all the orbiters would be devoted to building the ISS. This mission carried the student experiment called "EarthKAM," a digital camera formerly known as "KidSat," into space for the sixth time. Janet explained:

> "EarthKam is something that we've set up as an educational program with middle schools throughout the country. We had a 35-millimeter camera in the overhead window of the shuttle, and we let students throughout the country select sites that they wanted to photograph and study. Those sites were programmed into a computer, and the camera automatically took pictures of these particular areas at the appropriate time. The students could access the data, in a very timely manner, while we were still up in space. They could get their images back and then talk about what they saw, and do comparisons of earlier images of the same area."

This project, started by Sally Ride when she was professor at the University of California, San Diego, and managed by the engineers of the NASA Jet Propulsion Laboratory, was very successful and resulted in 2715 digital photos for students from schools in the US and around the world. The main goal of the SRTM mission was to generate the most accurate and complete high-resolution digital topographic database of Earth. Janet explains:

> "This mission was to accomplish a three-dimensional topographical map of about eighty percent of the landmass on the Earth. We had small sections of the planet that were mapped with aircraft and satellites, but we did not have a three-dimensional map of the surface of the Earth. We had one type of radar in the payload bay, and a long 200-foot boom that extended out from the payload bay from a canister. On the end of this boom we had a second set of radars. We did simultaneous radar measurements, thereby getting the three-dimensional aspect of the Earth surface. The mission mapped more that 47 million miles of Earth's land surface."

Following this second mission, Janet worked in the Astronaut Office Robotic Branch, where she trained in both the Shuttle arm and the space station robotic manipulator systems. The following year, Janet flew on the STS-104 mission. The primary objective of this seventh ISS assembly flight was to install the Joint Airlock named Quest. This module was designed to allow spacewalks from the ISS with both American Extravehicular Mobility Unit (EMU) spacesuits and Russian Orlan spacesuits. Before the installation of Quest, spacewalks could only be done from the Russian Zvezda service module using Orlan suits, and American spacewalks using EMUs were only possible while a Space Shuttle was docked.

Janet had also trained in the Neutral Buoyancy Laboratory for a contingency extravehicular activity (EVA) should it have been required. As loadmaster, she was responsible for the relocation and logistics of the payload from the Shuttle to the ISS and for used items transferred to the Shuttle for return to Earth.

Since her third and final mission, Janet's career has been firmly rooted on the ground. She initially served as Payloads Branch Chief, then as Chief for the International Space Station Branch in the Astronaut Office. She later became the Director of Flight Crew Operations at the NASA JSC, Houston, where she was responsible for the NASA Astronaut Corps and the Aircraft Operations Division at Ellington Field.

She was in that position on February 1, 2003, when the Space Shuttle *Columbia* STS-107 disintegrated on return from a 14-day science mission, killing all seven astronauts on board. This was an accident that hit Janet particularly hard because three of *Columbia*'s crew were part of her 1995 astronaut class:

> "It was all the more painful because you trained with them so closely for a couple of years. It is a family-type bond. You know their families well because you have family gatherings, and your kids play with their kids, and your spouses get to know their spouses; it's a very close-knit community."

Janet was called to serve as the Lead Casualty and Assistance Calls Officer (CACO) for the families of the crew after the tragedy. She explains:

> "CACO is a military term. It's a person who takes care of the families of the fallen soldiers, or in this case, astronauts. The loss of a crewmate is like the loss of a family member. Also, the loss was so publically visible. Helping the families maintain some privacy was important. There were visits with the President of the United States, ceremonies at Arlington National Cemetery, and their own private memorials. Just dealing with the publicity, the aftermath of the accident, and the investigation was huge. It was a very traumatic event, but we took care of everyone as best as we could. Still to this day, we keep in close contact with the families."

From August 2014 to March 2015, Janet served as the Deputy Director of the Health and Human Performance Directorate at NASA's JSC in Houston, where she was responsible for the NASA flight surgeons and human research investigations on the ISS. Janet is currently the Director at NASA's Glenn Research Center in Cleveland, Ohio. The Glenn staff consists of more than 3200 civil service and support contractor employees and has an annual budget of approximately US$625 million. She says:

> "I am honored to lead the incredibly intelligent and inspired scientists and engineers who develop the technologies that make space flight possible and air travel safer and more energy efficient. My background in flight crew operations has helped to add an operational perspective to a facility that is primarily dedicated to the design, evaluation, and testing of aeronautical concepts and space flight hardware."

Dr. Kavandi is married to John Kavandi and they have two children.

References

Heidman, K. "Biography—Janet L. Kavandi", *nasa.gov* (March 17, 2015).

Official NASA biography of Janet Kavandi, *jsc.nasa.gov* (February 2011).

Personal contacts by e-mail with the Author in April 2016.

"Preflight Interview: Janet Kavandi", STS-99, *spaceflight.nasa.gov* (April 7, 2002).

"Preflight Interview: Janet Kavandi", STS-104, *spaceflight.nasa.gov* (April 7, 2002).

Shayler, D.J.; Moule, I. *Women in Space—Following Valentina*, pp. 284–285, 292, 295. Springer/Praxis Publishing, Chichester, UK (2005).

Woodmansee, L.S. *Women Astronauts*, pp. 110–111. Apogee Books, Burlington, Ontario, Canada (2002).

37

Julie Payette: "To Assemble a Ship in the Ocean During a Storm"

Credit: NASA

U. Cavallaro, *Women Spacefarers*, Springer Praxis Books, DOI 10.1007/978-3-319-34048-7_37

Mission	*Launch*	*Return*
STS-96	May 27, 1999	June 6, 1999
STS-127	July 15, 2009	July 31, 2009

A veteran of two spaceflights, Julie Payette was the first Canadian to visit the International Space Station (ISS). A professional engineer, she is fluent in English and French, and has conversational skills in Spanish, Italian, German, and Russian. She also plays the piano and has been a member of both the prestigious Montreal Symphony Orchestra and Toronto's early music ensemble, the Tafelmusik Chamber Choir, in which she sang as soprano.

Julie Payette was born in Montreal, Quebec, Canada, on October 20, 1963, the second of three children. Her ancestors, she says, settled in Montreal, coming from France in 1655. After watching at Apollo Moon landing, she decided she wanted to be an astronaut:

"I wanted to be an astronaut since I've been a little girl. I was growing up in Montreal, Canada, and during the Apollo missions I was watching that on TV, on French TV because that's the only language I spoke, and I was fascinated."

Julie began making scrapbooks of space missions and taped posters of astronauts to her bedroom door. She recalls:

"It didn't matter to me that I was the wrong nationality, the wrong gender, and spoke the wrong language. I had never been in an airplane, and most of my family had never been in an airplane or anywhere near an airplane, Canada didn't have a human space program and didn't have a single astronaut at the time. So … there were a couple of things that were a little far away from my reach at the time!"

At the beginning, when Julie said that one day she wanted to be an astronaut, people and friends would pat her on the back and smiled a little bit, hoping she would find a more down-to-earth job. But the family never discouraged her from having that dream and that aim. She says:

> "I was never told that it was silly. I was just encouraged to do: 'Well, that's what you want to do? Go after it then, but don't think it's going to be easy or won't require any effort. You better work, you better go to school, you better be good'."

When Julie was deciding on what she would specialize in high school and then what kind of subject she would study at college, since Canada at the time did not have astronauts, she picked electrical engineering, thinking:

> "I like science. I like math. And the chances that I will do this astronaut career are really, really slim; so I might pick something that I really like to do because I will most likely do this for the rest of my life. In the back of my mind there was: 'Hey, if there's a recruitment program one day, if there's an opportunity one day to apply, then I might as well put all the chance behind me,' and it was not a conscious decision."

So Julie enrolled in McGill University, where she completed a B.Eng. in Electrical Engineering in 1986, after which she completed with honors a M.A.Sc. in Computer Engineering at the University of Toronto in 1990:

> "Engineering is extremely useful for being an astronaut because it's extremely applied. What engineering teaches you in particular is to look at a problem, analyze that problem, look at what you've got available to solve that problem. That's exactly what we do in space."

After graduating, Julie was a visiting scientist at the IBM Research Laboratory in Zurich, Switzerland, in 1991. The following year—while, after returning to Canada, she was working as a research engineer at Bell Northern Research in Montreal, supporting research in computer systems, natural language processing, and automatic speech recognition—she saw an ad in the paper announcing that Canada was recruiting astronauts again for the second time in its history. She says:

> "I thought 'I'm twenty-eight years old, it's a little young. I only have two degrees, a Master's Degree. I really have no operational experience. I'm not a military person. I'm not a pilot, but this is what I've wanted to do all my life,' and I'm a strong believer in statistics. If I do apply, I have a chance. If I don't, I have a hundred percent chance of not being picked; so I applied."

In June 1992, Julie was selected by the Canadian Space Agency (CSA) as one of four Canadian astronauts from a field of 5330 applicants. She worked as a technical advisor on the Mobile Servicing System (MSS), the robotics system which is Canada's major contribution to the ISS.

In August 1996, Julie reported to the NASA Johnson Space Center (JSC) in Houston, Texas, and started her training as a NASA astronaut. After completing her basic training, she worked on technical issues in robotics for the Astronaut Office and, in April 1998, was assigned as a mission specialist to space mission STS-96, scheduled for launch in mid-1999: "I was the first one of my class to get an assignment after the graduation in April. And I was really surprised. I really didn't expect to be assigned so quickly."

Commemorative cover of mission STS-96, signed by Julie Payette (from the collection of Umberto Cavallaro)

Julie was the second Canadian female astronaut to fly into space after Roberta Bondar and was the first Canadian citizen to visit the ISS that, at that time, only consisted of two modules: the Russian Zarya and the American Unity. She told the interviewer before the launch:

"This mission means a lot to Canada. We've been players in the aerospace business since the beginning—we were actually the third nation to have sent a satellite in space—and we are providing a higher robotic technology to the International Space Station. But we are also providing human resources, and I am part of those human resources. We're all very excited about this; and I have the immense privilege to be the first one to represent my country aboard the International Space Station; that is something I take with much humility."

During the approach and rendezvous, Julie was in charge of documentation, including all of the photo survey to capture, before docking, the changes that occur to some parts of the station due to the hostile environment of space. She added:

"For us who grew up in French it's always very interesting to realize that NASA has used a French word to describe this absolutely incredible operation, which is the approach, the docking of two space vehicles in space. It's called a rendezvous and a docking. And this rendezvous is extremely complex because objects in space don't quite behave like they do on the ground and in two dimensions. We have orbital mechanics, we have definite orbits and speed, so it is almost just as tricky to fly a rendezvous with a space shuttle as it is to land an airplane on the carrier while the carrier is going up and down in the sea."

One of the goals of this mission, in which Ellen Ochoa and Tamara Jernigan also flew, was to bring to the station 4 tons of equipment and supplies to be stored in the Functional Cargo Block (the Russian Zarya module) waiting for the arrival of the first team who would permanently inhabit the station. Julie was heavily involved in the operations as backup loadmaster and primary stowage master on the ISS. The main task of the mission was to install two cranes: the very massive Russian crane to be mounted outside the Russian segment and the American ORU transfer device, a small crane mounted outside the Unity module to allow crewmembers to move big pieces of equipment:

> "Assembling the International Space Station in orbit is extraordinarily complicated. We're building an enormous infrastructure in a very hostile environment. It would be just very similar as if we wanted to assemble a full ship in the middle of the ocean during a storm. We don't have any infrastructure out there in the middle of the ocean. We have to bring everything with us, every single piece of material, every single bolt, every single cable. We have to make sure they fit somehow before you leave in the middle of the sea in the storm, because if things don't fit, and if we don't have the right bolt, we just can't go and walk to the store and buy it. The other difficulty about the International Space Station is in the word international. We have several different nations putting together pieces, developing and designing these pieces in their own country, sometimes under a different measurement system. And then having everything fit together in orbit for the first time because sometimes those pieces won't see each other on the ground before they get to space. And that is an extraordinary challenge to make sure that everything is going to be fitting together. That is a challenge that we've been tackling now for several years and we see, so far, that it's working quite fine."

There wasn't much spare time during the mission:

> "You have very little time to think about what it represents in terms of inner self or emotion. On your first flight, usually you don't have that much time even to enjoy weightlessness or this absolutely magnificent view of Earth. The reason is that if you want this to be your profession, then you know that you're under evaluation. How you do on that first flight is going to determine whether or not you fly again."

One of the STS-96 payloads was STARSHINE, the satellite designed to allow students to perform research on Earth's atmosphere.

From 1999 to 2002, Julie represented the Astronaut Corps at the European and Russian space agencies and supervised the development of procedures, verification of equipment, and processing space hardware for the ISS Program. From 2000 to 2007, she also held the position of Chief Astronaut of the CSA. In 2006, she served as CapCom (capsule communicator) at Mission Control in Houston during the Space Shuttle mission STS-121.

Julie flew her second space mission to the ISS in July 2009 on mission *Endeavor* STS-127 when the station had been inhabited for 9 years. "I was very privileged," she said before the launch, "to go on the station when it was at the very beginning of construction. I find I am extremely privileged to go and work on it again when it's nearing completion." She was responsible for operating three robotic arms during this space mission: the Shuttle's Canadarm, the ISS Canadarm 2, and a Japanese robotic arm on the ISS Japanese

Laboratory Kibō, which had been carried into space 1 year before by mission STS-124 and which she helped to install permanently on the ISS. Several other components were also installed to complete the station, such as the JEF (Japanese Exposed Facility) platform for scientific experiments in open space, the VCC (Vertical Cargo Carrier), as well as batteries and other spare parts. Five spacewalks were performed to complete the job in the 16 days of the mission, thus surpassing all previous records.

On the ISS, Julie met up with Canadian Astronaut Bob Thirsk, who was on a 6-month stay on the ISS. It marked the first time that two Canadians were in space at the same time. During this flight of the Shuttle, the ISS for the second time reached a record 13 occupants (from five different countries: the US, Russia, Japan, Belgium, and Canada): in addition to the six members of the ISS Expedition-20/21, the seven crewmembers of the Shuttle *Endeavour* arrived (as had already happened in 1995 when the crews of both Shuttle STS-65 and Soyuz TM-21 had visited Mir at the same time).

Julie logged more than 650 h (over 27 days) in space altogether. At the beginning of 2011, she began a fellowship at the Woodrow Wilson Center for International Scholars in Washington, DC. In October 2011, she was appointed scientific authority for Quebec in Washington for the Quebec Department of Economic Development, Innovation and Export Trade. She remains a member of the Canadian Astronaut Corps.

In mid-2014, after 20 years in the US, Julie returned to live in her hometown of Montreal, where she accepted the position of Chief Operating Officer for the Montreal Science Centre in the Old Port of Montreal—the most important museum in Quebec, with more than 750,000 annual visitors—and started a new exciting chapter of her life aimed at introducing the general public and youth to science, to stimulate them and give them the tools to better understand the many phenomena and technology. She says:

> "In my opinion, any big city must have flagship institutions, for example, a fine art museum, a philharmonic orchestra or a (legendary!) professional sports team. If a city is also a knowledge society, it becomes critical to have academic institutions, educational infrastructure and sufficient capacity in research and innovation. My wish is for the Science Centre to continue growing and contributing to the overall development of Montréal. Over the next four years, we will embark upon the ambitious project of reviewing all of our permanent exhibitions and creating interactive and engaging attractions for people of all ages. The teams at the Science Centre are constantly searching for creative ideas and activities to attract people to visit. The Science Centre has everything it needs to become one of the pillars in the knowledge culture of Québec. And as we say: Science is fun!"

References

Gibson, S. "Ms. Universe: Astronaut Julie Payette Prepares for Her Second Mission in Space", *magazine.utoronta.ca* (Winter 2009), 38–43.

Gueldenpfenning, S. *Women in Space Who Changed the World*, pp. 74–82. The Rosen Publishing Group, New York (2012).

Munroe, S. "Julie Payette", *canadaonline.about.com* (December 2012).

Official CSA biography of Julie Payette, *csa.gc.ca* (September 2013).

Official NASA biography of Julie Payette, *jsc.nasa.gov* (June 2012).
"Preflight Interview: Julie Payette", STS-96, *spaceflight.nasa.gov* (July 2002).
"Preflight Interview: Julie Payette, Mission Specialist", STS-127, *nasa.gov* (May 2009).
Shayler, D.J.; Moule, I. *Women in Space—Following Valentina,* pp. 290–292. Springer/Praxis Publishing, Chichester, UK (2005).

38

Pamela Melroy, the Second and Last Woman to Command a Shuttle

Credit: NASA

U. Cavallaro, *Women Spacefarers*, Springer Praxis Books, DOI 10.1007/978-3-319-34048-7_38

Mission	*Launch*	*Return*
STS-92	October 11, 2000	October 24, 2000
STS-112	October 7, 2002	October 18, 2002
STS-120	October 23, 2007	November 7, 2007

Pamela Melroy flew first as Shuttle pilot in two missions—STS-92 (2000) and STS-112 (2002)—before becoming the second (and last) woman commander of the Space Shuttle during the STS-120 mission in October 2007.

Pamela Ann Melroy was born in Palo Alto, California, on September 17, 1961, between two brothers. After moving to different places, following her father, who was a pioneer computer expert, her family settled in Rochester, New York. "That was pretty much the longest I'd ever lived in one place and gone to school to one school," Pamela says. "So, obviously I'm very rooted there and very connected, and I consider Rochester, N.Y., to be my hometown." She graduated from the catholic Bishop Kearney High School in 1979. She remembers that becoming astronaut was her dream when she was an adolescent, fascinated by the Apollo program, like many others at that time. She found lavish support from her father who, when she turned 18, took out a life insurance policy for her, telling that, when she became an astronaut, she might have trouble in getting one because of the hazards of the job.

Pamela decided that she wanted to major in Physics and Astronomy. After looking at different schools in the north-west, she "fell in love" with Wellesley College, a women's college just outside Boston, Massachusetts, with an "amazing observatory," she says. She earned a Bachelor of Arts degree in Physics and Astronomy in 1983 and a Master of Science degree in Earth and Planetary Sciences from Massachusetts Institute of Technology in 1984, conducting a theoretical study of the atmosphere of Neptune by observing the occultation of stars by the planet. Meanwhile, she was commissioned through the US Air Force ROTC (Reserve Officers' Training Corps) program. She was the only woman in her class when, the following year, she attended Undergraduate Pilot Training at Reese Air Force Base in Lubbock, Texas. She quickly discovered that inordinate teasing was part of the military culture. It took 2 years for her to get used to it, but she learned how to survive in the military environment, which was meanwhile slowly starting to integrate women:

> "I'm fortunate in my life that I came into the Air Force as a pilot just at the time that women were being integrated. They started in 1976. I went through pilot training in '85. Women were just achieving a position of middle management in the Air Force as pilots. There were still no senior women managers, of course. You didn't see the 20-year colonels. But military culture was already starting to change."

Assigned to Barksdale Air Force Base in Bossier City, Louisiana, Pamela flew the KC-10 for 6 years as a co-pilot, aircraft commander, and instructor pilot. In 1989, she took part in the Operation *Just Cause*, the invasion of Panama by the US between mid-December 1989 and late-January 1990, when the Panamanian dictator Manuel Noriega was deposed. At the beginning of 1990, she participated in the Operation *Desert Shield* and *Desert Storm*, and spent over 200 h in combat and combat support missions during the first *Gulf War*. In June 1991, she attended the Air Force Test Pilot School at Edwards Air Force Base, California. Upon her graduation, she was assigned to the C-17 Combined Test Force. She was the only third woman test pilot in the Air Force, and the second at Edwards Air Force Base, the legendary base of Chuck Yeager. She served as a test pilot until her selection for the astronaut program in NASA. Pamela retired from the Air Force in February 2007, with the rank of Colonel, after logging over 6000 h of flight time in over 50 different aircraft.

Pamela entered NASA with the 15th Astronauts Group and started her basic training in 1995, qualifying as a Shuttle pilot. She was assigned to support the Kennedy Space Center (KSC)—with her colleague Mike Anderson—in the role of "Cape Crusader," assigned as an ASP (Astronaut Support Personnel) and responsible for monitoring the orbiter status during preparations for the next flight, initially under the lead of Marsha Ivins. A "nerve-racking job," she recalls:

> "My first mission was the Tethered Satellite [STS-75]. One of the Pc meters told that the engines were dropped to like 50 % after they lifted off and they throttled back. It didn't throttle back up. The engine was actually working fine, but the meter failed. Mike Anderson and I were standing on the roof of the Launch Control Center and listening to this. We had both the same thought at the same time. We looked at each other. We're like, 'It wasn't something I did, was it?' We were both so scared, 'I was in the cockpit an hour ago, did I do something?' That was an amazing experience to do that."
>
> "I also loved the intimacy of spending the last couple moments on Earth with the crew. That was definitely my favorite part of that job, was that last 45 minutes before you close the hatch because the crew has got a million things on their mind. They actually rely on the Cape Crusader .... The same thing with bringing them home, being the first one into the cockpit and seeing them."
>
> "I went in shadow mode for multiple experiences with Marsha Ivins who taught me to be a Cape Crusader, before I could actually lead and be prime, as we call it. I primed four or five times. That was great. But hard. It's operationally very rigorous. So it was very comfortable for me as a military pilot in terms of the checklist following, although I was pretty stressed whenever I strapped a crew in, don't get me wrong. You're on a timeline."

Back in Houston, Pamela served for a while as CapCom (capsule communicator):

> "That's a really interesting experience, because you're sitting next to a flight director who is in charge and telling you what to say. But how to translate it to the crew and how to guess in your head what it is they're doing at that exact moment and in the shortest number of words can you give them the flavor of the situation. Also recognizing everyone in the world is listening on air-to-ground. It's a hectic thing on a Station assembly flight to be a CAPCOM. There's always drama during the spacewalks so you're really trying to rack your brain all the time. What are they doing? What do they need from us? How long should we talk about this?"

Pamela realized soon—as happened to her colleague, Eileen Collins—that, in the Astronaut Office, she was in a privileged position compared with the other colleagues; there was some kind of a status difference between pilots and mission specialists:

> "The one thing that was clear to me from the beginning was that the smarter people in the office—and believe me, astronauts are nothing if not smart—figured out pretty darn quickly that because I was a pilot, someday I was going to be a commander, likely, and potentially their boss. It really changes the dynamic in my opinion."

Mission *Discoverer* STS-92, the 100th Shuttle mission, lifted off as usual from the KSC. The mission carried a crucial part to the then incomplete International Space Station (ISS) in order to prepare it for its first resident crew: Z1 in fact, the first segment of the Integrated Truss Structure, where the non-pressurized components are mounted, like the solar panels, water cooling, and other devices. After 13 days, she landed at Edwards, where she had been at home for years.

Pamela's second mission, STS-112, delivered and installed, with the help of Expedition-5, the S1 Truss, the third component of the station's 11-piece Integrated Truss Structure. It took three spacewalks to outfit and activate the new component, during which Pamela acted as internal spacewalk choreographer. Mission STS-112 was the first mission to mount a camera on the Shuttle's external tank to capture *Atlantis*'s ascent to orbit and provide a live view of the launch to NASA TV viewers and mainly to flight controllers to capture images of the falling debris from the external tank and locate possible damage to the heat shield.

After the flight, Pamela served in different positions in the Astronaut Office and, the following year, she was involved in the *Columbia* investigation and also served in the Astronaut Office as chief for the Orion Branch, the next generation of NASA's spacecraft, which will take humans to the space station, to the Moon, and to Mars.

In 2007, Pamela commanded the STS-120 mission, thus becoming the second and last woman to command the Shuttle. Originally, the mission had been planned with the *Atlantis* orbiter but, due to the delay in the STS-117 mission, Shuttle *Discovery* was used instead. The mission docked with the ISS while, for the first time, another woman, Peggy Whitson, was the commander of the Station: for the first time, two women commanders shook hands through the hatch in orbit. Pamela said that this coincidence, caused by delays in the launch schedule, actually made a deeper impression on her than being the second female Shuttle commander: "I think to me," she said, "it was actually a bigger thing that Peggy

Whitson and I were flying at the same time in space and that no one had planned it that way" (Fig. 38.1).

The main payload of STS-120 was the Harmony module (also known as Node 2), the pressurized habitable module, built for NASA by Thales Alenia Space in Turin, Italy, as part of an agreement between NASA and the European Space Agency (ESA). This element opened up the capability for future international laboratories to be added to the station. Pamela says:

> "I enjoyed the experience much more than STS-112, because as the commander your job is really not to have too much on your schedule. It's to be with everyone and to make sure that you're keeping track of the timeline."

At least this was true before the accident caused by the solar panel.

During the mission, the P6 solar arrays were in fact relocated as a preparatory step to allow the installation of Node-2 Harmony. The arrays were on the center point of the station on top of Unity, and had to be moved completely to their final position at the far end of the Z1 truss, using both the Shuttle arm and the station arm—a challenging operation also due to the great distance to be reached to install this element, which was difficult to manage

Fig. 38.1 Historic handshake (October 25, 2007) between Space Shuttle commander Pam Melroy and the International Space Station's Expedition-16 commander Peggy Whitson. They became the first female spacecraft commanders to simultaneously lead Shuttle and station missions. Credit: NASA

because of its size (approximately 13 × 110 ft, or 4 × 35 m) and because, as Julie Payette highlighted, "it's like assembling a ship in the ocean during a storm" (see Chap. 37 on Julie Payette). While deploying it, two small tears were produced in one of the large Solar Array Wings. A further complication was due to the fact that the whole panel was electrified with 120 V DC and it was not possible to disconnect it. It was important to prevent the bad tears from continuing to extend, but the damaged array was hard to reach. Pamela says:

> "We talked to the chief of the Astronaut Office. We talked to our flight director. And also the Station crew helped. They've usually got their own assignments. They've got lots and lots of work to do on the Space Station, as we well know. They've got their own experiments and things going on. But we had to all come together as a single crew. I think it helped the fact that Peggy and I knew each other so well from having flown together. Everybody played a part."

Engineers worked through the night at Houston assessing a variety of options for possible repairs. It was a kind of "mini-Apollo 13." Finally, the mission managers decided to fix the damaged solar array by installing five hinge stabilizers, or "cufflinks." The teams worked for 2 days on the station. Using materials found on board—strips of aluminum, a hole punch, a bolt connector, and 66 ft of wire—the astronauts constructed the cufflinks to the specifications provided by the ground. To install them, a contingency Extra Vehicular Activity (EVA) was organized, conducted by Scott Parazysnki, who had a vast EVA experience (having already performed six spacewalks). Pamela says:

> "We had to wrap every tool and everything that was metal on Scott's suit. We used up every piece of Kapton tape on the Shuttle and about half of what the Station had. For me, my biggest concern was for Scott's safety."

Scott was suspended from the space station's robotic arm (SSRMS) in combination with the orbiter's boom (OBSS) that brought him within arm's length of the tears in the P6-4B solar array. After five-and-a-half hours, the five cufflinks were safely installed. Pamela summarizes:

> "This was a monster mission. We had altogether five spacewalks. Thank God we had Stephanie [Wilson] with us, Madam Robotics Expert. She and Dan Tani, they were real-time moving that arm to try to get that last little inch out of it."

Pamela has logged 924 h (38 days) in orbit. After serving as Chief of the Orion Branch in the Astronaut Office again from January 2008 to August, she left NASA in 2009, after 14 years, and joined Lockheed Martin Corporation where, until April 2011, she served as the Deputy Director of Orion Space Exploration Initiatives, overseeing over 320 engineers designing Orion. She then became, until the end of 2012, Acting Deputy Associate Administrator and Director of Field Operations in the Federal Aviation Administration's Office of Commercial Space Transportation. As acting Deputy Associate Administrator, she was responsible for developing human commercial spaceflight safety guidelines and oversaw interagency policy coordination with the White House, NASA, and the Department of Defense on space policy. As Director of Field Operations, she was responsible for overseeing and growing activities from three to six field offices supporting operational safety oversight, licensing, and inspection of commercial space activities. Since

January 2013, she has been Deputy Director of the Tactical Technology Office (TTO) of DARPA (Defense Advanced Research Projects Agency).

References

Anon. "Pamela A. Melroy, Senior Technical Advisor", *faa.gov* (November 1, 2011).

Anon. "Tactical Technology Office: Ms. Pamela Melroy Bio", *darpa.mil.*

Cavallaro, U. "Pamela Melroy: 'Try to be True to Yourself'", *AD*ASTRA, ASITAF Quarterly Journal*, 31 (December 2016), pp. 15–20.

Gibson, K.B. *Women in Space: 23 Stories of First Flights, Scientific Missions and Gravity-Breaking Adventures*, pp. 139–144. Chicago Review Press, Inc., Chicago (2014).

Kevles, T.H. *Almost Heaven: The Story of Women in Space*, pp. 178–182. The MIT Press, Cambridge, MA, and London, UK (2006).

Malik, T. "Female Space Commanders Set for Landmark Mission", *space.com* (September 13, 2007).

Moskowitz, C. "Last Female Space Shuttle Commander Leaves NASA", *space.com* (August 11, 2009).

Official NASA biography of Pamela Melroy, *jsc.nasa.gov* (April 2013).

Personal contacts by e-mail in May 2016.

Ross-Nazzal, J. "Pamela A. Melroy: Oral History Transcript", *nasa.gov* (November 16, 2011).

Shayler, D.J.; Moule, I. *Women in Space—Following Valentina*, pp. 293–294, 297. Springer/Praxis Publishing, Chichester, UK (2005).

"STS-120 Preflight Interview", *nasa.gov* (September 27, 2007).

39

Peggy Whitson: The First Woman Commander of the International Space Station

Credit: NASA

U. Cavallaro, *Women Spacefarers*, Springer Praxis Books, DOI 10.1007/978-3-319-34048-7_39

Mission	*Launch*	*Return*
STS-111	June 5, 2002	
STS-113		December 7, 2002
Soyuz TMA-11	October 10, 2007	April 19, 2008

Peggy Whitson was the first, and so far only, woman Chief of the NASA Astronaut Office and the first woman commander of the International Space Station (ISS). A veteran of two long-duration space missions aboard the ISS, where she spent over 376 days altogether, in 2008, broke the record by both a US astronaut and a woman astronaut for cumulative time in space. With a total of six spacewalks totaling 39 h, 46 min, Peggy also set new records for the most spacewalks and most time spent in outer space by a woman (eventually superseded by Sunita Williams). She is NASA's most experienced woman astronaut and has now been selected for the crew of Expedition-50, scheduled to start in November 2016.

Peggy Annette Whitson was born on February 9, 1960, in Beaconsfield, Mount Ayr, Iowa, and grew up on a farm in a very rural area with only a post office and a church. "The closest town," she says, "had only 32 people living in it, so my high school was consolidated for the whole county." From this small environment and from the example of her parents whom she describes as "the hardest-working people I've ever met," working on the farm from sunup to sunset, she inherited a double dose of the dedication and stubbornness gene, which contributes to her success.

After graduating from the Mount Ayr Community High School in 1978, Peggy double-majored in Biology and Chemistry in 1981 from the Iowa Wesleyan College, a small college in Mount Pleasant, in south-eastern Iowa. "Biology was kind of my love," she explains. "I found chemistry challenging and the aspects of both of those very interesting together, which is why I got my Ph.D. in biochemistry from Rice University." It was for her "a huge culture shock" to move to Houston, she confesses. After her doctoral degree in 1985, she continued at Rice as a post-doctoral fellow until October 1986, when she began working at Johnson Space Center (JSC) in Houston, Texas, as a National Research Council Resident Research Associate. "I've never had a real job because I've always done what I wanted to do, which was work at NASA," she jokes. In fact, this had been set as her goal since she was a young girl:

"As a kid of 9 years old, I was inspired by the men who walked on the moon. It really didn't become a goal to me until I graduated from high school, which was coincidentally the same year they picked the first set of female astronauts. At that point I thought: this is going to be something I'm going to try and do, and so I chose my goals in education to be consistent with working at NASA, even as a scientist."

That was possible thanks to the biochemist Shannon Lucid, Peggy would explain later. At that moment, "it became more than just a dream." And she applied for the Astronaut Corps. She would then apply every year. "I think my talent, if I have any, is perseverance," she would say later. Hired by KRUG International, one of the medical sciences contractors at NASA-JSC, Peggy served as the supervisor for the Biochemistry Research Group from April 1988 until September 1989, when she was hired directly by NASA to lead the biochemistry section at the JSC. There, she started to be involved in the US–Soviet joint scientific activities, and traveled for the first time to Moscow, where she conducted joint biomedical research with the Russians. Because of her experience with the new partners, she was asked to help to develop and lead the entire science program during the Shuttle–Mir program. From 1991 to 1992, she was the Payload Element Developer for the Bone Cell Research Experiment (E10) aboard Spacelab-J (STS-47) and was a member of the US–USSR Joint Working Group in Space Medicine and Biology. In 1992, she was named the Project Scientist of the Shuttle–Mir Program (STS-60, STS-63, STS-71, Mir 18, and Mir 19) and served in this capacity until the conclusion of the Shuttle–Mir Phase One in 1995. "I was put in charge of developing the entire U.S. and Russian joint science program," she explains, "and this gave me a lot of visibility and maybe was a big contributor to why I was selected to be an astronaut." From 1993 until her selection as an astronaut in 1996, she also held the additional responsibilities of the Deputy Chief of the Medical Sciences Division at NASA-JSC.

Peggy is the co-inventor of a patented "simple, portable, relatively inexpensive apparatus for accurately and efficiently collecting, separating, testing, and even storing blood samples" that does not require a fridge to preserve the sample. It was invented for space use, but has great applications for field workers in remote areas where fridges are not available.

From 1991 to 1997, Peggy was invited to be an Adjunct Assistant Professor in the Department of Internal Medicine and the Department of Human Biological Chemistry and Genetics at the University of Texas Medical Branch, Galveston, Texas, and also earned a position as Adjunct Assistant Professor at Rice University in the Maybee Laboratory for Biochemical and Genetic Engineering.

Selected as a NASA astronaut candidate in Group 16 in 1996, Peggy completed 2 years of training and evaluation, was assigned technical duties in the Astronaut Office Operations Planning Branch, and served as "Russian Crusader," leading the Crew Test Support Team in Star City from 1998 to 1999. She then started training for her long-duration mission on board the ISS. She lifted off on June 5, 2002, aboard Shuttle *Endeavour* STS-111 and docked with the ISS 2 days later to begin the Expedition-5 together with the Russian cosmonauts, Valery G. Korzun and Sergei Y. Treshchov, so becoming the second woman resident and the first science officer to inhabit the ISS. During her stay on the station, she was the US expert, trained to manage and, if necessary, repair every US system including communication, life support, guidance, and navigation. As robotic arm operator, she helped to transfer the supplies delivered to the station through the Leonardo MPLM module that—provided to NASA by ASI (the Italian Space Agency) and built by Thales Alenia Space Italy in Turin—was flown in space for the second time, full of supplies and experiments. The Expedition-5 altogether was visited by three Shuttle missions, two Progresses and one Soyuz, and Peggy, as robotics operator, contributed to the installation of the Mobile Base System for Canadarm2, then the S1 truss segment (which was delivered to the station by STS-112 in October 2002), and finally the P1 truss segment (delivered in November 2002 by STS-113). As the first NASA science officer, she also conducted 21 investigations in human life sciences and microgravity sciences, including one in which Peggy was the primary investigator on the prevention of renal or kidney stones in astronauts living in orbit. She explains:

> "We are interested in trying to reduce that risk of stone formation. We've had crewmembers form stones after flight, and there's one case where a Russian mission was aborted because of a crewmember who during flight formed a stone that moved."

What little spare time she had, Peggy used to look out of the window. She says:

> "One of the most beautiful sights is when the rim of the Earth is bright on one side, and you see this defined line of the atmosphere. You see how close and thin it is. We've got to be careful. We've got to take care of this planet."

On August 16, 2002, Peggy performed a 4-h and 25-min spacewalk in a Russian Orlan spacesuit to install micrometeoroid shielding on the Zvezda Service Module, thus becoming the seventh woman to perform an extravehicular activity (EVA). She says:

> "Outside on a spacewalk takes it up another notch. You are traveling 17,500 miles an hour across the planet. You are looking down with views going past you. It's like being a bird maybe, the perspective of flying over the Earth."

After more than 184 days in space, Peggy returned to Earth on December 7, 2002, aboard STS-113. While, after 6 months in space most people would be ready to go home, she found very difficult leaving the station. In June 2003, she served as the commander of the fifth NEEMO mission (NASA Extreme Environment Mission Operations), living and working in the Aquarius underwater laboratory for 14 days. From November 2003, she served as Deputy Chief of the Astronaut Office and, in this position, she was in 2004 a member of the selection group for the Group 19 astronauts. From March 2005, she was named Chief of the Station Operations Branch, Astronaut Office, until she began training as backup ISS commander for Expedition-14 from November 2005.

Named commander of Expedition-16, she launched on October 10, 2007, on board the Soyuz TMA-11, few days after the 50th anniversary of the launch of Sputnik. Commenting on the event, Peggy said:

> "It's interesting to me that so much of the early Space Race was a competition, and it has evolved into being a huge international project. I think the legacy of the International Space Station will be the fact that it is a peaceful international project that we have conducted together."

During Expedition-16, Peggy surpassed Sunita Williams as the woman with the most spacewalks. She was the first woman to command the ISS and in fact the first mission specialist to become a commander: all of her predecessors, until then, had in fact been pilots. Hers was a very aggressive mission, which received the visit of ATV-1 "Jules Verne" (the European Automated Transfer Vehicle), two Russian Progresses, and three different Shuttle flights (namely STS-120, STS-122, and STS-123). In her pre-flight interview, Peggy describes the ATV-1, built in Turin, Italy, by Thales Alenia Space:

> "ATV has got a three to four times the mass of a Progress vehicle. It can bring up pressurized gases as well as water for transfer, and it has obviously a lot of volume to carry up food, clothing and other crew provisions. The Europeans wanted to try out a new rendezvous and docking system, so that'll be a new test. Yuri Malenchenko and I have been trained on it. It's all automated, but we have to monitor to make sure that the system, which will be tested for the very first time, is actually working correctly. We'll have two demonstration days where we'll approach the station to different distances and check and make sure the abort systems all work properly, before we actually approach the station to actually dock."

This was a very exciting mission during which many new key components were installed on the station. While ISS commander, Peggy oversaw the first expansion of the working and living areas of the station in its 6 years of life. In October 2007, the Shuttle STS-120 carried the Node-2 Harmony, built in Turin, Italy, by Thales Alenia Space. In February 2008, the Shuttle STS-122 carried to the station Columbus, the European Laboratory, whose structure had been built, again, in Turin. In March 2008, the first of the three components of Kibō—the Japanese Space Agency's contribution—arrived at the station on board Space Shuttle STS-123. Kibō is the largest pressurized module of the ISS: in the initial phase of the project, this was one of the smallest but later the US and Europe decided to reduce the size of their modules, while the Japanese component remained unchanged. The Shuttle also brought "Dextre," the Canadian "robotic hand" to be connected to the Canadarm2 robotic arm to allow finer manipulations. For assembling all those components, Peggy led five spacewalks, thus reaching her personal record of six spacewalks performed by a woman, and also setting the new record for time spent by a woman on activities in open space: 39 h and 46 min (it has since been surpassed by astronaut Sunita Williams).

One of Peggy's most intense moments happened when the crew was rearranging solar arrays to prepare for the installation of Node-2 and one solar panel tore, as already referred to in the previous chapter (see Chap. 38). If they jettisoned the ripped array, then they wouldn't have enough power to continue the next mission. Under the guidance of the Mission Control Center, the astronauts on board managed to work with the materials at hand and prepared precise "cuff links" to repair the rip. "It was intense up there. It was our Apollo 13 moment!" Peggy says.

During the mission, much interest was also devoted to research and Peggy, as a biologist, was on the front line. As she explained in a pre-flight interview:

"We have some integrated immune studies where we're looking at the effects of spaceflight on the immune function in crew members; we're looking at the nutritional levels and how that impacts the crew members' health on orbit. One in particular that most people can relate to is the Vitamin D levels that are very important for bone density. Obviously during spaceflight, we have seen in the past decreased bone density and we're trying to correlate Vitamin D levels to that process, to get a better feel for how that's happening and whether or not higher levels might help prevent that loss."

The crew also explored different solutions of iron in a magnetic field, which one day could be used on suspension bridges and earthquake-resistant structures. After 192 days in space, Peggy returned to Earth on April 19, 2008, together with the cosmonaut Yuri Malenchenko and the South Korean astronaut Yi So-Yeon. The spacecraft landed 260 miles (more than 400 km) west of the expected landing site in Kazakhstan, after undergoing for up to 1 min what is called a "ballistic descent" at eight times the force of Earth-normal gravity, or 8 Gs (while it is usually no more than 4.5 Gs). Peggy compared the 60-s bumpy ride and rough landing to a "rolling car crash": "It was pretty dramatic. Gravity is not really my friend right now and 8 Gs was especially not my friend, but it didn't last too long." It was the second ballistic descent in a few months: a not uncommon problem with the Soyuz. This return flight is also remembered as the first spaceflight in history to carry more women than men.

After her return in October 2009, Peggy was appointed the 13th Chief of the NASA Astronaut Office and the first woman to hold the position that in the old days of Mercury–Gemini–Apollo was held by the legendary and indecipherable Deke Slayton. In this capacity, she led the selections board for new astronauts in 2009 and acted as deputy chief of the selection board in 2013. "It wasn't until I was on the selection board that I realized how lucky I was. We had 8400, and we picked eight," she says. Those were challenging years when NASA suffered several budget cuts and flight opportunities for American astronauts dropped drastically due to the retirement of the Space Shuttles and cancellation of NASA's human spaceflight programs under the Obama administration. She had to face the challenge of a drastically reduced Astronaut Corps, competing for a handful of slots on the ISS and flying there on Russian Soyuz capsules. Peggy publicly declared: "We hope we will overcome this hurdle and continue to explore."

In July 2012, Peggy resigned as Chief of the Astronaut Office to get back in line for a mission assignment. She was selected from a group of 43 active astronauts for her third long-duration mission on the ISS. She was thrilled by the chance to return on the ISS, as part of Expedition-50/51, launched on November 17, 2016. The space station has grown in size with several additions since Peggy was there in 2007. She says:

"The U.S. has scientific facilities in both the European and the Japanese laboratory so it'll give me a lot more places to do different science. Most importantly, the cupola was added on, since I was there last. I'm looking forward to the view from the cupola."

The mission will add new records to the many she already holds.

References

Azriel, M. "Peggy Whitson Steps Down as NASA Astronaut Chief", *spacesafetymagazine.com* (August 4, 2012).

Chang, K. "For Astronauts, a Vanishing Frontier: As Program Cut, Exploring Takes on New Meaning", *boston.com* (April 24, 2011).

Firth, S. "Peggy Whitson, First Woman to Command the International Space Station", *findingdulcinea.com* (February 9, 2010).

Gibson, K.B. *Women in Space: 23 Stories of First Flights, Scientific Missions and Gravity-Breaking Adventures*, pp. 145–150. Chicago Review Press, Inc., Chicago (2014).

Kauderer, A. "Peggy Whitson, Preflight Interview", *nasa.gov* (October 28, 2010).

Kilen, M. "Age Is No Barrier for Astronaut Peggy Whitson", *usatoday.com* (March 17, 2015).

"Method and Apparatus for the Collection, Storage, and Real Time Analysis of Blood and Other Bodily Fluids", US Patent Nos 5665238 (September 9, 1997) and 5866007 (February 2, 1999), Inventors: Peggy A. Whitson, Vaughan L. Clift; Assegnee: NASA.

Official biography of Peggy Whitson, *jsc.nasa.gov* (October 2012).

Parson, A. "Scientist at Work: Peggy Whitson—Testing Limits, 220 Miles above Earth", *nytimes.com* (September 5, 2006).

Contacts by e-mail with the Author in April 2016.

Peterson, L. "Shaky Soyuz descents Plague Astronauts' Only Ride Home", *Houston Chronicle* (April 26, 2008).

Petty, J.I. "Peggy Whitson, Preflight Interview", *nasa.gov* (September 28, 2007).

Shayler, D.J.; Moule, I. *Women in Space—Following Valentina*, pp. 297, 334–336. Springer/Praxis Publishing, Chichester, UK (2005).

Woodmansee, L.S. *Women Astronauts*, pp. 114–115. Apogee Books, Burlington, Ontario, Canada (2002).

40

Sandra Magnus: Soaring to New Heights

Credit: NASA

U. Cavallaro, *Women Spacefarers*, Springer Praxis Books, DOI 10.1007/978-3-319-34048-7_40

Mission	*Launch*	*Return*
STS-112	October 7, 2002	October 18, 2002
STS-126	November 14, 2008	
STS-119		March 28, 2009
STS-135	July 8, 2011	July 21, 2011

Sandra Magnus decided that she wanted to become an astronaut when she was at middle school in Belleville West High School:

> "I've always been interested in why things work and how things work and, and here's this real complicated world that seems to work and go explore it. Just the whole idea of exploring and learning new things just grabbed me and space was the place to do it."

Since then, Sandra started to organize her life to reach that goal, even if, she explains:

> "I was putting this plan together in ignorance because, when I was that age, and really all the way through high school, I really didn't know anything about engineering. I was never exposed to it. I thought engineers were people who drive trains."

She urges students to pursue what she learned in person: "The path you end up on may not be what you planned, but you have nothing to lose if you do your best," she says.

Sandra Hall Magnus, "Sandy" among her friends, was born on October 30, 1964, in Belleville, on the Illinois side of the Mississippi River, across from St. Louis. She explains:

> "It's a kind of an extended suburb where a lot of the farming country starts: a perfect place to grow up. I had the best of both worlds. I had a city within 30 minutes with baseball teams and airports and things like that that are very comfortable to have when you want to participate, and then also it was a small place on the edge of the country where you know a lot of people and don't have a lot problems that you associate with the big city."

After graduating from Belleville West High School in 1982, Sandra went to the University of Missouri-Rolla (now known as the Missouri University of Science and Technology) and earned degrees in Physics and Electrical Engineering. She explains:

> "Physics is my first love because it answers the questions why, and you learn, you learn how to derive the equations that engineers use, so you really do understand what's going on. But while I was doing physics, I discovered engineering that looked interesting to me. So I took a few electrical engineering classes as an undergraduate, just for fun, to kind of see what it was all about."

As she was "tired of school and a little burned out," in 1986, Sandra started to work as a stealth engineer for McDonnell Douglas Aircraft Company, working for 5 years on internal research and development and then on the Navy's A-12 Attack Aircraft program, studying the effectiveness of radar signature reduction techniques, until the program was cancelled. She crossed over into Engineering and did her master's degree at night in Electrical Engineering. While working on stealth engineering, she became particularly interested in new materials, because, she explains, "materials drive everything when you're trying to put airplanes together—working with airplane design, and materials, specifically, and how they function, interact with electromagnetic fields, and so I got interested in materials." So she decided to go back to university and, in 1996, she earned a Ph.D. from the School of Material Science and Engineering at the Georgia Institute of Technology with a dissertation that was supported by NASA Lewis Research Center through a Graduate Student Fellowship and involved investigations on materials of interest for "Scandate" thermionic cathodes: "At that point I felt that I was ready to apply to the NASA Astronaut Office, and see what happens, and what happened was I was lucky enough to get chosen."

Sandra was selected as an astronaut candidate in 1996 with the 16th NASA Astronauts Group and qualified as mission specialist. From January 1997 to May 1998, she worked in the Astronaut Office Payloads/Habitability branch. Her duties involved working with the European Space Agency (ESA), the National Space Development Agency of Japan (JAXA), and Brazil on science freezers, glove boxes, and other facility-type payloads. In May 1998, she was assigned as a "Russian Crusader," which involved travelling to Russia in support of hardware testing and operational products development. In August 2000, she served as a capsule communicator (CapCom) for the International Space Station (ISS). "It's a fun job," Sandra says. "I get to learn a lot of new things every day. No day is the same, and so I'm very lucky."

Sandra is a veteran of three space missions. Her first space mission, STS-112, was the third out of 11 Space Shuttle missions entirely dedicated to the assembly of the ISS. Shuttle pilot was her colleague, Pamela Melroy. With the cooperation of the members of Expedition-5, the "S-1" truss, the third piece of the ISS 11-piece integrated truss structure, was installed. Sandra was mission specialist and loadmaster in charge of logistic transfers to and from the orbiter. She operated the station robotic arm during the three spacewalks required to outfit and activate the new component. STS-112 was the first Shuttle mission to use a camera on the external tank, providing a live view of the launch to flight controllers and NASA TV viewers. After the mission, she was assigned to work with the Canadian Space Agency (CSA) to prepare the Special Dexterous Manipulator robot for installation on the space station. She was also involved in "Return to Flight" activities, leading the Astronaut Office team in that effort after the *Columbia* tragedy in February 2003.

From September 16–22, 2006, Sandra served as the commander of the 11th NASA NEEMO mission (NASA Extreme Environment Mission Operations), an undersea expedition at the National Oceanic and Atmospheric Administration (NOAA) Aquarius laboratory located off the coast of Florida where NASA equipment and techniques for future space exploration are tested.

In July 2005, Sandy was assigned as flight engineer and science officer to the long-duration Expedition-18 and started her training. She flew to the ISS on November 16, 2008, aboard the Shuttle *Endeavour* STS-126 and spent four-and-a-half months on the station, where she completed experiments and other work vital to the health of the orbiting laboratory complex. The mission brought into space, on its fifth spaceflight, the "Leonardo" Multi-Purpose Logistics Module (MPLM), which had been provided by ASI, the Italian Space Agency, and built in Turin, Italy. Leonardo held over 14,000 lb (over 6.3 tons) of supplies and equipment, including the components necessary to expand the station and to enable it to accommodate a crew of six persons: two new crew-quarters racks, a second kitchen for the Destiny laboratory, a second Waste and Hygiene Compartment (WHC), and two water-reclamation racks. After 133 days in space, Sandra returned to Earth in March 2009 aboard the Shuttle *Discovery* STS-119.

After this mission, Sandra served for 6 months at NASA Headquarters in the Exploration Systems Mission Directorate in Washington. In July 2011, she was mission specialist during the final Shuttle mission, STS-135, which for the last time brought into space the logistic MPLM module "Raffaello." The prime job of the mission was to deliver tons of logistics to the space station while the huge cargo-carrying capacity of the Shuttle was still available. Another task was to bring down the pump module that, a few months earlier, had failed a little bit earlier than expected; NASA wanted to learn what happened and how to improve its engineering designs. Sandy was loadmaster, in charge of transferring almost 4 tons of supplies, water, and equipment to the station, and loaded into the MPLM over 2.2 tons of return items that were coming home, including the foam and packing materials, while having to consider the many constraints and dependencies in moving items to and from: "A giant three-dimensional puzzle!" she said. This was a challenging mission, with only four astronauts. She explains:

> "Really the driver for that was the fact that our rescue scenario was a little bit different than normal. Ever since Columbia, we've been mandated to have a shuttle on the pad ready to launch in case the crew has an issue with the orbiter and they need to be rescued. Because we were the last orbiter, there was not an orbiter there waiting for us so our rescue scenario involved the Soyuz capsules which we're flying to station via the Russians, and on the Soyuz capsules only one person could come down at a time. With a crew of four it had taken a year to get everybody down. if we had six or seven people up there it had taken close to two years to get everybody down."

The emblem of the last Shuttle mission, STS-135, features the Shuttle during lift-off over elements of the NASA logo, framed on the patch by omega, the last letter in the Greek alphabet. Sandra explains:

> "The omega came to mind immediately as it is the last letter of the Greek alphabet and we are the last shuttle mission we wanted to highlight that this was the end of the shuttle program. It's not just something that affects the shuttle program but it

> affects all of NASA we felt like having part of the NASA symbol on our patch was appropriate as well."

In September 2012, Sandra was appointed Deputy Chief of the Astronaut Office. She left NASA after 1 month and accepted the position of Executive Director at the American Institute of Aeronautics and Astronautics (AIAA), the main American private professional society in the field of aerospace engineering, with 35,000 members in 79 countries worldwide. So she started her new adventure:

> "I look forward to working with the Board of Directors, members, and staff to expand the relevance and reach of this distinguished organization. The aerospace industry is important to our country's future. AIAA, with its broad base of talented members and their depth of experience, will continue to play a key role not only in creating new opportunities and pushing the boundaries of technology, but also in recruiting the next generation of scientists, engineers, and technologists. I am thrilled to be a part of such a dynamic, vibrant organization!"

References

Official NASA biography of Sandra Magnus, *jsc.nasa.gov* (May 2004).

Contacts by e-mail with the Author in May 2016.

"Sandra Magnus: Preflight Interview", *nasa.gov* (September 24, 2008).

"Sandra Magnus: Preflight Interview", *nasa.gov* (June 3, 2011).

Shayler, D.J.; Moule, I. *Women in Space—Following Valentina*, p. 297. Springer/Praxis Publishing, Chichester, UK (2005).

Woodmansee, L.S. *Women Astronauts*, pp. 117–118. Apogee Books, Burlington, Ontario, Canada (2002).

41

Laurel B. Clark: From Deep Oceans to the Stars

Credit: NASA

U. Cavallaro, *Women Spacefarers*, Springer Praxis Books, DOI 10.1007/978-3-319-34048-7_41

Mission	*Launch*	*Return*
STS-107	January 16, 2003	–

US Navy Captain Laurel Clark died at age 41, with six of her crewmates, on February 1, 2003, over the southern US when Space Shuttle *Columbia* and the crew perished during re-entry, 16 min prior to the scheduled landing. Before going into space, she was an undersea medical officer and then a flight surgeon.

Laurel Blair Salton Clark, known as "Laurie" to her family and friends, was born on March 10, 1961—the oldest of four children—in Ames, Iowa, where her father, Robert Salton, attended graduate school at the Iowa State University. But she didn't remember Ames because the family left when she was just 2 years old and moved to her parents' hometown: the tiny hamlet of Delhi, New York, with a population of just 2000. She attended elementary school at the Delaware Academy until 1975. Then the family moved again to Albuquerque, New Mexico, and, after her parents got divorced, Laurel ended up settling with her mother and her siblings in Racine, Wisconsin, which she considered to be her hometown. She enjoyed scuba-diving, hiking, camping, biking, parachuting, flying, and traveling. She said in an interview:

> "I was interested in the Moon landings just about the same as everyone else of my generation. But, I never really thought about being an astronaut or working in space myself. I was very interested in environment and ecosystems and animals."

"A boring straight-A student, without many hobbies," she described herself in an interview, but she was on the swim team and in the ski club. Those skills became important when she decided to go through the very demanding Navy diving program.

After graduating in 1979 from William Horlick High School, Racine, Wisconsin, Laurel received in 1983 a Bachelor of Science degree in Zoology from the University of Wisconsin–Madison. Eventually, she decided to pursue medicine and, in 1987, earned a doctorate in Medicine from the same university. During medical school, to help pay for her college education, she did active duty training with the Diving Medicine Department at the Naval Experimental Diving Unit in March 1987. After completing medical school, she underwent post-graduate medical education in Pediatrics from 1987 to 1988 at Naval Hospital Bethesda, Maryland. The following year, she became a Navy undersea medical officer and was involved with submarines and with divers:

> "Submarine crews, like astronauts, are selected from a pool of people who turn out to be very healthy. If you have medical problems, then you're not allowed to continue in the submarine service because you're out at sea for long periods of time."

Laurel was then assigned as the Submarine Squadron Fourteen Medical Department Head in the Holy Loch in Scotland, and dove with US Navy divers and Naval Special Warfare Unit Two Seals, and performed numerous medical evacuations from US submarines. "People get appendicitis and can get infections," she explains. "And there were certain times when I had to be involved in getting people off the submarine and getting them to hospitals for further medical care." After 2 years of operational experience, she was designated as a Naval Submarine Medical Officer and Diving Medical Officer, and did her job with interest and passion:

> "If you're trying to get someone who's sick with a fever off of a submarine and it's cold and raining outside, and then you've got to get them off of the submarine (they're not able to walk), and the only way in and out of a submarine, generally, is through a fairly narrow hatch. So, you have to be able to transport them without hurting them or anyone else who's trying to move them off of the submarine. And then, once you get them off the submarine, you still have to get them onto another ship, then to land. And you're doing all of this in a different country, with a different medical system."

Laurel underwent then 6 months of aeromedical training at the Naval Aerospace Medical Institute in Pensacola, Florida, and made numerous deployments, practiced medicine in austere environments, and flew on multiple aircraft. Her squadron won the Marine Attack Squadron of the year for its successful deployment. She was then assigned as the Group Flight Surgeon for the Marine Aircraft Group MAG 13 and served as a flight surgeon for the Naval Flight Officer advanced training squadron VT-86 in Pensacola, Florida. She was a Basic Life Support Instructor, Advanced Cardiac Life Support Provider, Advanced Trauma Life Support Provider, and Hyperbaric Chamber Advisor:

> "In the Navy I was exposed to a lot of different operational environments, working on submarines and working in tight quarters on ships, and learning about radiation medicine. And it was really just sort of a natural progression when I learned about

NASA and what astronauts do, and the type of things that they are expected to do, that I thought about the things I had done so far and became more interested in that as a career."

Laurel was selected as a NASA astronaut in April 1996. She had already applied in 1994, but she didn't make the program on the first tryout. She tried again 2 years later, when she was 5 months' pregnant with her son Iain, and she got in. As she said during a pre-flight interview: "Motherhood has been incredible, and I tell my son all the time that my most important job is being his mother."

Prior to receiving her first flight assignment, Laurel worked in the Astronaut Office Payloads/Habitability Branch. Originally, it seemed that she had to be assigned to a long-term expedition to the International Space Station (ISS)—which would make her the first woman ever to participate in an ISS long-duration mission—and went through the severe Russian winter and water-survival training courses and got familiar with the Soyuz spacecraft. She was then assigned as a mission specialist on the ill-fated STS-107 mission in 2000. The launch was planned for June 2001. But a series of unrelated delays postponed Laurel's flight again and again. She always referred to this mission with passion:

"STS-107 is very exciting. It's the first multidisciplinary science mission we've done since STS-95, and the longest mission that we've done since STS-90. And we're doing a multitude of different scientific experiments on orbit, using also ourselves as test subjects. As a physician the life science research that we're doing is extremely exciting. It's just a great feeling to be part of the team of researchers and investigators that have been working for years to bring this all to fruition."

During the 16-day mission, working 24 h a day in two alternating shifts, the crew successfully conducted approximately 80 experiments—almost a record. Laurel was mainly involved with bioscience experiments, including the OSTEO (Osteoporosis Experiment in Orbit). On January 27, on the 11th day of the mission, the team commemorated the Apollo 1 accident in which, in 1967, exactly 40 years before, Virgil Grissom, Roger Chaffee, and Edward White had been killed. The next day, they commemorated the *Challenger* disaster in which, 17 years before, on January 28, 1986, seven astronauts were killed, including two women. It is curious to note that all three of the tragedies of the American space program happened, within just 6 days on the calendar, at the end of January. The day before, Laurel sent to family and friends her last e-mail:

"Hello from above our magnificent planet Earth. The perspective is truly awe-inspiring. This is a terrific mission and we are very busy doing science round the clock. Just getting a moment to type e-mail is precious so this will be short. I have seen some incredible sights: lightning spreading over the Pacific, the Aurora Australis lighting up the entire visible horizon with the city glow of Australia below, the crescent moon setting over the limb of the Earth, the vast plains of Africa and the dunes on Cape Horn, rivers breaking through tall mountain passes, the scars of humanity, the continuous line of life extending from North America, through Central America, and into South America, a crescent moon setting over the limb of our blue planet. Mount Fuji looks like a small bump from up here, but it does stand out as a very distinct landmark. Magically, the very first day we flew over Lake Michigan and I saw Wind Point, Wisconsin clearly. Haven't been so lucky since. Every orbit we go over a slightly different part of the Earth. Of course, much of the time I'm working back in Spacehab and don't see any of it. Whenever I do get to look out, it is glorious. Even the stars have a special brightness. I have seen my 'friend' Orion several times. Taking photos of the earth is a real challenge, but a steep learning curve. I think I have finally gotten some beautiful shots the last 2 days. Keeping my fingers crossed that they're in sharp focus. My near vision has gotten a little worse up here so you may have seen pics/video of me wearing glasses. I feel blessed to be here representing our country and carrying out the research of scientists around the world. (…) The food is great and I am feeling very comfortable in this new, totally different environment. It still takes a while to eat as gravity doesn't help pull food down your esophagus. It is also a constant challenge to stay adequately hydrated. Since our body fluids are shifted toward our heads our sense of thirst is almost non-existent. Thanks to many of you who have supported me and

Fig. 41.1 Laurel Clark and her family. Credit: Jonathan Clark/NASA

my adventures throughout the years. This was definitely one to beat all. I hope you could feel the positive energy that beamed to the whole planet as we glided over our shared planet.

"Love to all, Laurel."

Her 8-year-old son, Iain, who was very close to her mum, that Saturday morning had wanted to go with his father Jonathan ("Jon") Clark and wait for his mum to return from space; he eagerly awaited the sonic booms that would herald *Columbia*'s approach. He, his parents, and the family dog had survived a harrowing crash just 6 weeks earlier in December, when Jon's single-engine Beech Bonanza airplane hit a violent downdraft while trying to land during a storm. No one was injured, but the plane was destroyed. The experience haunted young Iain and he begged his mother not to fly on the Shuttle. In a family video conference during *Columbia*'s flight, Iain asked his mother "Why did you go?" She traveled often and had even been sent to Russia, so it wasn't just the separation between mother and son that bothered the child—Jon recalled: "He was very worried about her. He had some very moving premonitions that something bad was going to happen, and he didn't want her to go." Once it was clear that there had been trouble, the families were hustled to crew quarters, where they got the grim news. The joy and the longing of Iain to see his mum return from space turned quickly into anguish (Fig. 41.1).

References

Chien, P. Columbia: Final Voyage—The Last Flight of NASA's First Space Shuttle, pp. 69–76, Springer/Praxis, New York (2006).

Dismukes, K. "Laurel Clark: Preflight Interview", *nasa.gov* (November 12, 2002).

Gibson, K.B. *Women in Space: 23 Stories of First Flights, Scientific Missions and Gravity-Breaking Adventures*, pp. 194–195. Chicago Review Press, Inc., Chicago (2014).

Official biography of Laurel B. Clark, *jsc.nasa.gov* (October 2012).

"Racine Remembers a Hero", *racine.wi.net* (January 31, 2003).

Shayler, D.J.; Moule, I. *Women in Space—Following Valentina*, pp. 298–299. Springer/Praxis Publishing, Chichester, UK (2005).

Woodmansee, L.S. *Women Astronauts*, pp. 116–117. Apogee Books, Burlington, Ontario, Canada (2002).

42

Stephanie Wilson: "Madam Robotics Expert"

Credit: NASA

U. Cavallaro, *Women Spacefarers*, Springer Praxis Books, DOI 10.1007/978-3-319-34048-7_42

Mission	*Launch*	*Return*
STS-121	July 4, 2006	July 17, 2006
STS-120	October 23, 2007	November 7, 2007
STS-131	April 5, 2010	April 20, 2010

Stephanie was the second African American woman to travel into space after Mae Jemison.

"I was very fortunate to have a good friend who witnessed to me. I accepted Christ through her witness, and I've been striving to have a closer walk with the Lord ever since. My faith has played an essential role in my career as well as other areas of my life. I hope that my faith governs the decisions that I make in all areas."

Stephanie enjoys snow skiing, music, traveling, and stamp collecting: "I mostly collect stamps off letters that I receive. It's interesting to me to see the designs of the different stamps from the various countries."

Stephanie D. Wilson was born on September 27, 1966, in Pittsfield, Massachusetts, near Boston, where she grew up. She took the decision to explore the unknown when she was only 13 years old and, as part of a school assignment during the career week at the Crosby Junior High School, had to interview an adult whose career reflected specific student interests. Stephanie, who had an avid interest in astronomy, spoke to a local area astronomer, Professor Jay M. Pasachoff of the Williams College. She recalls:

"I was hanging on his every word. Pasachoff was passionate about his work. As he talked with personal enthusiasm about the incredible discoveries he worked on and the exploration of the unknown, I knew that I wanted to be a part of that, too."

But when she went to high school, Stephanie started looking at careers and opportunities and began to rethink her priorities:

"I thought that engineers basically had more opportunities. As a mechanical engineer, I could work on automobiles if the bottom fell out of aerospace. I could work building designs. I could work on city planning. I always felt like engineering was a good career move."

After graduating in 1984 from the Taconic High School, Pittsfield, in 1988, Stephanie earned a Bachelor of Science degree in Engineering Science from the Harvard University.

And she found that engineering could also lead her to the stars: "It was always something in the back of my mind," she said. Stephanie spent 2 years employed at the former Martin Marietta Astronautics Group, Denver, Colorado, where, as a loads and dynamics engineer, she was responsible for performing loads analyses for the Titan IV rocket.

In 1990, Stephanie left Martin Marietta to attend graduate school, focusing her research sponsored by the NASA Langley Research Center on the control and modeling of large flexible space structures and, in 1992, she earned a Master of Science degree in Aerospace Engineering from the University of Texas. Following the completion of her graduate work, Stephanie was hired by the NASA Jet Propulsion Laboratory (JPL), Pasadena, as a member of the Attitude and Articulation Control Subsystem for Galileo spacecraft, responsible for assessing attitude-controller performance, science platform pointing accuracy, antenna pointing accuracy, and spin rate accuracy during the Galileo mission to Jupiter. After Galileo, she also supported the Interferometry Technology Programme. After 6 years of experience in the field, including the four spent at the JPL, she decided to apply to be an astronaut. "That for me was all," she said later, "a natural progression from working on launch vehicles and robotic spacecraft to now flying on the shuttle." One important thing was still missing, however: astronauts spend a lot of time training underwater so, during their first month in the program, they're required to pass a swimming test. "I couldn't swim. So the hardest part was learning how to swim," she says. "I was very lucky that the California Institute of Technology coach agreed to teach me to swim. But it was barely enough time."

Stephanie was selected as mission specialist for the 16th NASA Astronaut Group: "When I entered in 1996 there were more than 2500 applicants, and only 35 new trainees were accepted, so I felt extremely fortunate." She was initially assigned technical duties in the Astronaut Office Space Station Operations Branch to work with space station payload displays and procedures. She then served in the Astronaut Office CAPCOM Branch, working in Mission Control with on-orbit crews during several missions. She was prime communicator the day of the *Columbia* accident:

> "I was the lead capsule communicator for that mission, so I was the voice of the mission control team talking to the crew. It was definitely a sad day for NASA and a sad day for the world. As a result of accidents, though, we learn a great deal. We're able to make safety improvements. Personally I handled the Columbia tragedy by making sure that their memories are kept alive. I did what I could to help the families through their losses, and I tried to remember that God is in control. I might not understand His plan, but He does have one."

Despite the risk involved, Stephanie remained committed to space exploration. She says:

> "I do believe I have found my purpose in life. I believe that this is what I was destined to do. The Space Program is important. There are many things that transfer in technology such as computers and medical equipment that are results of the space station that apply to our everyday lives and make our lives better. I do put my trust in God for protection. That helps me to have confidence in all the things that I do."

Stephanie was eventually assigned technical duties in the Astronaut Office Shuttle Operations Branch involving the Space Shuttle main engines, external tank, and solid rocket boosters. She participated in three Space Shuttle missions: STS-121 (2006), STS-120 (2007), and STS-131 (2010), logging more than 42 days in space. She flew all of the three missions with Shuttle *Discovery* and all of the three missions that docked with the International Space Station (ISS).

In her first mission, STS-121, Stephanie was loadmaster and cargo specialist, splitting robotic-arm responsibilities with the other rooky astronaut, colleague Lisa Nowak. Stephanie was in charge of transferring to the station 5000 lb (approx 2.3 tons) of new equipment, cargo, and food, which were delivered to the space station using the Leonardo Italian logistic module. "This has been a dream of mine for a long time," Stephanie said. After her mission, she served in the Astronaut Office Robotics branch performing robotics procedure reviews and serving as a robotics mentor and instructor astronaut.

In October 2007, Stephanie flew as flight engineer in mission STS-120—where Pamela Melroy was commander—with the task of helping the flight deck crew, to assist the commander and the pilot with any malfunctions that may occur during the ascent phase and during re-entry. Once in orbit, she was assigned as the primary robotic-arm operator for vehicle inspection and spacewalk support, helping to replace the S-band antenna and to relocate the P6 solar array from the Z1 truss to the end of the Integrated Truss Segment. During the deployment of the solar array, the array panels snagged and were damaged. She recalls:

> "We had some trouble deploying a solar array and during the deploy the port six solar array tore. That was almost very similar to an Apollo 13–like moment. The teams on the ground, the extravehicular activity (EVA) teams, the robotic teams, our flight directors, our structural and mechanics personnel all came together to come up with a plan to repair this solar array only using the materials that we had on board. They sent up procedures for us to fabricate cuff links that a spacewalker would install to bridge over the tear to give the array some structural stiffness to complete the deploy. So, we put together with the help with our ground team a spacewalk and robotics procedures that we hadn't seen before. It was very complex, but very memorable."

Stephanie's great ability in operating the robotic arm was key for the successful solution: "Thank God we had Stephanie with us, Madam Robotics Expert. She and Dan Tani, they were real-time moving that arm to try to get that last little inch out of it," said Pamela Melroy, recalling the case (see Chap. 38 on Pamela Melroy). Stephanie explains:

> "We had to send a spacewalker out to the end of the robotic arm at the end of the space station to do this repair. It was definitely a moment when I was trusting in God. His life was in my hands and in the hands of another robotic arm operator."

Mission STS-131 brought three women into space, including two rookies: Dorothy Metcalf-Lindenburger (USA) and Naoko Yamazaki (Japan). It was not the first time: three female astronauts had already flown in 1984 (STS-41G), in 1991 (STS-40), and in 1999 (STS-96). But this time was memorable because, on arriving the station, they

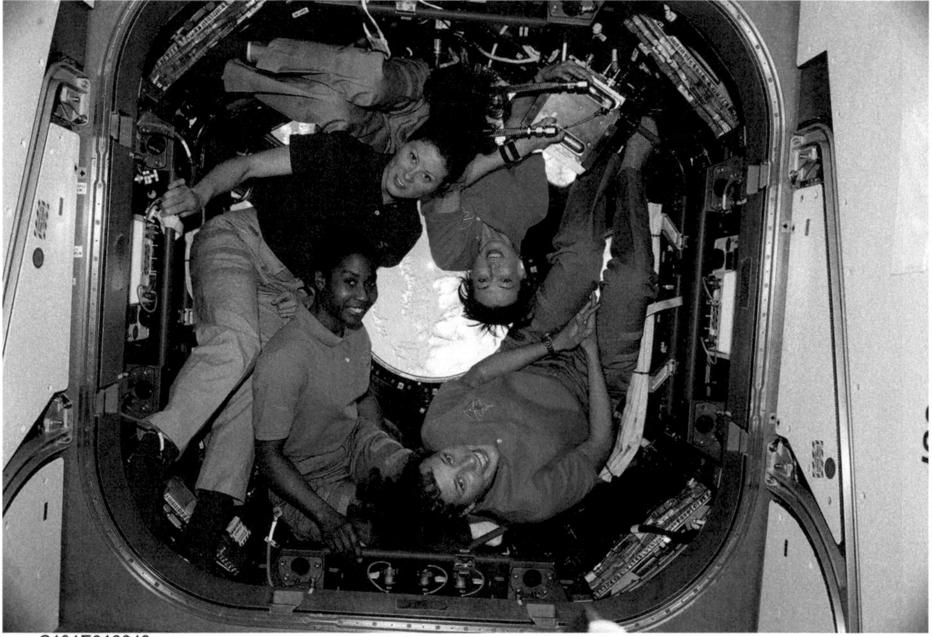

Fig. 42.1 For the first time in history, four women meet in space: Wilson, Dyson, Yamazaki, and Metcalf-Lindenburger in the ISS-Cupola. Credit: NASA

found on board her colleague, Tracy Caldwell Dyson, who a few days before had arrived at the station riding a Russian Soyuz to start her long-duration mission. It was the first time in history that four woman astronauts had met in space. One of the first goals of the mission was to deliver the Leonardo Multi-Purpose Logistics Module (MPLM) filled with supplies and equipment for the ISS and one new crew quarters rack, and to remove and replace an ammonia tank assembly outside the station on the S1 truss. Once more, Stephanie was responsible for moving Leonardo from the Shuttle bay using the robotic arm, and then to transfer 9 tons of material to and from the space station (Fig. 42.1).

References

Bush, S. "NASA Astronaut Stephanie D. Wilson: Faith in Abilities", in *iberkshires.com* (September 14, 2006).

Dean, B. "Stephanie Wilson: Becoming an Astronaut Kicking and Swimming", *nasa.gov* (November 20, 2006).

Evans, M. "Stephanie Wilson: God's Final Frontier", *cbn.com* (July 2009).

Kahn, J.P. "In Her Orbit", *The Boston Globe* (July 6, 2010).

Nevills, A. "Preflight Interview: Stephanie Wilson", *nasa.gov* (October 1, 2007).

Official NASA biography of Stephanie Wilson, *jsc.nasa.gov* (July 2013).
Pearlman, R. "Stephanie Wilson Wants Your Stamps …", *collectspace* (July 5, 2006).
Petty, J.I. "Preflight Interview: Stephanie Wilson", *nasa.gov* (August 11, 2005).
"Preflight Interview: Stephanie Wilson", *nasa.gov* (March 8, 2010).
Shayler, D.J.; Moule, I. *Women in Space—Following Valentina*, p. 300. Springer/Praxis Publishing, Chichester, UK (2005).
Woodmansee, L.S. *Women Astronauts*, pp. 131–132. Apogee Books, Burlington, Ontario, Canada (2002).

43

Lisa Nowak: The First Woman Astronaut Ever Dismissed from NASA

Credit: NASA

U. Cavallaro, *Women Spacefarers*, Springer Praxis Books, DOI 10.1007/978-3-319-34048-7_43

Mission	*Launch*	*Return*
STS-121	July 4, 2006	July 17, 2006

Lisa Nowak was one of NASA's rising stars until she gained international attention on February 5, 2007, when she was arrested in Orlando, Florida, and subsequently charged with the attempted kidnapping of a US Air Force captain.

Lisa Marie Caputo Nowak, the oldest of three girls, was born in Washington on May 10, 1963, 1 month before the historical mission of Valentina Tereshkova. She grew up in Rockville, Maryland, near Washington, and graduated from C.W. Woodward High School, Rockville, Maryland, in 1981, the year of the first Shuttle mission, *Columbia* STS-1. Though she watched the development of the Space Shuttle Program with interest, becoming an astronaut wasn't in her plan, but she enjoyed studying math, science, and engineering. She also liked quiet hobbies such as reading and doing crossword puzzles, taking care of indoor plants, gourmet cooking, bike riding, and sailing. After graduation, she enlisted in the US Navy and joined the Naval Academy in Annapolis, where, in 1985, she majored in Aerospace Engineering and was sent on temporary duty for 6 months to Ellington Field in Houston, to provide engineering support for the Johnson Space Center (JSC)'s Shuttle Training Aircraft Branch. Then something changed. She recalls:

> "I got to meet everybody in the program not just the astronauts but everybody that works in all the different jobs. What impressed me was that all of the people were so into the mission and knew they were a big part of it. It seemed really exciting seeing it up close. I thought if there was a chance to be able to come here that I would love to do that."

After her temporary duty in NASA, with less than perfect eyesight, Lisa was not able to be a pilot for the Navy but she knew there were other ways that she could get to fly. "If something looks like, 'I can't do this' it doesn't mean it's the end of the road," she said, and went for 6 months to the Naval Flight School at Training Squadron VT-86 at NAS Pensacola, Florida, where, in June 1987, she earned her wings as a Naval Flight Officer. "That's the person in the Navy airplanes," she explains, "that does communications and weapons systems and navigation." In 1988, she married Richard Nowak, a colleague whom she had met at the Flight School. The two had a son in 1991.

After an assignment to Electronic Warfare Aggressor Squadron 34 (VAQ-34) at NAS Point Mugu, California, where she flew both the EA-7L and ERA-3B aircraft, supporting the US Pacific Fleet, Lisa joined the US Naval Postgraduate School in Monterey, California, where, in 1992, she earned both a Master of Science degree in Aeronautical Engineering and a degree in Aeronautical and Astronautical Engineering, and began working as an aircraft systems project officer at the Systems Engineering Test Directorate at Patuxent River, Maryland. She was then assigned to the Naval Air Systems Command, working on testing and acquisition of new aircrafts and new navigations and weapons systems for naval aircraft. She logged over 1500 h of flight in over 30 different aircraft during her career in the Navy; she obtained the rank of captain.

Selected by NASA in April 1996, Lisa reported to the JSC in August and began 2 years of training and evaluation, and qualified as a mission specialist. Initially assigned technical duties in the Astronaut Office Operations Planning Branch, she also served in the Astronaut Office Robotics Branch and in the CAPCOM Branch, working in Mission Control as prime communicator with on-orbit crews. In 2001, she had twin daughters.

In 2006, Lisa was assigned to Space Shuttle mission STS-121, the second "Return to Flight" Mission after the *Columbia* tragedy in 2003. She explains:

> "The first two flights instead of just bringing up supplies or assembling part of the Station, which are all very important, are dedicated to testing out some new repair methods that we might have if something did hit the Shuttle during a flight. We have a new boom that we're attaching to the robotic arm. That's something that we've never used before."

Lisa flew as flight engineer during the ascent phase and was in charge of the robotic arm. In a pre-flight interview, she explained:

> "From a robotics point of view that's one of the most interesting spacewalks we are doing. We're doing some things we've never done before. We'll be using the Shuttle's robotic arm to go and get that large boom, that's actually as long as the robotic arm itself. In fact it was built from pieces that were originally going to be another robotic arm. So, we're going to have a very long extended arm to work with and the idea is that with the extended piece we would then be able to inspect different places of the orbiter that we couldn't reach or see with just the robotic arm itself. But a new thing that we're looking at is putting an EVA person on the end of that arm with the idea that they might be able to repair a small piece of damage on the Shuttle somewhere. But we're going to test if that long extended arm and boom can handle the loads of a person on it moving around and, say, smoothing out an area that they had fixed. And we're going to move them to what we call a strong position, where we think is probably OK take those loads. Then we're going to stretch it to what we call a weak position where we're not sure."

The mission also delivered to the International Space Station (ISS) the Italy-built Leonardo Multi-Purpose Logistics Module (MPLM) with tons of supplies and pieces of equipment, but she explained:

> "We're not doing any assembly because we are a test flight. We will be bringing up some spare pieces to put up on a pallet that some later flights will use for assembly, but our main goal of this flight is to test the new safety features and repair techniques."

Shortly after the end of the mission, in January 2007, Lisa separated from her husband after 19 years together. Regarded at the time as one of the best NASA astronauts, with "all the right background credentials, skills and abilities," her career suddenly crashed when, on February 5, 2007—leaving absolutely stunned NASA officials, and anyone who knew her—she was arrested at Orlando International Airport on charges of attempted kidnapping and attempted first-degree murder, after confronting a women she believed was romantically involved with another astronaut whom Lisa loved and spraying pepper spray into her car. Her 2-year affair with William Oefelein, "Billy O" to his friends, an astronaut and Shuttle pilot who turned out to have been involved in a bizarre love triangle, surfaced. The state's attorney argued that the facts indicated a well-thought-out plan to kidnap and perhaps to injure her victim. After being confined for 2 days in jail, Lisa was released on US$25,500 bail and ordered to wear a monitoring device. NASA immediately grounded her, removing her from mission activities, and put her on a 30-day leave, together with Oefelein, and, since she was US Navy officer, sent her back to the US Navy. Lisa and Billy O became the first astronauts ever to be dismissed from NASA. After this case, NASA announced that it would review the astronauts' psychological screening process "to determine if any modifications are needed."

On November 10, 2009, Lisa was sentenced to a year's probation and the 2 days already served in jail, with no additional jail time. Following this trial, a US Navy administrative panel decided to give her a discharge of "other than honorable" and to downgrade her rank from captain to commander. Jon Clark, a former NASA flight surgeon who lost his wife, astronaut Laurel Clark, in the *Columbia* disaster in 2003 (see Chap. 41 on Laurel Clark) said that Lisa had supported his family then and he wanted to support her now:

> "She was a mother before she was an astronaut. She really was into family life, and what's happened has just been totally a shock. She is a really wonderful, good, caring person. You have to find forgiveness and love in your heart to get her through this."

References

Curtis, H.P. "Lisa Nowak: Records Sealed in NASA Astronaut's Love-Triangle Arrest", *orlandosentinel.com* (January 18, 2012).

McLaughlin, M.E. "Shuttle Astronaut Visits Stone Ridge", *web.archive.org* (October 26, 2006).

NASA Press Release. "Statement Regarding the Status of Lisa Nowak", *nasa.gov* (March 7, 2007).

Official NASA biography of Lisa Nowak, *jsc.nasa.gov* (March 2007).

Petty, J.I. "Lisa Nowak: Preflight Interview", *nasa.gov* (August 11, 2005).

The Associated Press, "Astronaut Flown to NASA for Evaluation: Weeks Ago Nowak and Her Husband Separated after 19 Years", *nbcnews.com* (February 7, 2007).

Woodmansee, L.S. *Women Astronauts* pp. 126–127. Apogee Books, Burlington, Ontario, Canada (2002).

44

Heidemarie Stefanyshyn-Piper: From Diver to Astronaut

Credit: NASA

U. Cavallaro, *Women Spacefarers*, Springer Praxis Books, DOI 10.1007/978-3-319-34048-7_44

Mission	*Launch*	*Return*
STS-115	September 9, 2006	September 21, 2006
STS-126	November 14, 2008	November 30, 2008

Heidemarie Stefanyshyn-Piper says:

> "To me exploring space is just a natural progression of where humans are going. As we become more advanced and we have more technology to go farther. Thousands of years ago people would just go beyond the next hill, go over the mountain, go across the river. Then it led to going across the oceans. And, then it was 'OK, let's go into the skies.' We now have airplanes. We can fly. We have submarines and submersibles; we can go into the waters. So looking into the skies and looking at the stars and at the planets and thinking, what's out there …. We're curious. We, as humans always want to know what's out there. To me it just seems natural that we've looked around here and we're just going to go look out farther. We're still developing the means to go out there farther. But that's just where we're going to go next. To me, exploration makes sense because we're always looking at what's the next thing out there, what else can we learn, and how can we go there. Maybe we can learn something that we can bring back here and help solve some of the problems we have on Earth."

Captain Heidemarie Martha Stefanyshyn-Piper was born in St. Paul, Minnesota, on February 7, 1963, second in line with four brothers in a typical Midwestern, middle-class family. Both of her parents were Europeans who migrated to the US after the World War II: her father came from Ukraine and her mother from Germany. Heide was raised in the Ukrainian cultural community of St. Paul, Minnesota, where she learned to speak Ukrainian and became a member of Plast—a Ukrainian scouting organization, where she learned how to use maps and compasses, and how to paddle a canoe and hike in the woods. She said later:

> "The biggest part of scouting that I use as an astronaut is being part of a team. In scouting, you learn to work as a team to accomplish a goal. As an astronaut, you are a small part of a very large team to put people in space."

As a small girl, Heide's favorite movie was *Heidi* because she thought it was named after her. "My parents wanted all of us to have a good education," she says, "school was very important. That's why they sent us all to Catholic schools thinking that we would get a better education there." In 1980, she graduated from the all-girls Derham Hall High School, St. Paul. Very active in sports, she practiced running, roller blading, ice skating, swimming, and scuba-diving:

> "One of the things I learned after not living in Minnesota for a while is that I really like four seasons. It gets very cold in Minnesota in the wintertime, but that never hindered us as kids. You always went out and played in the snow, and if it was cold you just put on another sweater or another scarf or an extra pair of mittens. It was nice to have different seasons. In the summertime it was actually hot, but in Minnesota there are lots of lakes and rivers. I learned to swim, which I guess helped when I became a diver."

Since math and science were her stronger subjects in high school, and there was a big push to get a lot more girls to go into engineering, Heide decided to go to MIT (Massachusetts Institute of Technology) for Mechanical Engineering. She applied as a Navy pilot and took the US Navy Reserve Officer Training Corps (ROTC) scholarship, to help pay for college. She says:

> "I've always had a fascination with flying. As my mother came from Germany, every couple of years she and my father would send one or two of us, usually two of us, to Germany. I remember when I was 4 years old going and flying in an airplane and I thought that that was the neatest thing. So I've always had this bug in the back of me that says 'I really want to fly.' I was going in the Navy, and I decided I was going to try to fly for the Navy. After I graduated, I decided that, or it was decided for me that, because I didn't pass the eye exam, I wasn't going to become a pilot. And so I became a Navy diver."

Heide received a Bachelor of Science degree in Mechanical Engineering in 1984 and a Master of Science degree in Mechanical Engineering in 1985, and was commissioned an Ensign in the Engineering Duty Officer Community in 1985. After completing her training at the Naval Diving and Salvage Training Center in Panama City, Florida, as a Navy Basic Diving Officer and Salvage Officer, she served at the Pearl Harbor Naval Shipyard, Shore Intermediate Maintenance Activity Pearl Harbor, Commander Naval Surface Force, US Atlantic Fleet, and Naval Sea Systems Command Supervisor of Salvage and Diving: "Most of my time in the Navy was spent doing ship repair. I was at the shipyard in Pearl Harbor, a shore maintenance in Pearl Harbor, in Little Creek or Norfolk, Va., at a tech commander staff doing maintenance."

Heide qualified as a Surface Warfare Officer on board USS *GRAPPLE* (ARS 53), becoming the first woman to reach this position in the US Navy. In that role as experienced diving and salvage officer, she took part in major salvage projects that attracted high attention in the media, including the development of a plan for the Peruvian Navy salvage of the submarine *Pacocha* in 1988 and the de-stranding of the tanker *Exxon Houston*, off Barber's Point, Oahu, Hawaii, in 1989.

It was at that time that Heide first thought of becoming an astronaut. She says:

> "When I was growing up I didn't really think much of being an astronaut. I was pretty much removed of the space program so I just didn't think that I could become an astronaut. It wasn't until when I was in college and I was in the ROTC program. We had an astronaut come talk to us and, even at that time, I thought, 'OK. That's pretty neat.' but I never really put two and two together that that would be me. It wasn't until seven or eight years after that I was in the Navy and a fellow officer that was stationed with my husband had applied to the astronaut program. I was talking to him and I looked up what NASA was doing and they were talking about building space station Freedom. I looked at what they were planning and some of their training that we're doing and I thought, 'You know, that looks to me more like diving than flying.' I was fixing ships for the Navy, doing underwater ship repair and I thought, 'You know, if I could fix ships underwater, I can build a space station in space.' I thought I had a lot to offer, to help build the space station. That's how I applied to the program."

In 1996, Heide was selected as a NASA astronaut for the 13th Group. She says:

> "I remember that morning because I was scheduled to take my physical fitness test for the Navy. They happened to have caught me right before I was leaving to take the test so and my secretary called me and said, 'You have a phone call from NASA' and I said, 'Oh, I better take this.' So I went over and I took it and it was Bob Cabana who at the time was the head of the astronaut office. You always knew that if you got the phone call early in the morning by the head of the astronaut office that was good news as opposed to getting the call later in the day, saying that you can try again in a couple of years. He said, 'We'd like you to come down and be part of the astronaut program.' I was very excited but, I figured, 'OK, I'm in a professional office here. I can't start screaming and jumping up and down' I had to take my physical fitness test and my husband, who was also stationed at the Naval Sea Systems Command, was taking his test the same time so he was there and he could tell that something was up because I just walked in and he said I had the biggest smile on my face. And, so I told him."

Heide reported to the Johnson Space Center (JSC) in August 1996. After 2 years of training and evaluation, she qualified for flight assignment as a mission specialist. Initially assigned to astronaut support duties for launch and landing, she also served as lead Astronaut Office Representative for Payloads and in the Astronaut Office's extravehicular activity (EVA) branch.

During her two Shuttle missions—STS-115 and STS-126—Heide had the chance to make the best of her diver's skills for maintenance of the International Space Station (ISS). She knew of her assignment to her first mission, STS-115, on the day of her birthday. "It was a great birthday present, the best birthday present that I had!" she recalls. And she started her long training:

> "We've stayed together as a crew for four years. We don't need that much time to train for a shuttle mission. When the Columbia accident happened, we were about 3,5 months from flight. After the accident our training obviously was put on hold, or it wasn't an intense, daily training. Some of us did do some other technical assignments. For example I was doing some of the EMU recovery. They had a problem on board station that they couldn't get cooling to the EVA suits, the EMUs. So I spent a year and a half working the technical issues, that was something that helped me understand more about the suits, so now I know more about the suit and I know more about station systems."

The STS-115 mission finally was launched in September 2006. This was the mission during which NASA returned to build the ISS after the 3-year break following the *Columbia* accident. In a pre-flight interview, Heide explained:

> "The main goal of our flight is to deliver the P3/P4 Truss element to the station. On P4, there's a set of solar arrays, so that, after our mission station will have the capability to generate more electricity. And by having more electrical capability, you can add the partner modules. You can bring up the European laboratory and the Japanese modules, and now the partners are able to have their components on orbit and we can also increase the crew size so that we are able to utilize the space station for what it was built for."

To complete the installation of the truss and activate the solar arrays, Heide participated in two of the mission's three EVAs for a total of 12 h, which made her the second most experienced female spacewalker. With astronaut Joe Tanner, she conducted initial installation of the P3/P4 truss onto the space station, connected power cables on the truss, installed a GPS antenna on the Japanese Logistic Module, and lubricated the Port SARJ (the joints on a solar array). The following year, from May 7 to 18, 2007, Heide was commander of the 12th NEEMO (NASA Extreme Environment Mission Operations) expedition, a NASA program for studying human survival in the Aquarius underwater laboratory in preparation for future space exploration.

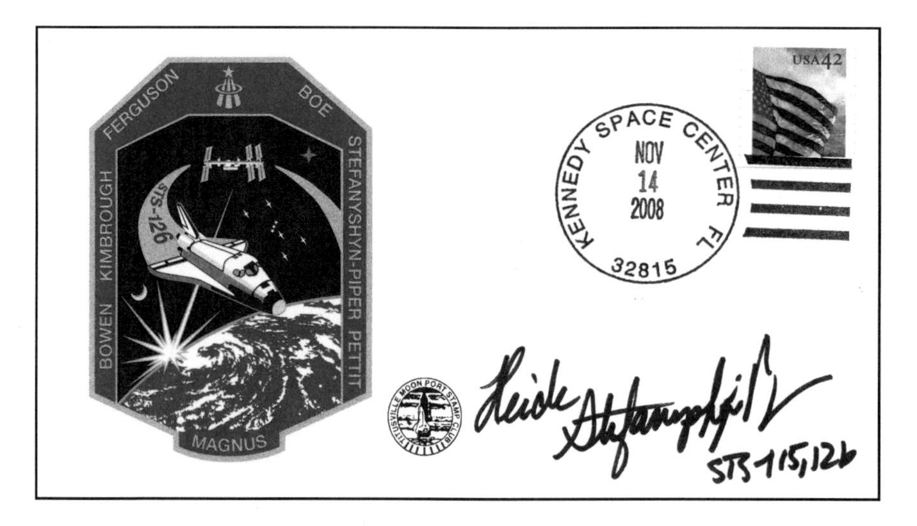

Commemorative cover of mission STS-126, signed by Heidemarie M. Stefanyshyn-Piper (from the collection of Umberto Cavallaro)

STS-126 was the ULF-2 mission (second Utilization Logistics Flight) that delivered to the ISS an extra bathroom, a kitchen, two bedrooms, an exercise machine, a water-recycling system, and all that was required to expand the living quarters of the space station to eventually house six crewmembers. Before the mission, Heide explained:

> "Our mission will be a ULF or Utilization Logistics Flight and we are a logistics flight. We're not bringing up a new module to space station so we're not really changing the outside of station, or the inside. But we're going to add a tremendous capability to the inside of space station. We're taking up the largest MPLM that they've had. It's packed full, with more equipment, more spares up to space station so that we can enhance the capability. We're bringing up some crew quarters that will be needed for a year from now when we double the size of space station and go from three to six crew members. For that we're bringing up another space toilet and another space galley. More important is we're bringing up, what they're calling the ECLSS, the Regenerative Environment Control and Life Support System."

Sandy Magnus was also flying on this mission and remained on board the ISS as a member of the long-duration Expedition-18. Heide participated as the lead spacewalker in three out of the four spacewalks that resulted in restoring full power-generation capability to the ISS. It was during her first spacewalk the infamous mishap occurred: on November 18, while Heide was trying to clean up grease that had inadvertently spurted from her grease gun, her US$100,000 toolbag drifted out of her reach and sailed off into space. The toolbag remained in orbit circling Earth and was visible from the ground for more than 8 months.

During her two missions, Heide has logged over 27 days and 15 h in space, including 33 h and 42 min of EVA in five spacewalks that put her at 25th in the all-time list of spacewalkers by duration. In August 2009, she left NASA and reported to the Naval Sea Systems Command as Naval Systems Engineering Chief Technology Officer. Two years later (on May 20, 2011), she was appointed commander of the Carderock Division of the Naval Surface Warfare Center in Maryland.

References

Hitt, D. "Dream of Flying", *nasa.gov* (August 29, 2006).

Official NASA biography of Heidemarie Stefanyshyn-Piper, *jsc.nasa.gov* (July 2009).

"Preflight Interview: Heidemarie Stefanyshyn-Piper", STS-126, *nasa.gov* (October 31, 2008).

"Preflight Interview: Heidemarie Stefanyshyn-Piper", STS-115, *nasa.gov* (July 19, 2006).

Woodmansee, L.S. *Women Astronauts*, pp. 118–119. Apogee Books, Burlington, Ontario, Canada (2002).

45

Anousheh Ansari: The First Iranian Spacewoman

Anousheh Ansari, the first female private space explorer.

Credit: NASA

U. Cavallaro, *Women Spacefarers*, Springer Praxis Books, DOI 10.1007/978-3-319-34048-7_45

Launch		*Return*	
Soyuz TMA-9	September 18, 2006	Soyuz TMA-8	September 29, 2006

Ever since she could remember, Anousheh Ansari wanted to be an astronaut. As a child, she would draw pictures of herself sitting in a spacecraft, blasting into orbit. Her dream, which she kept in her heart since she was a little girl in Iran, was fulfilled a few days after her 40th birthday, when, on September 18, 2006, she captured headlines around the world as the first Iranian astronaut, the world's first female Muslim in space, and the first-ever self-funded woman to fly to the International Space Station (ISS).

Anousheh Ansari was born Anousheh Raissian in Mashhad—Iran's second largest city—on September 12, 1966. When she was a young girl, her family moved to Teheran, where she attended the schools. She left her country, which had become unstable when, after political turmoil (culminating in the overthrow of the King of Persia and the Islamic Revolution in 1979), universities were closed for 2 years and it was no longer possible to pursue an advanced education. In 1982, as a teenager, without knowing a single word of English, Anousheh migrated with her family to the US, where a family member was living, in Virginia. "When I left Iran, I was 16 years old," she says, "and in class XI. But in the US, I was admitted to class IX." She was, however, determined not to waste those academic years. "In the holidays," she recalls, "I went to the nearby college and attended non-stop English classes from eight in the morning to eight in the evening."

In 1989, Anousheh earned a bachelor's degree in Electronics and Computer Engineering from George Mason University, Fairfax, Virginia. After graduation, she began working at MCI, where she met her future husband, Hamid Ansari. They married in 1991, and Anousheh became a US citizen. In 1992, she earned a master's degree in Electrical Engineering from George Washington University. A successful serial entrepreneur with a strong personality, in 1993, she co-founded with her husband, Hamid Ansari, and her brother-in-law, Amir Ansari, the Telecom Technologies Inc., a service provider that invented an innovative software switch technology, allowing integration between existing legacy telecom networks and next-generation networks, and enabling voice communications over the Internet. Assignee of three key US patents, in 2000, TTI successfully merged into Sonus Networks, Inc., a provider of IP-based voice infrastructure products, in a deal worth approximately US$750 million. Anousheh served as general manager and vice president of the Softswitch division.

In 2006, Anousheh co-founded and was chairwoman of Prodea Systems, a company headquartered in Dallas, Texas. Prodea provides the "Internet of Things"—managed service solutions for global operators and service providers in several verticals via their powerful Residential Operating System (ROS).

She recalls that she has always been fascinated with space:

> "I remember it has always been in my heart and a part of me. I don't know how it began or where it began. Maybe I was born with it. Maybe it's in my genes. I don't know. My husband sometimes jokes and says you know I think you're not from this planet. You may have come from another planet and you're just trying to get back home."

Endorsing the motto of her hero, Gandhi—"Be the change you want to see in the world"—Anousheh is an active proponent of world-changing technologies. With the support of her family, on May 5, 2004 (on the 43rd anniversary of the American suborbital first of Alan Shepard), she decided to contribute a US$10 million cash award for the first non-governmental organization to launch a reusable manned spacecraft into space twice within 2 weeks. The Ansari X-Prize, as it was renamed to honor the generous family, was won that same year by the legendary aircraft designer Burt Rutan with his SpaceShipOne. When Burt Rutan later announced that he was joining with Sir Richard Branson's Virginia-based Virgin Galactic Company to make a larger version for commercial suborbital flights by 2008, Anousheh immediately reserved her seat. She was the fourth overall self-funded space traveler on the Soyuz spacecraft after the American Dennis Tito (2001), the South-African Mark Shuttleworth (2002), and the American Greg Olsen (2005), under a deal arranged by Space Adventures with the Russian Federal Space Agency.

Originally, the world's fourth space tourist was supposed to be the Japanese businessman Daisuke Enomoto, scheduled for flying with Soyuz TMA-9, along with Russian cosmonaut Mikhail Tyurin and US astronaut Miguel Lopez-Alegria. Since Anousheh strongly wanted to fly into space, she accepted becoming the backup of Enomoto. It happened that, a few weeks from flying, on August 21, 2006, Enomoto was disqualified for medical reasons and Anousheh became a primary crewmember. She recalls:

> "I was actually going back to my room after finishing my day of training and I received a call from Space Adventures telling me that I've been moved up to become part of the primary crew. First I couldn't believe it. I thought they were joking with me and then, as I started believing them, I was in complete shock and total excitement and you know, I would've screamed if I wasn't so aware of the people around me."

On September 18, 2006, Anousheh blasted off on board the Sojuz TMA-9 and reached the ISS. She wanted to wear both the American and the Iranian flags on her spacesuit, to honor her two countries:

> "I was born in Iran and lived there until the age of 16 and then moved to the United States. So I have a lot of roots in Iran and feel very close to the Iranian people and the culture of the country."

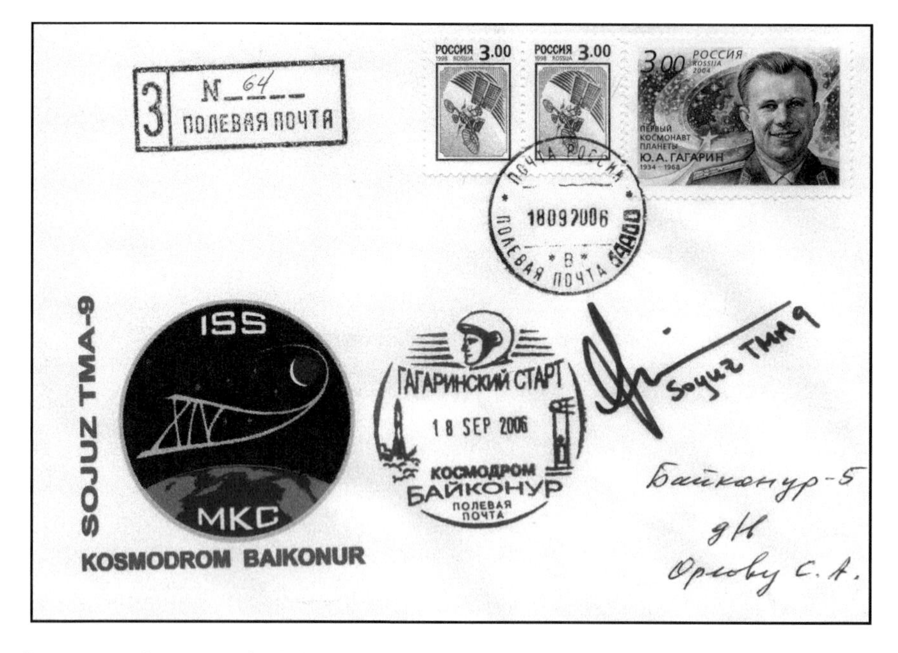

Commemorative cover of mission Soyuz TMA-9, signed by Anousheh Ansari (from the collection of Umberto Cavallaro)

At the insistence of the Russian and US governments, she did not wear the flag, but wore the Iranian colors instead and kept the Iranian flag on her flight patch.

Anousheh remembers the intense odor of a burnt almond cookie that she sensed when she entered from the Soyuz to the ISS. Later, she found out that it was because the thrusters that they fire when approaching for docking use a fuel that contains cyanide, which has an almond smell. The enclosed environment of the ISS made a lasting impression on her—"almost like a bachelor pad." "Imagine if you were stuck in an aircraft and the door wasn't opened for several years—it's something like that," she recalls. But finally she had accomplished something that many had told her she couldn't:

> "I was finally at the destination I always wanted to go to. I described it sometimes as being like when people go on a pilgrimage to Mecca. For me it was my pilgrimage and I was there, I finally made it, and I didn't want to leave."

During her 8-day stay on board the ISS, Anousheh performed a series of scientific experiments on behalf of the European Space Agency (ESA). She conducted four experiments, including researching the mechanisms behind anemia, how changes in muscles influence lower-back pain, and consequences of space radiation on ISS crewmembers and different species of microbes that have made a home for themselves on the Space Station. She also became the first person to publish a weblog from space. She explains:

> "I felt it was important to share this experience. I felt that I was very privileged, that I was very lucky to be there. I know that there are lots of people like me who dream about this, so I wanted to share this experience and take them on the journey with me."

She doesn't like the term "space tourist" and calls it an "over simplistic label to a complicated process." She prefers the title of "spaceflight participant":

> "In a way I take offense when they call me a tourist because it brings that image of someone with a camera around their neck. I've been training for it for six months. I think if it is to be compared to an experience on Earth it probably is closer to expeditions like people who go to Antarctica or people who climb Mount Everest. So I would probably compare it more to an expedition than to a touristy trip to another city."

Anousheh is a strong promoter of the participation of the private industry in space exploration and fosters the "privatization of space." Her family keeps helping the industry and making space travel more affordable and accessible to more people. She adds:

> "I hope to inspire everyone, especially young people, women, and young girls all over the world, and in Middle Eastern countries that do not provide women with the same opportunities as men, to not give up their dreams and to pursue them. It may seem impossible to them at times. But I believe they can realize their dreams if they keep it in their hearts, nurture it, and look for opportunities and make those opportunities happen."

On her website, *anoushehansari.com*, Anousheh declares that she wants to be a "Space Ambassador" to promote peace and understanding amongst nations:

> "Looking at it from up there you can't see any borders or any differentiation between different races or anything like that and all you see is one planet; one place that all of us have to take care of if we want to be able to live on it for a long time. Our current technologies and everything we have does not afford us the luxury of saying 'ok if we blow up this planet and make it inhabitable for ourselves we can pack up and live some place else.' So on one hand you look at your safe haven on Earth and then you turn around and then you look at the blackness of the universe and see that there is not a lot of habitable planets or moons around you. You sort of feel like you need to take care of the precious gift you've been given."

She adds:

> "I hope that more and more people will get to have this experience because it does give you a new perspective on life, and on everything else like how to live your life and interact with your environment. I've talked to different astronauts and cosmonauts and read their books, and think that it's a common theme that you hear from all of them. It does make a big difference. I am hoping that more and more people will be able to have that experience first hand and I think it may make our world a better place to live if more people flew to space."

She lives in Dallas, Texas, and has two honorary Doctor of Science degrees from Utah Valley University and International Space University. She is currently working toward a master's degree in Astronomy from Swinburne University. She has published her memoirs with Homer Hickam in the book *My Dream of Stars: From Daughter of Iran to Space Pioneer*, issued by St. Martin's Press in 2010.

References

Gibson, K.B. *Women in Space: 23 Stories of First Flights, Scientific Missions and Gravity-Breaking Adventures*, pp. 167–168. Chicago Review Press, Inc., Chicago (2014).

Goudarzi, S. "Anousheh Ansari: A Passion for Space Travel", *news.bbc.co.uk* (September 15, 2006).

Goudarzi, S. "Interview with Anousheh Ansari, the First Female Space Tourist", *space.com* (September 15, 2006).

Gupta, G. "The Dramatic Journey of Anousheh Ansari, the First Female Private Space Explorer", *dnaindia.com* (January 31, 2016).

Hollingham, R. "Space Tourist's Sick Trip to 'Bachelor Pad' in Space", *bbc.com* (November 18, 2014).

Iran Chamber Society, "Anousheh Ansari: The first Iranian Astronaut and First Female Space Tourist", *iranchamber.com* (February 26, 2016).

Official biography of Anousheh Ansari, *anoushehansari.com.*

Contacts by e-mail with the Author in April–June 2016.

46

Sunita Williams: A Marathon Runner in Space

Credit: NASA

U. Cavallaro, *Women Spacefarers*, Springer Praxis Books, DOI 10.1007/978-3-319-34048-7_46

Mission	*Launch*	*Return*
STS-116	December 9, 2006	
STS-117		June 22, 2007
Soyuz TMA-05M	July 15, 2012	November 18, 2012

During her two long-term stays on the International Space Station (ISS), Sunita Williams broke several records, including the one for longest continuous spaceflight by a woman (192 days), for total female spacewalks (seven), and the longest spacewalk time for a woman (50 h, 40 min). She was the second woman ever to command the ISS and the second astronaut of Indian descent to fly into space:

> "Records are for breaking. They are just little beacons out there that somebody else will want to do and they are measures of the accomplishments that we've made so far. I don't think too much about them because I was also just in the right place at the right time."

Sunita Lyn Pandya Williams, "Suni" to her family and friends, was born in Euclid, Ohio, on September 19, 1965, to a father from India who lectured on neuroscience at medical schools in Harvard and Boston University, and a mother of Yugoslavian descent who was working as an X-ray technician. Both of them were helicopter pilots. When Suni was a young girl, the youngest of three siblings, her family moved to Needham, Massachusetts, near Boston, which she considers to be her hometown. There she went through elementary, middle, and high school, and practiced many sports, including swimming, triathlon, biking, baseball, football, basketball, and hockey. She graduated from Needham High School in 1983. As a little girl, she contemplated becoming a veterinarian, but her way of life led her in quite another direction.

Following the steps and the suggestions of her first brother, who went to the Naval Academy, Suni decided to pursue Physical Science in the US Naval Academy. The

Academy offered a lot of opportunities and, among other things, she became a swim team captain, and learned leadership and followership, and a sense of camaraderie. After receiving a Bachelor of Science degree in 1987, she was commissioned in the US Navy in May 1987 and, after training, was designated a Basic Diving Officer. She next reported to the Naval Air Training Command, where she become a Naval Aviator in July 1989. She would want to fly jets, but "there weren't a lot of women in combat aircraft at the time," she explains, "so the opportunities were really limited. So I went into helicopters." She was assigned to Helicopter Combat Support Squadron 8 (HC-8) in Norfolk, Virginia, with which she made overseas deployments to the Mediterranean, Red Sea, and the Persian Gulf for Operation *Desert Shield* and Operation *Provide Comfort*, to provide humanitarian aid.

In September 1992, Suni was the Officer-in-Charge of an H-46 detachment for Hurricane Andrew relief operations aboard USS *Sylvania*. In January 1993, she began training at the US Naval Test Pilot School. She graduated in December, and was assigned to the Rotary Wing Aircraft Test Directorate as a chase pilot. She has logged more than 3000 flight hours in more than 30 aircraft types.

While attending the Test Pilot School, Suni got in contact with NASA.

She said in an interview:

> "There are some people in our Astronaut Office who knew from the very beginning they wanted to be astronauts. Not me. I really just didn't know anything about becoming an astronaut or really too much about what the program was pursuing. It wasn't really until later on, when I was established in my career. In my mid-20s, when I was a test pilot, I had the opportunity to go to Johnson Space Center and meet astronaut John Young and understand that he landed on the moon in some type of vertical apparatus—that sounds like helicopters, and I was a helicopter pilot—so it seemed like it might fit. Then, I started looking at what I needed to do to become an astronaut. I needed a master's degree."

After earning a Master of Science degree in Engineering Management from Florida Institute of Technology, in December 1995, she decided to apply to become a NASA astronaut and went back to the Naval Test Pilot School as an instructor in the Rotary Wing Department. After 11 years in the US Navy, she entered NASA with the 18th Astronauts Group in June 1998.

After intensive training in Shuttle and ISS systems, Suni worked in Moscow with the Russian Space Agency to coordinate the Russian contribution to the space station and to support, as Russian Crusader, the first expedition crew. Following the return of Expedition-1, she returned to Houston and worked within the robotics branch on the station's robotic arm and the follow-on Special Purpose Dexterous Manipulator.

From May 13 to 20, 2002, Suni lived underwater in the Aquarius habitat for 9 days as a crewmember of the NEEMO-2 expedition (NASA Extreme Environment Mission Operations)—the NASA program aimed at investigating survival in a hostile, alien place for humans to live. During the expedition, she performed extravehicular activities (EVAs) to simulate underwater spacewalks. She participated in two long-duration expeditions on board the ISS.

Suni's first trip into space was Expedition-14/15. Launched aboard the STS-116, she participated in the 20th Space Shuttle mission to the station. The launch was originally scheduled for December 7, 2006, but was delayed because of bad weather conditions and left 2 years later. She recalled:

> "My first mission was six-and-a-half months. We weren't exactly sure how long it was going to be because I went up and back on the space shuttle which was dependent on weather for launch and landing. So you might have to say goodbye a couple of times and you might get excited to come home and then have to wait. It was an emotional rollercoaster, particularly because it was my first space flight."

Suni took along with her to the space station a copy of *Bhagavad Gītā* (the traditional Hindu "spiritual dictionary") and an idol of Lord Ganesha. She said:

> "I knew Ganesha was looking after me. When you are thinking about going away for a long duration mission, it has to be part of your mindset that you're leaving your family, but it's for the right reasons, for good reasons, and hopefully helping humanity. So you settle yourself on that but you also have to prepare yourself. I call it tying up the ends of your life before you go because you never know what's going to happen. You want to make sure that when you leave, you feel at peace with everything."

During the mission, Suni set the record for the longest continuous single spaceflight by a woman: 192 days. With four EVAs, she also established another female record, beaten only in 2008 by Peggy Whitson, who made five "spacewalks." She explains:

> "You don't just 'go outside.' Usually that is the fun and easy part of the entire thing. The days leading up to the EVA are the intense days with battery charging, METOX (CO^2 removal cartridge) regenerations, suit sizing, tool gathering and preparation, equipment gathering and preparations, studying new procedures, reviewing and talking through how to get us suited and how to get the airlock depressed, reviewing the tasks we will do with each other and with the robotic arm, talking about cleaning up, and then talking through a plan to get back into the airlock, and any emergencies that can come up—loss of communications, suit issues, etc."

Even return may involve complex decontamination procedures before entering into the ISS. In particular, in her last two spacewalks, Suni was in contact with the ammonia used in the air-conditioning system of the station and had to undergo a long cleaning process to avoid taking the ammonia—which is toxic—inside the space station. As all the astronauts who have been in space for some time used to say, from up there, you look at Earth and its problems with different eyes:

> "When you're flying in space some of the things down on Earth seem trivial. Things like politics leave your mind. I didn't feel like I was a person from the United States, I felt like I was lucky enough to be a person from Earth. For me, most news wasn't important but people are important, so when you hear about natural disasters like hurricanes and fires, that makes you miss home and wonder how everybody's coping. But I would also look back at the planet and think 'gosh it's a pretty little place,

everybody's going for a walk on the beach or something like that, they must be enjoying life down there'. If you are having a bad day, you can go to the cupola window and see a part of the Earth. It makes you smile."

Cupola was her preferred place: "the crowning jewel of the space station," as she named it. "It's like being on a spacewalk," she adds, "but it's a little bit more comfortable; you're in just in T-shirt and shorts and you can turn 360°. It's just spectacular": a great place to have a pretty much heads-on view and to manipulate the robotic arm. Like many astronauts after their extraterrestrial experience, she suggests that we should broaden our vision: "It is hard to believe that we have borders on our planet. In space, all we see are oceans. I believe that we are all citizens of this universe."

Suni made headlines when, while she was on the ISS, on April 16, 2007, she ran the famed Boston Marathon, the best marathon in the world, which happened to be in her home town. With her race number 14,000 posted on the space station's treadmill, she successfully crossed the metaphorical finish line in about 4 h and 24 min, while circling Earth at least twice, running as fast as 8 miles/h but flying more than 5 miles/s. She ran in tandem with fellow astronaut Karen Nyberg, who was running on Earth, among the 24,000 other runners participating in the marathon. Despite having to contend with the monotony of running 42.2 km (26.2 miles) on a treadmill without crowds to cheer her on, Suni ran under better weather conditions than her Boston counterparts: in Boston, it was 48° with some rain, mist, and wind that day. "I did it," she said, "to encourage kids to start making physical fitness part of their daily lives. I thought a big goal like a marathon would help get this message out there." Media sang the praises of the marathoner "pioneering new frontiers in the running world." Suni—who had already participated in the Boston Marathon once before as a teenager—had qualified for this 2007 race during the 2006 Houston Marathon with a time of 3 h, 29 min, and 57 s, finishing among the top 100 women.

Apart from this event, Suni ran at least four times a week on the station: two longer runs and two shorter runs. Exercise is indeed essential to counteract the effects of long-duration weightlessness on astronauts' bones and muscles: "In microgravity," Suni explains, "both of these things start to go away because we don't use our legs to walk around and don't need the bones and muscles to hold us up under the force of gravity."

Back from her mission, Suni was appointed Deputy Director of the Astronaut Office. In July 2012, she was assigned again to a long-duration mission: Expedition-32/33. She lifted off from Baikonur Cosmodrome aboard Soyuz TMA-05M, with the Russian Commander Jury Malenchenko and the Japanese flight engineer Akihiko Hoshide. Also in September 2012, she became the first person to do a triathlon in space, which coincided with the Nautica Malibu Triathlon held in Southern California. She used the space station's own treadmill and stationary bike and, for the swimming portion of the race, she used the Advanced Resistive Exercise Device (ARED) to do weightlifting and resistance exercises that approximate swimming in microgravity. After "swimming" half a mile (0.8 km), biking 18 miles (29 km), and running 4 miles (6.4 km), Suni finished with a time of 1 h, 48 min, and 33 s. She ran this triathlon in tandem with doctor Sanjay Gupta concurring on Earth.

During this mission, Suni performed six EVAs, the first of them being far longer and less productive than planned. Goal of the mission of August 30, 2012, was to replace a failed power distributor, one of the four Main Bus Switching Units (MBSUs) installed in the S-zero truss that connects a box that routes 25 % of the power coming to the station from the solar arrays. They had to give up because of a "sticky" bolt that—after being exposed for 10 years to 16 day/night heat cycles per day—didn't want to unscrew. After spending more than 8 h—making it the third longest spacewalk in history—Suni and her Japanese colleague Akihiko Hoshide returned aboard the ISS without completing the task:

> "The lesson we learned is that you can't get married to a plan. something you thought was going to be difficult turns out to be easy, and something you thought was going to be easy turns out to be hard."

A returned spacewalk was required on September 5 to fix the problem. With this EVA, Suni's sixth spacewalk, which lasted for 6 h and 28 min, she broke Peggy Whitson's record for longest total time spent on an EVA by a woman, receiving live congratulations by Peggy, who at that time was her boss, as head of the NASA Astronaut Office. Peggy sent her a message via Mission Control: "It's an honor to hand off the record to someone as talented as you! You go girl."

Ten days after breaking this record, Suni took over as commander of the ISS and, from mid-September to November 2012, was the second female commander in its history after Peggy Whitson. It was a very busy time: within a 17-day period, nine spacecraft docked to the ISS.

Fig. 46.1 Sunita Williams is the only woman in NASA's Commercial Crew Astronaut Corps eligible for the maiden test flights on board the Crew Dragon capsules. Credit: NASA

Suni has so far logged 322 days off the planet. She is the woman astronaut who, after Peggy Whitson, has spent more time in space and holds a number of records (although her record for the longest stay in space was overtaken in June 2015 by Samantha Cristoforetti) and is likely to establish more in the near future. She was the second female astronaut of Indian descent to fly in space after Kalpana Chawla, who was tragically killed in the *Columbia* STS-107 accident in 2003. In July 2015, Sunita Williams was selected with three other experienced astronauts—Bob Behnken, Eric Boe, and Doug Hurley: all of them former military test pilots—to work closely with the Boeing Company and SpaceX to develop their crew transportation systems and provide crew transportation services to and from the ISS. She is now in NASA's Commercial Crew Astronaut Corps, eligible for the maiden test flights on board the Boeing CST-100 and Crew Dragon astronaut capsules, scheduled for launch in 2017. This operation, on the one hand, will relieve the US's reliance on Russia's Soyuz and regain some of the capability lost with the retirement of the Space Shuttle and, on the other hand, as NASA's Administrator Charles Bolden has stated, will put Suni and her three veteran colleagues on "a trail that will one day land them in the history books" (Fig. 46.1).

References

Ahmad, N. "Working Woman: Sunita Williams", *niralimagazine.com* (October 1, 2004).

Gibson, K.B. *Women in Space: 23 Stories of First Flights, Scientific Missions and Gravity-Breaking Adventures*, pp. 151–157. Chicago Review Press, Inc., Chicago (2014).

Gladden, B. "Interview with Astronaut Sunita Williams", *resources.alljobopenings.com*.

Jeavans, C. "Astronaut Suni Williams: You Need to Make Peace before Leaving Earth", *bbc.com* (July 25, 2013).

NASA Press Release. "Astronaut to Run Boston Marathon in Space", *nasa.gov* (March 29, 2007).

Official NASA biography of Sunita Williams, *jsc.nasa.gov* (November 2012).

Pearlman, R. "NASA Assigns 4 Astronauts to Commercial Boeing, SpaceX Test Flights", *collectspace.com* (July 9, 2015).

Petty, J.I. "Preflight Interview: Suni Williams", *nasa.gov* (September 8, 2006).

"Preflight Interview: Suni Williams", *nasa.gov* (June 4, 2012).

Sardesai, R. "Sunita Williams on Why She Loves Samosas!", *ibnlive.in.com* (October 4, 2007).

Valentine E. "Race from Space Coincides with Race on Earth", *nasa.gov* (April 16, 2007).

Williams, S. "Space Is Busy, Active and Unkind", *fragileoasis.org* (September 3, 2012).

Woodmansee, L.S. *Women Astronauts*, pp. 130–131. Apogee Books, Burlington, Ontario, Canada (2002).

47

Joan Higginbotham: An Unplanned Adventure

Credit: NASA

U. Cavallaro, *Women Spacefarers*, Springer Praxis Books, DOI 10.1007/978-3-319-34048-7_47

Mission	*Launch*	*Return*
STS-116	December 9, 2006	December 22, 2006

Joan Higginbotham is the third African American woman to fly into space, after Mae Jemison and Stephanie Wilson. When she is not fulfilling duties for NASA, she enjoys training and competing as a bodybuilder, weightlifting, cycling, motivational speaking, and music. She also speaks Spanish and Russian.

Joan Elizabeth Higginbotham was born in Chicago, Illinois, on August 3, 1964. After graduating from the Whitney M. Young Magnet High School, Chicago, in 1982, she entered the Southern Illinois University at Carbondale, where she received a Bachelor of Science degree in Electrical Engineering in 1987. During college, she had interned for 2 years with IBM. Her original career plan was to continue with this company at the end of her study. She comments:

> "It seemed like a natural fit at the time, because I had interned with them for two years in college; they were a good company; and they thought I was a good employee. However, at the time I was graduating from college, they had a hiring freeze on engineers. So they offered me a position as a sales associate and would move me over to engineering once that hiring freeze came off. In the interim someone from NASA called me and had gotten a hold of my résumé and thought I'd be a real good fit in two positions they had."

When this phone call arrived from NASA, Joan's first reaction wasn't really a positive one. She said:

> "It was a huge step for me because I lived in Chicago where I was born and raised. That's where my friends and family are. This guy on the phone was asking me to move a thousand miles away for some company that had a very bad accident. This was in'87 and the Space Shuttle Challenger explosion of 1986 was still fresh on everyone's mind at the time. So I didn't know if that was the wisest thing to do, and I didn't really know much about NASA to be honest. I was not a space junkie or anything like that. So it took a little convincing."

Joan made up her mind to accept the offer after seeing the launch pad: "It just looked like something out of Star Wars," she says. She flew to the Kennedy Space Center (KSC) in Florida 2 weeks after graduating, and was hired as Payload Electrical Engineer and assigned to the Electrical and Telecommunications Systems Division. Within 6 months, she became the lead for the Orbiter Experiments (OEX) on OV-102, the Space Shuttle *Columbia*. She says:

> "I had the background to do this, but it wasn't necessarily my goal. I think the message to kids is to just prepare yourself. Have goals. Have dreams. But they don't necessarily have to be set in stone. As long as you're prepared … I think you'll have a lot of opportunities open to you."

In her 9 years spent at the KSC, Joan took on increasingly challenging positions and participated in 53 successful Space Shuttle launches and landings. She reconfigured the electrical wiring and performed electrical compatibility tests for payloads and experiments flown aboard all Space Shuttles; she then undertook several special assignments in which she served as the Executive Staff Assistant to the Director of Shuttle Operations and Management, led a team of engineers in performing critical analysis for the Space Shuttle flow in support of a simulation model tool, and worked on an interactive display detailing the Space Shuttle processing procedures at the KSC's Visitors Center. Eventually, she served as backup orbiter project engineer for OV-104, Space Shuttle *Atlantis*, in which she participated in the integration of the orbiter docking station (ODS) into the Space Shuttle used during Shuttle–Mir docking missions. Two years later, she was promoted to lead orbiter project engineer for OV-102, Space Shuttle *Columbia*.

While working full time, 3 years into her NASA career, Joan decided to go back and get an advanced degree and, in 1992, earned a Masters of Management Science degree from Florida Institute of Technology. "I worked essentially night shift so that I could go to the school during the day and get my second degree. But obviously it paid off," she says. At the behest of her then boss, she applied for the Astronaut Corps in 1994 for the'95 class, together with 6000 other candidates. She explains:

> "I was one of the lucky ones that got interviewed (there were only 122 of us), and ultimately I was not one of the 15 selected. After talking to some board members, they suggested I go back and get a more technical advanced."

Joan therefore went back to Florida Tech and got a master's degree in Space Systems. She reapplied for the Astronauts' Corps in 1995 and was selected for the '96 class. With this class—the largest astronaut class ever—44 new Ascans (astronaut candidates) joined the Corps. The class was nicknamed "The Sardines," humorously implying that it was such a large class that their training sessions would be as tightly packed as sardines in a can.

Joan was initially assigned technical duties in the Payloads & Habitability Branch of the Astronaut Office. She then supported the SAIL (Shuttle Avionics & Integration Laboratory) and finally joined the Operations Support Branch at the KSC, where she tested various modules of the International Space Station (ISS) for operability, compatibility, and functionality prior to launch. She was originally assigned to the STS-117 mission that had to be launched in 2003:

> "I do remember the phone call, because it came the day before my birthday in the year 2002, and we were slated to fly in September of 2003. Due to the accident, of course, our mission was pushed back. But I continued to train with that crew for about two years. Then one day, just out of the blue, I got a phone call from the chief of the astronaut office, asking me to report to his office. As I was walking up the stairs, I was trying to figure out why I was being called into the principal's office and what I had done wrong. When I got there, he told me that they had been looking at some of the crews, and they were realigning the crews, and that I had been moved up to the crew of STS-116. It was basically a shock because I had been with my other crewmates for two years and I thought that's pretty much how we're going to launch in that crew configuration."

Joan took part in mission STS-116 with Sunita Williams, who eventually stayed on the space station for her long-duration mission, while Joan returned to Earth after 12 days. She recalls:

"My prime task was to be the Space Station robotic arm operator. There's also a robotic arm on the space shuttle. But I was one of the operators of the robotic arms on the Space Station. The piece that we carried up was called the P5 truss. We used the arm to robotically place P5 next to the rest of the structure. Then two of my crewmates went outside, did a spacewalk, and they physically bolted that piece that we brought up to the Space Station and did some electrical connections."

Joan's own mission job was a feat of daring-do. While on the Shuttle side, operating the Shuttle robotic arm, Nick Patrick grappled P5 and pulled it out of the payload bay. Joan, together with Sunita Williams, grappled the P5 truss with the robotic arm of the space station (the Canadarm2) and put within inches of its new position. So, at one time, there were two arms grappled to the truss. Joan explained:

"The arm operations are really complex. We have very tight tolerances between the arm and different structures. For example, on our mission, as we put the P5 truss into position, we had to come within inches of a box. That's unheard of."

One of the largest tasks of the mission was then to electrically reconfigure the space station, to prepare it to support the addition of European and Japanese laboratory modules by future Shuttle crews. Another goal of the mission was to pick up the European crewmember Thomas Reiter, who had been on the ISS for 6 months. After this mission, Joan was assigned to the STS-126 mission targeted for launch in September 2008. However, in November 2007, she retired from 20-year distinguished career with NASA in order to pursue a career in the private sector.

Joan logged over 308 h in space and over 2000 h piloting the T-38 high-performance jet. She served for 4 years in Marathon Oil Corporation, an oil and natural gas exploration and production international company based in Houston, Texas, where she spent 2 years as Senior Technical Consultant in charge of company strategies and eventually became Corporate Social Responsibility Manager. She managed, among others, a project to successfully reduce the rate of transmission and deaths due to malaria in Equatorial Guinea, Africa.

In October 2011, Joan accepted the position of Director Community Relations at Lowe's Home Improvement, the second-largest chain of retail home improvement and appliance stores, with over 1840 branches in the US, Canada, and Mexico, and with philanthropic corporate contributions totaling more than US$30 million annually. Joan is responsible for conceptualizing and implementing the multicultural marketing strategy to increase affinity for the company's brand, develops and executes ongoing philanthropic strategy, directs the company's natural disaster response program, and manages key cause-related marketing partnerships (Habitat for Humanity, American Red Cross). She says:

"I've been incredibly blessed as an individual, and I had wonderful parents and family and friends who just encouraged me to be the best that I could. I think that's why I am the person I am today and where I am today. I just feel a sense of responsibility to do the same for people who are coming up. I think nowadays there are a lot of children who weren't as blessed as I am. They don't come from homes where families encourage them to do things. I think if I can maybe help them and encourage them to do whatever it is—not necessarily become an astronaut—just encourage them to do their best and expect nothing but the best from themselves."

References

Mansfield, C.L. "Her Time for Discovery", *nasa.gov* (November 15, 2006).
Official NASA biography of Joan Higginbotham, *jsc.nasa.gov* (November 2007).
"Preflight Interview: Joan Higginbotham", *nasa.gov* (November 3, 2006).
Williams, J.A. "Joan Higginbotham: Soaring to New Heights in Space Exploration", *thefreelibrary.com* (July 2008).
Woodmansee, L.S. *Women Astronauts*, pp. 122–123. Apogee Books, Burlington, Ontario, Canada (2002).

48

Tracy Dyson: The Lead Vocalist for the All-Astronaut Band *Max Q*

Credit: NASA

U. Cavallaro, *Women Spacefarers*, Springer Praxis Books, DOI 10.1007/978-3-319-34048-7_48

Mission	*Launch*	*Return*
STS-118	August 8, 2007	August 21, 2007
Soyuz TMA-18	April 2, 2010	September 25, 2010

"I became serious about wanting to be an astronaut when I was a junior in high school. Astronauts were test pilots to me, they were the guys with 'the right stuff,' the Apollo mission astronauts, and I didn't have any interest in being in the military. It was 1986 and there was a lot of talk in newspaper and TV: it was the year that Christa McAuliffe was launching into space, the first teacher to be in space. I thought 'wow, teachers impact my life every day, and not just one teacher but I have six teachers; my coach on the track team was a teacher, my basketball coach is a teacher.' You start to realize that I've got something in common with this person, and if a teacher has 'the right stuff' and a teacher's teaching me all day long (I spend more time with my teachers than I do with my parents) then it became very interesting to me, so I started to look more into what these astronauts were and I found out that they weren't just pilots, they were called mission specialists, and these mission specialists were engineers and they were scientists, and I didn't consider myself much of a scientist at that age, but I was interested in science. I really enjoyed, of all things, chemistry and all the things that it answered like why does water boil, why is the sky blue, what makes a bush a bush, a tree a tree, all sorts of things, and I also learned that at that time it was transitioning from the notion of a Space Station Freedom to an International Space Station, and the fact that astronauts had to keep in shape, and they were going to be building this space station. There were all these aspects of this that fit my life-style. I was about to graduate from high school, I was entering my senior year, and I didn't know that I wanted to do later on in life. My parents had suggested 'why don't you write a list of all the things you like to do?'. So on that list were things like: I want to be athletic, I want to keep working out; I want to do science; I want to work with tools (I was an electrician at a pretty young age working for my father and I loved working with tools), I was learning languages (started Spanish, I knew sign language), I was pretty motivated to learn a different culture, how to communicate. When I looked at that list and then I looked at what astronauts were doing, this seemed like a pretty good match, and that was the diving board that I sprang from into college and graduate school and the path that I took to become an astronaut."

Tracy Caldwell Dyson was born in Arcadia, California, on August 14, 1969, the younger of two girls, and moved in different places in southern California. She was "the first astronaut who was born after Apollo 11." She enjoys hiking, running, weight training, softball, and basketball. She also inherited from her father, who was an electrician, a love for craftsmanship and working with tools. "Christa McAuliffe," she says, "inspired me to search and reach for a goal that I thought was unreachable."

Tracy received a Bachelor of Science degree in Chemistry from California State University, Fullerton (CSUF), in 1993. During her time at the university, she competed as both a sprinter and a long jumper. As an undergraduate researcher at CSUF, she designed and implemented electronics and hardware associated with a laser-ionization, time-of-flight mass spectrometer for studying atmospherically relevant gas-phase chemistry. During that time, she also worked as an electrician/inside wireman for her father's company doing commercial and light industrial construction. At the University of California at Davis (UCD), she taught general chemistry in the laboratory and began her graduate research. Her dissertation work focused on investigating molecular-level surface reactivity and kinetics of metal surfaces using electron spectroscopy, laser desorption, and Fourier transform mass-spectrometry techniques. She also designed and built peripheral components for a variable-temperature, ultra-high-vacuum scanning tunneling microscopy system and developed methods of chemical ionization for spectral interpretation of trace compounds. Her work was published in scientific journals and presented in many papers at technical conferences. She earned her Ph.D. in Chemistry from UCD in 1997. She recalls:

> "Upon the completion of graduate school, I decided it was time to apply. At the same time I submitted my astronaut application, I began a post-doctoral fellowship in chemistry at the University of California at Irvine. I waited about a year for a call back from NASA and the chance to interview. I didn't know what to expect. The only jobs I had had before were for my father, student jobs, and a research assistant job. I was nervous to say the least, but I decided that the best course of action was to just be myself; it was my only chance at capturing my dream. Several months later, while at work in my lab at UC Irvine, I received a phone call. I was completely frozen. The Chief of the Astronaut Office was on the line and asked if I wanted to begin training as an astronaut for NASA. I was so shocked that I hyperventilated and was unable to speak, so he then asked me to consider the position and that he would call back later

> to get my final decision. This gave me plenty of time to come back to reality and run ecstatically around the science building, embracing my friends and sharing the great news with everyone. Later, when he called back, he asked if I would only share the news with my immediate family. Unfortunately, it was a little too late for that!"

Selected for NASA's Group 17 in June 1998, Tracy reported for training in August 1998. She was initially assigned to the SAIL (Shuttle Avionics Integration Laboratory) in charge of checking the Shuttle flight software. The following year, she was assigned to the Astronaut Office ISS Operations Branch as a "Russian Crusader," participating in the testing and integration of Russian hardware and software products developed for the International Space Station (ISS). In 2000, she was assigned prime Crew Support Astronaut for the ISS Expedition-5 crew, serving as their representative on technical and operational issues throughout the training and on-orbit phase of their mission. During ISS Expeditions-4 through -6, she served as an ISS spacecraft communicator (CapCom) in Mission Control, and also worked supporting launch and landing operations at Kennedy Space Center, Florida. She was then the lead CapCom for Expedition-11.

In August 2007, Tracy participated as mission specialist in Shuttle *Endeavour* mission STS-118, which, in its 22nd flight to the space station, delivered 6000 lb (2.7 tons) of cargo, equipment, and scientific experiments contained in the Spacehab Logistics Single Module, built in Italy, that returned to Earth with some 4000 lb of hardware and equipment that was no longer needed. She flew with her colleague Barbara Morgan, who, in 1986, had been the backup for Christa McAuliffe. Christa, who was killed during the *Columbia* accident, had been the "role model" who inspired Tracy and attracted her towards the astronaut job. During the mission, the crew successfully added truss segment S5 and a new gyroscope. As a flight engineer, Tracy assisted in flight deck operations on ascent and also aided in rendezvous/docking operations with the ISS. As mission specialist, she operated *Endeavour*'s robotic arm to maneuver the Orbiter Boom Sensor System (OBSS) and handover the S5 truss segment to the space station. She served as the "internal spacewalk choreographer," directing the four spacewalks performed by three crewmembers.

On April 4, 2010, Tracy joined the Expedition-23 crew aboard the ISS. She lifted off from the Baikonur cosmodrome aboard the Russian Soyuz TMA-18, on which she served as flight engineer. She performed three successful contingency spacewalks to remove and replace the failed pump module on the station, logging 22 h and 49 min of spacewalk time. After 176 days' duty as part of the Expedition-23/24 crew, she returned to Earth with the Soyuz TMA-18 landing unit. Together with Commander Aleksandr Skvortsov and Flight Engineer Mikhail Korniyenko, Tracy landed in Kazakhstan on September 25, 2010:

> "During my time in space, one of my favorite things I was able to do was make a video for the deaf community to tell them about the International Space Station and life onboard. On the space station, the whole crew was ecstatic about the video and fascinated to learn some sign language. Each crewmate had a speaking role in the video. I was thrilled to have the opportunity to invite everyone on board, no matter what language they spoke."

While she was on the station, at the beginning of April 2010, the ISS was visited by Shuttle *Discovery* STS-131. Three of seven astronauts on *Discovery* were women: the two American astronauts Stephanie Wilson and Dorothy Metcalf-Lindenburger and the Japanese Naoko Yamazaki. It was not the first time that three female astronauts were flying at the same time: it had already happened at least three times during the STS-41G mission in 1984, STS-40 in 1991, and STS-96 in 1999. But this time was memorable because, on arriving at the station, they found Tracy, who had arrived at the ISS a few days before. This formed the largest gathering of women in space in history. After spending 176 days in space, Tracy re-entered to Kazakstan on September 25, 2010. She logged over 188 days in orbit altogether (Fig. 48.1).

Tracy is married to naval aviator George Dyson. She has been the lead vocalist for the all-astronaut band *Max Q*.

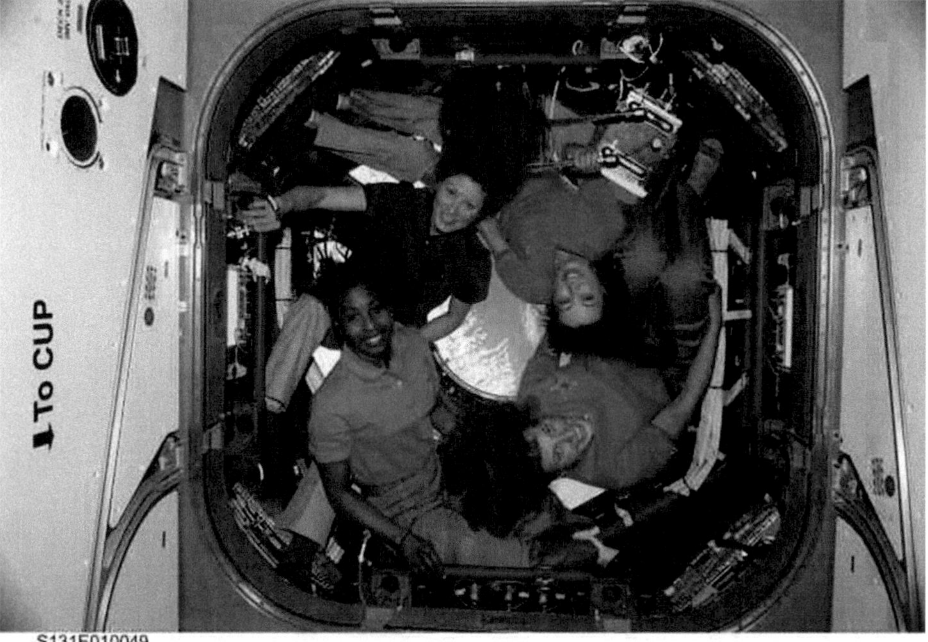

Fig. 48.1 For the first time in history, four women meet in space: Wilson, Dyson, Yamazaki, and Metcalf-Lindenburger in the ISS-Cupola. Credit: NASA

References

Antoun, C.; Antoun, M. "Tracy Caldwell Dyson" interview, *women.nasa.gov* (July 1, 2014).
Bosker, B. "NASA Astronaut Tracy Caldwell Dyson on the Challenges Women Face in Space", *huffingtonpost.com* (July 19, 2011).
Official NASA biography of Tracy Dyson, *jsc.nasa.gov* (February 2011).
"Preflight Interview: Tracy Caldwell Dyson—Expedition 23", *nasa.gov* (March 23, 2010).
Woodmansee, L.S. *Women Astronauts*, pp. 121–122. Apogee Books, Burlington, Ontario, Canada (2002).

49

Barbara Morgan: "I'll Fly with the Eyes, Ears, the Heart and Mind of a Teacher"

Credit: NASA

U. Cavallaro, *Women Spacefarers*, Springer Praxis Books, DOI 10.1007/978-3-319-34048-7_49

Mission	*Launch*	*Return*
STS-118	August 8, 2007	August 21, 2007

None of the astronauts has, perhaps, personally experienced the tragedy of the Challenger STS-51L more than Barbara Morgan did. As the backup of Christa McAuliffe, who became, for the media, the champion of that flight as the "Teacher in Space," Barbara trained intensely alongside her for 6 months. Although she had witnessed two Shuttle accidents, with the deaths of colleagues and friends, she continued to believe with tenacity that, despite the risks involved, space exploration is important (Fig. 49.1):

> "If we put our head in the sand and don't accept any risk at all, we're not going anywhere. We can never predict the future, but we can help shape the future. And if we want that future to be bright and open-ended and be one of lifelong learning, we've got to keep reaching for the stars. There is so much that we don't know, and

Fig. 49.1 Christa McAuliffe and Barbara Morgan. Credit: NASA

space exploration just provides so much. It truly motivates our young people and it's exploration, it's discovery, it's experimentation—it's all those things that make humans, humans. And it's all the things that help us gain more knowledge and help make the world a better place. And it's also important that we show our young people that there are risks that should be taken."

Barbara Radding Morgan was born in Fresno, California, on November 28, 1951. She is a classical flutist and enjoys jazz, reading, hiking, swimming, cross-country skiing, and spending time with her family. She graduated from Hoover High School, Fresno, in 1969, the year of the first Moon landing. Following graduation, she was accepted at Stanford University in Palo Alto, California, where in 1973 she earned a Bachelor of Arts degree with distinction in Human Biology. She recalls:

"At the age I was through high school and college years, basically the only thing that seemed like girls did was either became teachers or nurses. I really didn't like those limitations. But in my studies in college in human biology, one of my classes that really fascinated me a lot was on the brain. At the same time I was also taking a psychology class on learning theories and memory. At some point, I was walking around a bookstore and, I don't know why, I just got drawn to the education section and happened to pick up a book about somebody that I knew nothing about. It was Maria Montessori, who it turned out was a very famous and influential educator. Putting all those things together, I thought, 'If these are the things I'm interested in ...,' and it reawakened that desire when I was a little kid of wanting to be a teacher."

Barbara obtained her teaching credential from Notre Dame de Namur University in nearby Belmont in 1974 and began her teaching career on the Flathead Indian Reservation at Arlee Elementary School in Arlee, Montana, where she taught remedial reading and math. After marrying, from 1975 to 1978, she taught remedial reading/math and second grade at McCall-Donnelly Elementary School in McCall, Idaho, which was the home town of her husband, the writer Clay Morgan. From 1978, Barbara taught English and science to third-graders at Colegio Americano de Quito in Ecuador. The following year, she returned to McCall-Donnelly Elementary School, where she started to teach second, third, and fourth grades. She says:

"When the Teacher in Space program was started, I was sitting at home. It was after school, it was the five o'clock news, and the President came on the news and announced that they were going to send a teacher in space. I shot straight up and said, 'Wow!' What a great opportunity! Because as teachers, we're always looking for opportunities to bring the world to our classroom, to gain more experiences, gain more knowledge about our world so that we can make our classroom a better place for our kids. And, it was a tremendous opportunity. And, as all teachers, we don't pass up those opportunities."

Barbara applied together with some other 11,000 American teachers. She says:

"One of the best parts of that program was being able to meet teachers from all over the country who were doing a dynamite job. Every single one of us knew we were just representing hundreds and thousands of great teachers across the country."

Barbara was selected as the backup candidate for NASA's "Teacher in Space" project on July 19, 1985. For almost 6 months, she trained with Christa McAuliffe and the Space Shuttle *Challenger* crew at NASA's Johnson Space Center (JSC), Houston, Texas. Following McAuliffe's death in the *Challenger* disaster, NASA asked Barbara to stay on as the "Teacher in Space Designee," which included public speaking, educational consulting, and curriculum design: "One of my favorite things that I got to do during that period was serve on the National Science Foundation's Federal Task Force for Women and Minorities in Science and Engineering."

In the fall of 1986, Barbara returned to McCall-Donnelly Elementary School to resume her teaching career and resumed teaching second, third, and fourth grades—which she did for almost 20 years—while continuing to work with NASA's Education Division of the Office of Human Resources and Education. The "Teacher in Space" program was halted by NASA in 1990 amid concerns surrounding the risk of sending civilians to space. The "Educator Astronaut" project was created instead. The program carries on the objectives of the "Teacher in Space" program, seeking to elevate teaching as a profession and inspire students. Unlike in the previous program, however, educator astronauts leave their job and enter NASA Astronaut Corps to become full-time astronauts: fully trained to do the same jobs and duties and to fly as crewmembers with critical mission responsibilities, as well as education-related goals; they are "fully fledged" astronauts as opposed to "spaceflight participants."

Barbara was selected as mission specialist of NASA Astronaut Group 17 in January 1998, 12 years after McAuliffe's death. In August that year, she reported to the JSC to begin training and become a full-time astronaut. Following the completion of 2 years of training and evaluation, she was assigned technical duties in the Astronaut Office Space Station Operations Branch. She served in the Astronaut Office CAPCOM Branch, working in Mission Control as the prime communicator with on-orbit crews—she was CapCom also during the *Columbia* tragedy—and took this experience very seriously: "I can't imagine flying in space without having that job, where we really learn who all the folks are on the ground and how that all works on the ground, and the best ways to communicate back and forth."

After waiting to fly in space for half of her life with perseverance and patience, in December 2002, 1 month before the *Columbia* STS-107 accident, Barbara was assigned to mission STS-118, originally scheduled to be flown in November 2003 by Columbia Orbiter. However, the disaster altered the planned flight schedules and the mission had to be rescheduled for *Endeavour* almost 4 years later:

> "I'd like to say you're never waiting for a flight assignment. You're working and working very hard. That's really what you do as an astronaut, the spaceflight part is the ultimate, but it's such a small part time wise and really intensity wise of everything that you do. And, the other jobs that you do are fascinating, and they're really important jobs. They help support everybody else flying. They help support the entire program. So we have folks whose technical jobs is helping to work with the design of the next exploration vehicle, the CEV that we'll be using for station, for moon and Mars, etc. My particular assignments while I was quote/unquote 'waiting,' which wasn't waiting, it was working—every single one of those assignments I really loved. I felt like I was contributing, which I was; and I was also learning. It's the best kind of on-the-job training that you can get."

Barbara happened to fly on the Shuttle *Endeavour*. She feels very close to this "Orbiter." "Endeavour," she explains, "has special meaning to me as a schoolteacher, because Endeavour was our replacement orbiter for Challenger, and it was named by schoolchildren all over this country." In fact, *Endeavour* was built to replace *Challenger*, which broke apart on January 28, 1986, 73 s into its flight, leading to the deaths of its seven crewmembers, including Barbara's colleague and close friend, the "Teacher in Space" Christa McAuliffe. But, on top of this, the name of the new orbiter was chosen through a competition that was participated in by more than 6000 US schools. Students at McCall-Donnelly Elementary School, where Barbara taught for a total of 22 years before being selected as an astronaut in 1998, submitted the name "Endeavour" (Students could not suggest just any name: it had to be that of an exploratory or research sea vessel, be appropriate for a spacecraft, capture the spirit of America's mission in space, and be easy to pronounce; and "Endeavour" was the most popular entry, accounting for almost one-third of the state-level winners). When students at McCall-Donnelly Elementary School suggested a name for NASA's newest orbiter, little did they know they were helping to name the Shuttle that one day would carry into space one of their teachers. The mission emblem of her mission STS-118 features the "torch of knowledge" that represents the importance of education. She explains (Fig. 49.2):

"If you look closely at the patch, you'll see the trajectory or the orbit that the shuttle is taking as it circles around the astronaut symbol that's going up to the International Space Station and beyond. That orbit emanates from that flame of knowledge.

Fig. 49.2 The "torch of knowledge" in the mission emblem represents the importance of education. Credit: NASA

It's education, it's great education, that propels all of what we're doing in space exploration and as we learn more about our universe. That's so important because to us one of the primary purposes of all these missions is gaining knowledge, and it's gaining knowledge through exploration. The other thing that's near and dear to our hearts is that flame of knowledge really is there to honor teachers and students everywhere."

Due to the presence of Barbara and due to the fact that the official STS-118 mission patch included the torch of knowledge, NASA press releases and media briefing documents improperly stated that STS-118 was the first flight of a mission specialist educator. However, while it is true that tip of the flames touched Morgan's name on the patch, Barbara did not train in the Educator Astronaut Project and she flew with technical responsibilities like every astronaut of the team. As NASA Administrator Michael D. Griffin clarified in a post-mission press conference, Barbara was not considered a mission specialist educator, but rather a standard mission specialist who had once been a teacher, even if Barbara had made her point that she would "share that experience through a teacher's perspective and through the eyes, ears, the heart and mind of a teacher" (Fig. 49.3).

As mission specialist, Barbara was very busy. In charge of the Shuttle's robotic arm, she helped in coordinating the unloading of 5000 lb (over 2250 kg) of cargo contained in the Spacehab, and moving supplies and equipment from the Shuttle to the space station; she also served as *Endeavour*'s prime Shuttle robotic arm operator during the flight's three spacewalks that were planned to attach a new truss segment, relocate a stowage platform,

Fig. 49.3 Commemorative cover of launch of mission STS-118, signed by Barbara Morgan (from the collection of Umberto Cavallaro). The STS-118 mission patch features the trajectory or the orbit that the Shuttle is taking as it circles around the astronaut symbol that is going up to the International Space Station (ISS). "That orbit emanates from that flame of knowledge," Barbara says

and replace a gyroscope that helps to control the station's orientation. She faced it with a special perspective, however: "For me, especially from a schoolteacher's point of view, it is applied mathematics. If you love geometry, it's geometry."

Barbara brought into space 10 million basil seeds that were placed outside the station to test how they withstand the harsh environment of space. Some of them remained on the station to be grown in zero gravity, while the rest were returned to Earth, sorted, and placed into small packets, each containing approximately 50 space-flown seeds. Together with control packets of seeds that have not flown, they were eventually distributed to students to grow in small greenhouses that they had designed. More than 40,000 classrooms in all 50 states and 30 foreign countries participated in the program. Tracy Caldwell, who had been inspired to become an astronaut following in the footsteps of Christa McAuliffe, also flew in that same mission.

Even in this mission, a piece of insulation foam came off the external tank during lift-off, though the impact caused little damage and was not in a critical area. The mission landed successfully at Kennedy Space Center on August 21, a day ahead of schedule due to concerns about Hurricane Dean. Finally, Barbara had achieved her dream at age 55. She logged over 209 h (more than 8 days) in space.

Being an astronaut was always considered by Barbara as "a long while lateral move" and her plan has always been "going back to teaching when all is said and done here," as she said in an interview while at Houston. After this mission, she left the Astronaut Corps in 2008 to become a Distinguished Educator in Residence at Idaho's Boise State University, where she taught in the STEM (science, technology, engineering, mathematics) area. "Education," Barbara says, "is important, just like exploration. They are very much the same. It's about learning, it's about exploring, it's about discovering, it's about sharing, and building a future." In August 2008, the "Barbara R. Morgan Elementary School" opened in McCall, the small town where she taught for 22 years.

References

"Educator Astronaut Project", *en.wikipedia.org*.

Gibson, K.B. *Women in Space: 23 Stories of First Flights, Scientific Missions and Gravity-Breaking Adventures*, pp. 158–164. Chicago Review Press, Inc., Chicago (2014).

"In Their Own Words: Barbara Morgan", *nasa.gov* (December 14, 2011).

Kevles, T.H. *Almost Heaven: The Story of Women in Space*, pp. 105, 118, 133. The MIT Press, Cambridge, MA, and London, UK (2006).

Moskowitz, C. "Challenger Remembered: Q&A with Teacher Astronaut Barbara Morgan", *space.com* (January 25 2011).

Nevills, A. "Preflight Interview: Barbara Morgan", *nasa.gov* (November 23, 2007).

Official biography of Barbara Morgan, *nasa.gov* (July 2010).

Smith, H.R. "The Naming of Space Shuttle Endeavour," *nasa.gov* (February 21, 2008).

Woodmansee, L.S. *Women Astronauts*, pp. 125–126. Apogee Books, Burlington, Ontario, Canada (2002).

50

Yi So-Yeon: Korean "Spaceflight Participant"

Credit: KARI (Korean Aerospace Research Institute)

U. Cavallaro, *Women Spacefarers*, Springer Praxis Books, DOI 10.1007/978-3-319-34048-7_50

Launch		*Return*	
Soyuz TMA-12	April 8, 2008	Soyuz TMA-11	April 19, 2008

The flight of Dr. Yi So-yeon, the first South Korean astronaut to fly in space, was not only a personal accomplishment, but a national one, as this spaceflight in 2008 made South Korea the 35th country to reach space and one of only three nations to date to have a woman as its first astronaut. As had happened to the Briton Helen Sharman and the Iranian Anousheh Ansari, she was in fact not only the first woman, but also the first citizen from her country to achieve spaceflight, and the second Asian female astronaut, after the Japanese Chiaki Mukai. And, to date, she still is the only South Korean astronaut in a country that has a hard time allowing women to fit in and climb its social, political, and business ranks. Moreover, at age 29, she also beat Sally Ride, who, until then, had held the record for the youngest astronaut.

So-yeon "Lee" Yi was born in Gwangj, South Korea, on June 2, 1978. Since she was a child, she dreamed about being an astronaut or flying in the universe while watching science-fiction movies and animations. She says:

> "When I watched sci-fi movies as a kid, with the cool astronauts flying the spaceship there was always one female scientist who was always smart, thin, pretty, and always blonde; and if something happened, she always explained everything well, and then, when the men started fighting, she calmed them all down and solved things. So watching these cool women, I thought, 'Ahh! I want to be a cool scientist like that!' It was totally the same with other ordinary kids pretending to fly into space. No more and no less. As I grew up, I realized that it was impossible to have those kinds of jobs in Korea. Maybe only Russians or Americans."

But Yi continued to like technology and she recalls that, from primary-school age, she was attracted to mechanics, supported by her parents. She says:

> "Although I grew up in Korea, my parents are not typical 'conservative' Korean parents. For my generation, parents were more likely to encourage sons to go into the hard sciences and engineering. I never felt that engineering was a 'guys' thing … my parents raised me to believe that I could choose any field. Especially my daddy, who always made me his 'assistant' on his fix-up projects … working on the cars, boilers, and other things. I was his No. 1 assistant whenever he did technical things!"

After graduating from Gwangju Science High School, Yi chose a scientific career and earned a bachelor's degree in Mechanical Engineering at the Korea Advanced Institute of Science and Technology (KAIST) in Daejeon, was the first research-oriented science and engineering institution in South Korea, and then a master's degree. She recalls:

> "For my Ph.D., I transferred to the Bio Systems department. Even though I transferred departments, I did almost the same research, since I was working with BioMEMS (micro-electro mechanical systems) for my Masters in Engineering. My Ph.D. thesis was about how to develop micro-machines to separate DNA molecules by their sizes."

While she was working at her dissertation at the university, Yi occasionally saw the advertisement by KARI (Korean Aerospace Research Institute) and knew about the space program. She says:

> "When I read an article about astronaut program, I thought it would be cool, but I couldn't even imagine that I would make it. However, I thought, at least I want to try. Of course, I was sure that I would be eliminated after few rounds."

On December 25, 2006, Yi was instead selected as one of the two Korean astronaut candidates, beating more than 36,000 other applicants, and was sent to the Gagarin Cosmonaut Training Center at Star City, near Moscow, Russia, where she trained from March 2007 through April 2008 as backup for the "prime" astronaut Ko San, an artificial-intelligence expert and roboticist. Yi says:

> "I had never been in Russia at all, and it was not a familiar country to me. I didn't even know the Russian alphabet when I left, so it was a totally new world for me, even besides the training. One of the toughest things was to learn Russian. At first, the culture and the reactions of people were unusual to me. Most of all, the military base was a totally strange place: as a woman, I'd never been on a military base before."

Due to her training commitments in Russia, Yi was unable to be present at the ceremony at KAIST when, on 29 February, 2008, her doctorate in Biotech Systems was conferred and she had to delegate her mother to attend her doctoral graduation ceremony in her place.

Just weeks before the scheduled launch, the Russian Federal Space Agency asked for a replacement of Ko San who, reportedly, on several occasions had violated regulations of the training center and security protocols by removing without permission sensitive training materials from the Russian facility and mailing one classified document back to Korea. Even though he later denied and explained that he was trying to understand how Soyuz's systems worked so that he could participate in the mission safely, Ko was sent back to his country and the glory of becoming South Korea's first astronaut, in a reversal of roles, went to Yi, who succeeded him in the prime crew.

Yi lifted off from Baikonur on board Soyuz TMA-12 on April 8, 2008, thus becoming the first Korean and the 49th woman to visit space. She flew as a spaceflight participant together with the cosmonauts Sergei Volkov and Oleg Kononenko, who, respectively, would replace Commander Peggy Whitson and the flight engineer Jury Malenchenko, who had finished their 6-month rotations on the International Space Station (ISS). Peggy had just completed her tour as first commander of the ISS. Reportedly, for the "commercial agreement" and for Yi's ticket to space, the Korean government had paid to Russia US$20 million (25 million according to some sources)—something that has been called a "matter of national pride." The mission was part of an ambitious South Korean initiative into space exploration, including the planned completion of a US$265 million space research center.

During her 11 days on the ISS, Yi conducted on behalf of KARI 18 scientific experiments, including one that monitored the effects of microgravity on 1000 fruit flies that she transported into space. Other experiments involved the growth of plants in space, the study of the behavior of her heart, and the effects of gravity change on the pressure in her eye and shape of her face. With a specially designed three-dimensional Samsung camera, Yi took six shots of her face every day to see how it swelled in the different gravity. She also observed Earth, in particular the movement of dust storms from China to Korea, and also measured the noise levels on board the ISS. She says:

> "That was a lot. Everyone told me I didn't have to complete all of them, that it wasn't expected of me. But I knew everyone was watching me, so I gave up meals and sleep and completed all 18 experiments. It's a very Korean thing to do."

Yi felt that it was important that she should do a good job in representing South Korea and worked very hard: "The most fantastic thing is that I cannot feel my weight, and I can fly around like Peter Pan." In honor of the first Russian in space, Yuri Gagarin, on April 12, which Russia celebrates as Cosmonautics Day, Yi shared with the crew—with some hearty laughs—a traditional kimchi dinner with instant noodles, chili paste, and other traditional Korean recipes. "It was very delicious," diplomatically commented the veteran Sergej Volkov, Commander of the Space Station, during a live in-flight call, "despite the fact that you want things spicy when you get up here, it was a bit more spicy than we expected or required." Three top government research institutes had worked to create "space kimchi" because "If a Korean goes to space, kimchi must go there, too." When the Korean government finally decided to finance the space trip, they wanted the Korean spacefarer to be well prepared for this momentous journey, which meant that she had to take kimchi with her. After millions of dollars and years of research, South Korean scientists successfully engineered Korean space recipes. When the Russian space authorities approved the plan, the South Korean food companies that participated in the research took out full-page newspaper ads. Not only did Yi share the traditional Korean food with the crew, but she also left some for the next expeditions: "I am not certain if they will become common fare in space," she said, "but others enjoyed the food I brought, and if there are any left I plan to leave some behind for others to enjoy."

The return of Soyuz TMA-11 to Earth on April 19 is reported as the first spacecraft ever on which there were more females than males on board: Yi So-yeon, Peggy Whitson, and Jury Malenchenko. It wasn't a smooth return. The space capsule missed its mark and the re-entry started later than expected, as the Soyuz descent module did not separate properly from the rest of the spacecraft. Soyuz then entered a "ballistic trajectory"—a pre-programmed maneuver when the re-entry module does not separate as planned. It landed 260 miles (more than 400 km) away from its planned touchdown point—a highly unusual distance given how precisely engineers plan for such landings—near some shepherds in northern Kazakhstan. "They thought at first we were aliens," Yi said in an interview.

During this rough re-entry and landing, the crew was subjected to severe G-forces (more than 8 Gs, i.e. eight times the force of gravity) and arrived around 20 min later than scheduled. The capsule hit the ground so hard that it bounced. Despite the "hard landing," the entire crew ended up largely uninjured. Some controversy was sparked when the Roscosmos Chief Anatoli Preminov during the post-flight press conference referred to a naval superstition that having women aboard a ship was bad luck: "In Russia, we have a sort of omen regarding such occasions but thank God, everything ended well. Certainly we will try to somehow avoid a prevalence of females on a crew, though I don't think it will be mandatory." It is reported that, after her return to Korea, Yi was hospitalized at an Air Force facility in Cheongju, some 130 km south-east of Seoul, where she was treated for mild dislocation and bruising of some vertebrae.

To fulfill her dream "in a still male-dominated world," Yi had to overcome many obstacles. "Men and women are different," she comments, "and I believe that it used to be that people didn't distinguish the 'differences' from 'abilities'. Being different is not the same as being less capable." There are big obstacles in her country: "Confucian culture," she says, "is very chauvinist, but I think Korean society is changing." But she adds: "What I've seen in Russia is that the women cosmonauts are not treated like the other 'guys' in the program either."

Requested to comment on the mission of Liu Yang, she said:

> "I have not met Liu Yang yet, and I am not sure yet whether she is in a 'show window' for the international awareness of China's progress in space and science. I've read that one of the prerequisites in China was that they be married! In my view, married or single cannot be the criteria to evaluate whether a woman is 'mentally stable' or capable of managing the requirements of the job. As I understand it, this requirement was discredited as a misunderstanding rather than discrimination."

After her mission, Yi became officially a senior researcher in KARI and traveled a lot as Korea's space ambassador, not only in her country, but also abroad to give public lectures, and she participated in educational programs to inspire students. Right after the flight, she had a public lecture in schools almost every day, then her speaking engagements decreased to approximately one per week. She said:

> "I have a dream that one day, an engineer or scientist or winner of a Nobel Prize would get asked 'What makes you to do this?' And he or she might answer, 'When I was a kid, I watched TV and the first Korean astronaut was flying on the ISS and doing experiments there; it was really exciting! I thought only Russians or Americans

could fly to space, but she changed my mind. So I made up my mind I should be a person who does those things … and finally I made it'."

Yi left the state-run research space center in August 2014 by sending a letter of resignation by mail "for personal reasons," so bringing to an end the country's first astronaut program and sparking criticism of government waste. In an interview, she explained later that her retirement was because she was about to marry an American man and receive American citizenship; moreover, she had decided to add an MBA to her Ph.D. in Biotech Systems. She chose the University of California–Berkley's Haas School of Business because of its proximity to Silicon Valley and its distance from Korea, where she's a celebrity with little privacy. "California is an easy place to be an Asian woman engineer," she says. "It feels like home."

References

Bonilla, E.L. "International Brief: Dr. Soyeon Yi", *appel.nasa.gov* (May 24, 2012).
"First Korean Astronaut Yi So-yeon", *blog.sciencewomen.com* (March 11, 2008).
Gibson, K.B. *Women in Space: 23 Stories of First Flights, Scientific Missions and Gravity-Breaking Adventures*, pp. 187–190. Chicago Review Press, Inc., Chicago (2014).
Lopez-Alegria, D.; Yi, S. "Space Ambassador, Science Advocate", *spacebridges.com* (June 25, 2012).
Malik, T. "South Korean Astronaut Shares Laughs, Space Food with Station Crew", *space.com* (April 15, 2008).
Newcomb, A. "Why South Korea's Only Astronaut Quit", *abcnews.go.com* (August 13, 2014).
O'Neill, I. "Soyuz Capsule Hatch Nearly Burned Up and Crew's Lives Were on a 'Razor's Edge'", *universetoday.com* (April 22, 2008).
"Russian Spacecraft Returns Off-Course", *cbsnews.com* (April 19, 2008).
Sang-Hun, C. "Kimchi Goes to Space, along with First Korean Astronaut", *nytimes.com* (February 22, 2008).

51

Karen Nyberg: The Marathoner Engineer Who Competes with the Space Station

Credit: NASA

U. Cavallaro, *Women Spacefarers*, Springer Praxis Books, DOI 10.1007/978-3-319-34048-7_51

Mission	*Launch*	*Return*
STS-124	May 31, 2008	June 14, 2008
Soyuz TMA-19	April 2, 2013	November 10, 2013

Karen LuJean Nyberg was born, the fifth of six children, on October 7, 1969, in Vining, a very small town (fewer than 100 inhabitants) in central Minnesota, where she grew up in a house outside of town on a lake in the country. She says:

> "The town that I went to school in is a neighboring town, Henning, and its one school was kindergarten through twelfth grade. So, I went to school with pretty much the same people my entire life. There are a couple drawbacks. Education-wise, it was a good education and it all ended up okay. The good side about being in a town like that, in a school that size, is I was able to participate in everything that I wanted to. I was playing sports. I was in the band, in the choir, student council and being a part of a team in that way. It made me a more well-rounded person. I think if I had gone to a bigger school, like my sports' abilities, I was OK but I never would have been able to play in a big school."

Although she chose a career dedicated to science and technology, Karen still has varied interests. Her recreational interests include piano, running, drawing and painting, and sewing. She says:

> "My mom and dad are both very creative people and made a lot for all of us kids: everything from snowmobile suits to prom dresses. My mom taught me to sew when I was about five or six years old. I would sew all day every day if I could, I love it that much."

Karen especially enjoys quilting and appliqué work and used those skills to create a lot of the décor for her son's nursery before he was born. Drawing and painting is another of her hobbies that she inherited from her father, Ken Nyberg—of Norwegian descent—who creates painted steel sculptures made from scrap metal. She also packed a sketch book and pencils when preparing for her long-duration mission in orbit. Before leaving, she said:

> "I'm really hoping to spend some of my free time drawing, I used to mostly draw portraits, and gave them to friends, but I haven't done it in a long time. I am hoping I can get back to some of that while I am in space."

Karen started running as a graduate student while at the University of Texas and developed a love for long-distance running. She participated in nine marathons and made headlines when, in 2007, she completed the Boston Marathon in 3 h, 32 min, and 9 s, in tandem with fellow astronaut Sunita Williams, who ran the marathon while in orbit on the International Space Station (ISS) (see page 295). She graduated from Henning Public High School, Henning, Minnesota, in 1988. She wanted to be an astronaut since she was a little kid:

> "I can't pinpoint an event or a person or anything that made me decide that. I just decided that that's what I wanted to do and I kept that with me and most of my friends in high school knew that's what I wanted to do. They just called me 'the rocket scientist'."

Karen graduated summa cum laude with a degree in Mechanical Engineering from the University of North Dakota in 1994. She explains:

> "It's kind of funny how I chose mechanicals because this was before the time of computer-aided drawing and computer-aided modeling, and I like to draw so it was like drafting sounded fun. I do not think I knew enough about engineering to know that that's what I wanted: at this time I knew that 'astronaut' was my ultimate goal."

Meanwhile, in 1991, Karen got into the Cooperative Education Program at the NASA Johnson Space Center (JSC) and worked as a co-op student in a variety of areas: in the robotics on the first semester, in the MOD (Mission Operations Directorate) for another semester, and then with Crew and Thermal Systems Division. In 1994, she received a patent for her work on robot-friendly probe and socket assembly. "It extended my graduation by a while," she says, "but I think it was very valuable. I learned a lot about engineering and I got my foot in the door at JSC." She decided to continue her studies at the University of Texas at Austin, where, at the Austin BioHeat Transfer Laboratory, she investigated human thermoregulation and experimental metabolic testing and control, specifically related to the control of thermal neutrality in spacesuits, which was NASA research. This work led to her doctorate in 1998 and she was hired at JSC with the Crew and Thermal Systems Division, working as an environmental control systems engineer in charge of improving the thermal control systems of the spacesuits and designing thermal systems for the future Lunar and Martian missions. That summer, she applied for the astronaut program. "I thought it is time," she says. "I have my Ph.D., now I will put in my application. I never thought that I would be selected. Luckily, for some reason, I was selected." She joined the 18th Group of NASA Astronauts in July 2000 and, after 2 years of basic training and evaluation, she qualified as mission specialist and was assigned for technical duties in the Astronaut Office Station Operations Branch. She was crew support astronaut for the Expedition-6 crew during their 6-month mission on the ISS and then was chief of the Robotics Branch.

From July 22 to 28, 2006, Karen lived and worked underwater for 7 days in NEEMO-10 (NASA Extreme Environment Mission Operations), a deep-sea training and simulation expedition at the Aquarius underwater laboratory, to help prepare for the return of astronauts to the Moon and manned missions to Mars.

Karen flew for the first time into space in 2008 with Shuttle mission STS-124, the second of three missions that contributed to assembling and installing the Japanese Laboratory Kibō ("Hope", in Japanese) and the Japanese robotic arm. In a pre-flight interview, she said:

> "That's one of the biggest payloads the space shuttle has ever flown, so there's not a lot else that we're taking up, other than that module. So that's the main goal, to deliver that to the space station, install it, activate it, get it up and running."

Karen, together with the Japanese astronaut Akihiko Hoshide, was responsible for the installation of Kibō and worked heavily on robotics, being the lead for all the robotic-arm operations: she was the first astronaut ever to operate all three robotic arms that were on station at the time: the Shuttle Remote Manipulator System (SRMS), also known as Canadarm-1; the Space Station Remote Manipulator System (SSRMS), known as Canadarm2; and the Japanese Experiment Module Remote Manipulator System (JEMRMS), used to transfer the experiments from an airlock that the Japanese Kibō module has to the "Terrace" located outside, where they are exposed to the vacuum of space. One remarkable aspect of this mission was that it helped to highlight the international nature of the ISS program:

> "We flew in an American space shuttle, carried a Japanese laboratory, used a Canadian robotic arm to install it to a module that was built in Italy, and we did all this traveling 17,000 miles an hour. These modules never having been mated on Earth to see if they fit, and it's that international aspect of it that I think we have learned a lot."

Commemorative cover of Expedition-36/37 signed onboard the ISS by the Crew, including Karen Nyberg, on the day of the 50th anniversary of the flight of Vaentina Tereshkova (from the collection of Umberto Cavallaro)

This mission started the final assembly phase of the ISS. In a pre-flight interview, Karen highlighted:

> "After this mission the space station is going to be very close to its final configuration, so that it can be used to do the things that it was originally meant to be for, all the science that originally it was intended for. And I hope that it can be used to look at the effects of microgravity on humans which is very important for the future of what we plan on doing in sending humans back to the moon and even on farther where we're really going to be in microgravity for very long periods of time."

In May 2009, Karen was assigned to mission STS-132, scheduled for launch in 2010, but had to be replaced a few months later due to a temporary medical condition. She flew to the ISS in May 2013 as flight engineer of the Soyuz TMA-09M mission, the second-ever express trip to the ISS, and for 5 months she was part of Expedition-36/37, along with the Russian cosmonaut Commander Fyodor Yurchikhin, who carried into space the Olympic torch for the 2014 Winter Games in Sochi, and the Italian astronaut Luca Parmitano. She was one of only two women in space on June 16, 2013, on the 50th anniversary of the flight of Valentina Tereshkova, the other being Wang Yaping aboard the Tiangong-1 on the Shenzhou-10 mission. Karen was involved in a number of experiments on ocular health, osteoporosis, and cardiovascular health:

> "We've discovered that there are quite a few astronauts coming home with decay in their eyesight. Luca Parmitano and I have been involved in numerous tests. We're doing tonometry—we are looking at the pressure of the eye. We are doing ultrasounds to look at the morphology of the eye, we are doing fundoscopy to take images of the retina, vision tests. We are hoping that we can determine exactly what is causing this and hopefully mitigate the problem, especially if we start longer duration missions going to Mars. And I'm confident there will be some type of an Earth application that will come from this: that we could contribute to the success of solving some earthbound eye diseases. Another great one to talk about is bone density: there's great potential for using that and applying it to earthbound osteoporosis."

One of the fun pictures she sent during her mission was the "Made in space!" dinosaur, as she wrote on the caption, which she created for her 3-year-old son, Jack. "I made this dinosaur for my son last Sunday, September 22," she says. "It is made out of velcro-like fabric that lines the Russian food containers that are found here on the International Space Station (Fig. 51.1)."

During Karen's mission, as the Shuttles weren't coming to the station anymore after their retirement in 2011, the ISS was visited by four different transfer vehicles, prepared by ISS partners to deliver supplies and equipment: the Russian PROGRESS, the European ATV (built in Italy), the Japanese HTV, and the first demonstration vehicle of the Orbital Science Corporation, CYGNUS, whose pressurized module was built in Italy by Thales Alenia Space. She played a major part in grabbing most of them with the robotic arm.

She is married with astronaut Douglas Hurley in 2009 and they have one son.

Fig. 51.1 The "Made in space!" dinosaur that Karen made for her son: it is made out of Velcro-like fabric that lines the Russian food containers that she found on the International Space Station (ISS). Credit: NASA (CollectSpace)

References

Anderson, B. "Space Chat with Astronaut Karen Nyberg", *cnn.com* (February 20, 2014).

Binkley, M. "Finding Minnesota: Vining Sculptures in Otter Tail County", *minnesota. cbslocal.com* (March 27, 2011).

Official NASA biography of Karen Nyberg, *nasa.gov* (November 2013).

Pearlman, R. "'Made in Space!': Astronaut Sews Dinosaur Toy from Space Station Scraps", *collectspace.com* (September 27, 2013).

"Preflight Interview: Karen Nyberg", *nasa.gov* (April 4, 2013).

"Preflight Interview: Karen Nyberg, Mission Specialist—(STS-124)", *nasa.gov* (April 29, 2008).

"The Softer Side of Space: A Profile of Astronaut Karen Nyberg", *nasa.gov* (October 30, 2013).

Valentine, E. "Race from Space Coincides with Race on Earth", *nasa.gov* (April 16, 2007).

Woodmansee, L.S. *Women Astronauts*, pp. 127–128. Apogee Books, Burlington, Ontario, Canada (2002).

52

Megan McArthur: An Aerospace Engineer Riding a Pedal-Powered Submarine

Credit: NASA

U. Cavallaro, *Women Spacefarers*, Springer Praxis Books, DOI 10.1007/978-3-319-34048-7_52

Mission	*Launch*	*Return*
STS-125	May 11, 2009	May 24, 2009

Commemorative cover of mission STS-125, signed by Megan McArthur (from the collection of Umberto Cavallaro)

Katherine Megan McArthur was born in Honolulu, Hawaii, on August 30, 1971, but she feels that she doesn't have a hometown, as she grew up around airplanes and airbases. Following her father, who, as a career naval officer based on the Moffet Field Naval Air Station, had temporary duty assignment in different countries, her family moved through California, which she considers her home state, but also to Japan, to England, and to Rhode Island. She graduated in 1989 from the Francis High School in Mountain View near Moffet Field, California. She says:

> "For a number of years we lived at Moffett Field Naval Air Station, which is also where the Ames Research Center is, when I was in high school, and we used to see astronauts come out there to do training in one of the simulators there and they'd park their T-38s out there on the ramp, and I thought that looks like a pretty neat job. But mostly it made me interested in the space program in general, because that seemed like kind of a long shot to ever get selected to be an astronaut. I studied aerospace engineering in college and got interested in some other things along the way, but always came back to the idea of being an astronaut. What appeals to me about it, I think, it's a challenging job. You are having to know lots of different things. You have to be a generalist as well as a specialist in some areas, and, of course, it's a lot of fun."

Megan went to the University of California in Los Angeles, where she earned a Bachelor of Science degree in Aerospace Engineering in 1993, and there she fell in love with the ocean after she was required to become scuba-certified in order to participate in human-powered submarine races with other engineering students who had built a small two-person pedal-powered submarine. She recalls:

> "Towards the end of my studies at UCLA I got interested in a project with some other aerospace engineering students that was called 'Human Powered Submarine Project' and basically we built a small two-person submarine, and raced it against some other colleges and, it's kind of funny that a bunch of aerospace engineering students got interested in this but a friend of mine Derek, who was going into the Navy as a submarine officer, had read about this project and decided he wanted to participate. And, as part of that project I ended up being the pilot because, as the only girl, I was the smallest and the only one that would fit in the spot that we had designated for the pilot. I had to get scuba certified in order to do this project. And so I got interested in ocean engineering at that time and ended up working for a few months, at an ocean engineering company that designs underwater robots and some manned submersibles as well. Then I went to Ireland actually, for a few months, and I worked in a dive shop. That was a lot of fun. I loved being around the ocean."

Megan set out to find a way to combine her new "ocean" passion with her engineering background. She found Hodgkiss' lab at Scripps Institution of Oceanography, a graduate school for oceanography that had a program called "Applied Ocean Sciences." There she conducted graduate research in near-shore underwater acoustic propagation and digital signal processing. Her research focused on determining geoacoustic models to describe very shallow water waveguides using measured transmission loss data in a genetic algorithm inversion technique. She served as chief scientist during at-sea data-collection operations, and planned and led diving operations during sea-floor instrument deployments and sediment-sample collections. She also actively cooperated in in-water instrument testing, deployment, and maintenance, and in the collection of plants, animals, and sediments.

While at Scripps, Megan also volunteered at the Birch Aquarium, conducting educational demonstrations for the public from inside a 70,000-gal exhibit tank of the California Kelp Forest, where she dived for in-water tank maintenance, animal feeding, and observations. She also served as a volunteer in CHiPS (Committe for Humanity and Public Services), founded in those years by students at the Scripps Institution of Oceanography for the purpose of encouraging and facilitating student involvement in community services like water-quality monitoring, plant salvage, revegetation, and bank stabilization. Before completing her doctorate, she also applied to become an astronaut. She says:

> "I did that for six years but always held on to this idea that being an astronaut would be pretty exciting and pretty challenging. And so at one point during my graduate career I went ahead and applied to NASA and just got really lucky and got interviewed and got hired."

Selected as a mission specialist by NASA in July 2000, Megan reported for training in August 2000. In 2002, she completed 2 years of training and evaluation, and also completed her Ph.D. and was ready to be assigned to the Astronaut Office Shuttle Operations

Branch working technical issues on Shuttle systems in the Shuttle Avionics Integration Laboratory (SAIL). In 2004, she served as the crew support for Expedition-9. During this mission, she spent 6 months in Russia, from April to October 2004. She also worked in the Mission Control Center as a CapCom (capsule communicator) during several International Space Station (ISS) and Space Shuttle missions. Her first flight into space was aboard Space Shuttle *Atlantis* STS-125. She says:

> "I was one of the last people to find out. That week I was actually working in Mission Control Center and I was working the night shift, so I wasn't going into my office at all. The chief of the Astronaut Office, Steve Lindsey had been calling me at my desk and, of course, I wasn't there. And so the word was getting out amongst the crew. It happened that I ran into our pilot on the crew, who I see occasionally, and he said "Have you talked to Scooter lately?" He is not somebody I would see very often or talk to on a regular basis. That was very suspicious to me, but I had to go and work a shift in Mission Control and hadn't time to ask more. But when I got home that night and my phone rang and it was Steve Lindsey, I finally knew what was going on."

STS 125 was the fifth and final Shuttle mission to the Hubble Space Telescope. During the ascent-and-return phase, Megan operated as flight engineer, but her primary responsibility as mission specialist was to grasp the Hubble Telescope and place it in the payload bay of the Shuttle, and then to operate the robotic arm to stabilize and assist astronauts servicing the Hubble during five spacewalks conducted by Andrew Feustel, Michael Good, John Grunsfeld, and Michael Massimino to "perform impossibly hard, never-before-attempted 'brain surgery' on this one-of-a-kind telescope, a telescope that will help unlock the secrets of the universe," as Megan wrote in an article. In a pre-flight interview, she explained:

> "This is the fifth servicing mission to the Hubble Space Telescope, and what we're going to do is basically extend the life of the telescope for another five years at least, hopefully. We're also going to repair it by capturing the telescope and placing it in the shuttle's payload bay, and then some of our crew members will do five different spacewalks to do this repair and upgrade work. I'll be driving the robotic arm for all of the EVAs. At the end we'll use the robotic arm to move it away from the orbiter and do a series of burns to get farther away from it."

The 19-year-old telescope spent 6 days in the Shuttle's cargo bay undergoing vital repairs, including some to equipment that was never designed to be fixed in space. The crew overcame frozen bolts, stripped screws, and stuck handrails to refurbish the Hubble Space Telescope with rejuvenated scientific instruments, two new instruments (the Wide Field Camera 3 and the Cosmic Origins Spectrograph), new batteries, new gyroscopes, and a new computer. Megan was the last person to have their hands on the Hubble Space Telescope. She has logged almost 13 days in space, so far. She hopes to visit the ISS soon. In the meantime, she works with two US commercial companies that have a contract with NASA to deliver unmanned cargo to the ISS: "I work with those companies to give them input on how the crews will use those vehicles and how they'll operate in them."

In 2012, Megan took part as a NASA representative in a Crew Equipment Interface Test inside the Dragon capsule at Cape Canaveral Air Force Station's Space Launch Complex-40 (SLC-40) located in Florida. "CEIT" tests are an activity that dates back to the Space Shuttle program, when it provided a training opportunity to assess the compatibility of partners' equipment and systems with the procedures to be used by the flight crew and flight controllers and to familiarize astronauts with the actual hardware that they would use in space.

Megan is married to fellow astronaut Robert L. Behnken and they have one son.

References

Cavallaro, U. "NASA-ESA Hubble Space Telescope: The Greatest Leap Forward in Astronomy since Galileo". *Astrophile*, **57**(1, #317), 13–19 (2015).

Chin, M.; Lin, J. "10 Questions for Bruin Astronaut Megan McArthur", *newsroom.ucla.edu* (May 6, 2009).

"K. Megan McArthur, 2010 Stellar Award Presenter", *2010 Rotary National Award for Space Achievement*, RNASA Houston (2010), p. 16.

Mansfield, C.L. "Astronaut Meets Dragon", *nasa.gov* (March 29, 2012).

McArthur, K.M. "What It Was Like to Capture the Hubble Space Telescope", *discovermagazine.com* (November 3, 2014), excerpt from *Infinite Worlds: The People and Places of Space Exploration*, by Michael Soluri, Simon & Schuster (2014).

Official NASA biography of Megan McArthur, *nasa.gov* (June 2009).

"Preflight Interview: Megan McArthur, Mission Specialist", *nasa.gov* (July 31, 2007).

Woodmansee, L.S. *Women Astronauts*, pp. 124–125. Apogee Books, Burlington, Ontario, Canada (2002).

53

Nicole Stott: A Steady Flying Passion

Credit: NASA

U. Cavallaro, *Women Spacefarers*, Springer Praxis Books, DOI 10.1007/978-3-319-34048-7_53

Mission	*Launch*	*Return*
STS-128	August 28, 2009	
STS-129		November 29, 2009
STS-133	February 24, 2011	March 9, 2011

Commemorative cover of mission STS-128, signed by Nicole Stott (from the collection of Umberto Cavallaro)

Nicole Stott is a veteran astronaut with two spaceflights and 104 days living and working in space on both the Space Shuttle and the International Space Station (ISS), including three Space Shuttle missions (STS128, STS129, and STS133), ISS Expedition-20/21, and one spacewalk. As she summarizes in her website:

> "Nicole brought a small watercolor kit with her on her mission to the ISS and is the first astronaut artist to paint while there. She is also a NASA Aquanaut and holds the Women's World Record for saturation diving following her 18 day mission with the NEEMO-9 crew on the Aquarius undersea habitat."

"I never pictured myself as an astronaut when I was growing up," Nicole says. "It wasn't my lifelong dream or anything I ever really thought could be possible for me." But, since she was a little girl, she was enchanted with flying—a fascination brought on by her dad: "I developed a passion for flying from what I saw in him and the passion that he had for it. It took me a long time to figure out I could make career out of doing something I love."

Nicole Marie Passonno Stott was born in Albany, New York, on November 19, 1962, the oldest of three sisters, and grew up in Clearwater, where she moved to at a young age—"a beautiful, beach town on the west coast of Florida," as she says in her blog. She enjoys flying, snow skiing, scuba-diving, woodworking, painting, and gardening. She spent much of her free time growing up with her family at the local airport. Her father, paint-company owner Fred Passonno, who loved flying and built small airplanes in his garage, shared that love with his family. She recalls helping him with the Plexiglass and fabric that went into the biplanes and then at the Clearwater Executive Airport, flying on aerobatic airplanes with people who were there in the small local club. "I'd go out in the garage or out to the airport and help him whenever he'd let me," she says.

But, in 1978, just days before Nicole's 16th birthday, her father was killed when an experimental airplane he was piloting crashed into a canal near Lake Tarpon, where he had built their nearby dream house. Her father's death didn't turn off her passion for flying. On the contrary, it inspired her to get her pilot's license and study for a career in aviation. "I was devastated when it happened of course," she says, "but I never thought about giving up flying. He was a guy whose whole life was spent flying and building planes and it was something he just loved to do."

After graduating from Clearwater High School in 1980, Nicole enrolled in the aviation administration program at the Clearwater and Tarpon Springs campuses of St. Petersburg Junior College. The program included earning her private pilot's license at St. Petersburg-Clearwater Airport, which she achieved at age 18. The program "was the greatest thing. The ground school was all handled by instructors at St. Pete College," she says. "I was able to take courses that fed right into the engineering degree that I ultimately worked on." In 1987, she earned a Bachelor of Science degree in Aeronautical Engineering from Embry-Riddle Aeronautical University. "Kennedy Space Center was the number one place I applied for, and really wanted to be there after growing up in Florida and seeing shuttles launch while I was at university," she says.

KSC had at that time stopped hiring, however, so Nicole took a job for over a year as a structural design engineer with Advanced Engines Group at Pratt & Whitney in West Palm Beach, performing structural analyses of advanced jet-engine component designs. She says:

> "I really enjoyed the people I worked with at Pratt & Whitney, and the projects were all very interesting and cutting edge. But it became clear to me that my talents were more associated with operations and hands-on engineering work than analysis."

Finally, Nicole was contacted by NASA at KSC: "I got a call from my original application to NASA at Kennedy Space Center, where they were finally off their hiring freeze, and needed to pick some people up for the shuttle ops group that they were forming." In 1988, she joined NASA at Kennedy Space Center (KSC) as an operations engineer in the Orbiter Processing Facility (OPF) and worked there for over 10 years, holding a variety of positions within the Shuttle processing division, until she became Shuttle Flow Director for *Endeavour* and then Orbiter Project Engineer for *Columbia*, and finally NASA convoy commander for Space Shuttle landings and vehicle operations engineer, preparing Space Shuttles for their next mission. She says:

> "I've worked in every part of space shuttle operations and I've had so much fun. I have so many friends who have moved or changed jobs every couple of years to find something they enjoy. I feel fortunate that I get the opportunity to try something new every few years within the Space Shuttle program and benefitted from seeing something a little different about the program each time."

In 1992, Nicole received her Master of Science degree in Engineering Management from the University of Central Florida. During her last 2 years at KSC, she worked in the Space Station Hardware Integration Office. Nicole met her husband Chris, a British citizen from the Isle of Man who was assigned to KSC for 3 months while he was working on his master's degree. They married and, when Chris got a job with McDonnell Douglas in California, Nicole moved with him and was relocated to Huntington Beach, where she served as the NASA project lead for the ISS truss elements under construction at the Boeing Space Station facility. People at NASA encouraged her to apply for the astronaut program and, in 1997, she decided to try. "Until the point of starting work at Kennedy Space Center with NASA," she says, "it never crossed my mind that being an astronaut was a possibility."

Nicole was not selected the first time, but she was offered to join the Aircraft Operations Division at the Johnson Space Center (JSC), where she served as a flight simulation engineer on the STA (Shuttle Training Aircraft)—a Gulfstream II modified to mimic the flight characteristics and instrumentation on the Shuttle—helping to train astronaut pilots to land the Space Shuttle. She reapplied for the astronaut candidate training program in 1999 and, this time, she made the cut. In 2000, she entered the NASA Astronaut Group 18, nicknamed "The Bugz" (the name vaguely recalling the "year 2000 and the millennium bug"). "We all received the phone call on July 20, 2000—the anniversary of the Apollo moon landing," she recalls in her blog. And she began her astronaut training as mission specialist, after which she was assigned to the International Space Station Operations, responsible for evaluating payloads. She also worked as a support astronaut and CapCom (capsule communicator) for the ISS Expedition-10 crew.

In April 2006, Nicole was a crewmember on the NASA Extreme Environment Mission Operations (NEEMO) 9 mission, where she lived and worked with a six-person crew for 18 days in the Aquarius undersea research laboratory. This mission—the longest expedition in the program—aimed at testing advanced spacesuit design concepts, robotic devices for surface-based exploration, construction and communication techniques, and advanced telemedicine hardware and techniques for future lunar operations. "It was the best preparation for going to space," she explains, as, once you are down there, you have to be very thoughtful of what you are doing, and you can't just open the hatch and walk out and go to the surface without a long decompression: "You learn the dynamics of living and interacting with a crew in a confined living space." They tested and trained on undersea "moonwalks" and robotic surgeries controlled by a doctor high and dry in Canada. Just before "splash-up" for returning to the surface, Nicole told her fellow aquanaut and astronaut classmate Ron Garan: "You know, if we never get to fly in space, this experience would be enough." Nicole still holds the Women's World Record for saturation diving with a total of 18 days spent in this mission.

Assigned to long-duration Expedition-20/21 to the ISS, Nicole moved to Moscow, Russia, for a full-immersion Russian-language class and successfully completed ISS systems training at each of the international partner training sites in Star City (Russia), Tsukuba (Japan), Cologne (Germany), and Montreal (Canada):

> "Flying in space is most certainly the goal of any astronaut, but what you have to accept is that the life of an astronaut is 99 percent not about flying in space. No matter how quickly you get assigned to your first space flight, you are ultimately going to spend the majority of your time as an astronaut working here on the planet. Fortunately it's all very cool. Everything we do, whether it's training for a spacewalk in the neutral buoyancy laboratory or training on the robotic arm or flying T-38 training aircraft, it is stuff I would never have the opportunity to be exposed to otherwise. I kind of look at it like spaceflight itself is this perk at the end that someday might happen, but all these things we're getting to do on the ground are unique and interesting on their own."

Nicole was launched from KSC Launch Complex 39A (LC-39A) on her first spaceflight as a flight engineer on board the Shuttle *Discovery* STS-128 (the ISS assembly flight 17A) in August 2009 and reached the ISS to begin her expedition. On the fourth day of the mission, she participated, with her colleague Danny Olivas, in the first spacewalk of the mission and remained in open space for 6 h and 39 min to prepare for the replacement of an empty ammonia tank on the station's port truss. They retrieved a materials processing experiment and a European science experiment mounted on the EuTEF (European Technology Exposure Facility) outside the Italy-built European Columbus laboratory and stowed them in *Discovery*'s cargo bay for their return to Earth. Nicole says that it was "neat" to see the station both from the outside and from the inside. "I had very high expectations of what this experience would be like, and I can honestly say that every expectation I had was exceeded," she says in her blog. Particularly vivid are the memories of our planet seen from the ISS:

> "You can't (or at least I have never been able to) look at a picture of the Earth from space and not feel a sense of awe. The Earth, our planet, is indescribably beautiful. It glows like a colorful light bulb. It is placed perfectly against the blackest black I have ever seen. On the ISS I have been blessed with the opportunity to see our planet

> from a totally different perspective. At the same time it appears blue and calm and peaceful, you can look in a different direction and it is very dynamic and dark and even sad and unpredictable. It is a vantage point that can most certainly lead you to believe that we all might just be insignificant little specs in the grand, universal scheme of things. I believe that what I'm seeing out the windows is this awesome Creation that has been put in the perfect place in the universe, in the perfect place in the solar system, at the perfect distance from the sun, giving us the perfect conditions to survive."

As Nicole recalls, hers was a fairly exciting mission during which the station was visited by many "amazingly beautiful" spacecraft:

> "The space shuttle, the space station, the Soyuz, the Japan-built HTV, and the Russian Progress—these are all the spacecraft I've had the opportunity to see while I've been here in space. You can't look at these vehicles without being impressed, sometimes overwhelmed by how impressive they are. And the impression is not just from the incredible engineering marvels that they all are or from their size, but it's also very simply from how incredibly beautiful they each are. There is a shiny, spectacular independence to each of them when you see them hanging so naturally in space, like they were meant to be there with the forces of nature holding them in their place. And as they approach and come into view—starting out first as only a pinpoint of light against the very blackness of space or the backdrop of our glowing, colorful planet and then gradually/quickly transforming into the magnificent, shining, beautiful spacecraft that they are. Awesome!"

Working with crewmate Frank De Winne, Nicole executed the first track and capture of the Japanese cargo vehicle HTV with the robotic arm. She told a NASA interviewer:

> "The vehicle comes up, gets to about ten meters from the space station, and then the two vehicles are flying along together at 28,000 km/h. We have 99 seconds to achieve the capture with the big robotic arm. That's going to be cool because it's a totally new vehicle and on station we've never done a track and capture like this before."

Nicole also conducted a wide variety of science and research activities, including studies on nutrition in space and on our immune system. "What's interesting," she comments, "is that everything we're doing on the ISS has a direct application to helping us explore further off our planet and to also improving life here on Earth." During the mission, she participated, together with her crewmate Jeff Williams, in the "First Tweet Up in Space." It was not the instant Twitter most of us use today; it was a much slower and more labor-intensive process that involved e-mailing down the tweets and it required the help of ground personnel to relay the information and tweet it out.

Nicole was the last ISS expedition crewmember to fly on a Space Shuttle when she returned to Earth aboard STS-129 in November 2009. Nicole completed her second spaceflight at the beginning of 2011 as mission specialist on STS-133, riding *Discovery* again, originally scheduled as the final flight of the Space Shuttle program. After the addition of the *Atlantis* STS-135 flight and the delay of the *Endeavour* STS-134 flight, STS-133 was the final flight of the Space Shuttle *Discovery*. She arrived on the station in the period when the ISS was commemorating the 10th anniversary of the ISS Expedition-1, the crew that started the continuous human presence in orbit. "The key objectives for STS-133," as Nicole explained, "were primarily positioning the station for future years to come, both inside and outside." The PMM (Permanent Multipurpose Module), derived from the Italy-built Multi-Purpose Logistics Module (MPLM) Raffaello and properly modified, was permanently integrated into the ISS. The mission also delivered to the ISS the fourth ELC (Express Logistics Carrier), a large unpressurized platform attached outside of the ISS to provide mechanical mounting surfaces, electrical power, command and data handling services, and science experiments that must be exposed in outer space. Nicole, together with her colleague Michael Barratt, operated the robotic arm to support the installation of both the ELC-4 and the PMM. She also contributed, as internal choreographer, to coordinate the extravehicular activities (EVAs) performed by her two colleagues.

In her two spaceflights, Nicole logged 104 days in space. After completion of the STS-133 mission, she returned to KSC for a 1-year assignment as the Astronaut Office representative to the Commercial Crew Program and the NASA selection of the companies chosen to build the next generation of US human-rated spacecraft. In 2012, she returned to the JSC, where she served as Chief of the Astronaut Office Space Station Integration Branch and then Chief of the Vehicle Integration Test Office, responsible for a team of engineers that manages the astronaut interface to existing and future flight hardware; she was also the lead astronaut representative to the Orion Landing and Recovery team, responsible for determining how we will recover astronauts from the Orion spacecraft after splashdown.

After almost 28 years of service, Nicole retired from NASA in June 2015, to stay with her family and pursue her next adventure as a full-time artist, painting her experiences with the spacecraft she flew and her interpretation of Earth observation images she was blessed to photograph from space, "using oils and acrylics and incorporation of physical mixed media like sea glass and sand," she explains. Inspired by her spaceflights, she is using her artwork as "a powerful tool to communicate the importance of space exploration and share the awesome beauty" that she experienced through the windows of the Space Shuttle and space station, and during her spacewalks. She is also a motivational speaker who actively supports science, technology, engineering, math, and art (STEM/ STEAM) education.

References

Church, J. "'Unusual Interests' Might Lead to the Stars", *sptimes.com* (*St. Peterburg Times*, March 3, 2002).

Houck, J. "Clearwater High Grad to Join Space Station", *tbo.com* (August 18, 2009).

Official biography of Nicole Stott, *nasa.gov* (March 2013).

Personal contacts by e-mail with the Author in March/April 2016.

Personal website: *www.npsdiscovery.com/*.

"Preflight Interview: Nicole Stott", STS-128, *nasa.gov* (August 13, 2009).

"Preflight Interview: Nicole Stott", STS-133, *nasa.gov* (October 7, 2010).

Rhian, J. "Insider Interview: Nicole Stott Talks Leaving NASA, Orbital Artistry", *spaceflightinsider.com* (July 26, 2015).
Stott, N. "Anniversaries & Memories of My 1st Space Adventure", *fragileoasis.org* (August 30, 2010).
Taylor, J.D. "Women in Space: Nicole Stott—more than 100 Days on Orbit", *spaceflight-insider.com* (April 29, 2015).
Woodmansee, L.S. *Women Astronauts*, pp. 128–129. Apogee Books, Burlington, Ontario, Canada (2002).

54

Dorothy Metcalf-Lindenburger: From Space Camp to Space Station

Credit: NASA

U. Cavallaro, *Women Spacefarers*, Springer Praxis Books, DOI 10.1007/978-3-319-34048-7_54

Mission	*Launch*	*Return*
STS-131	April 5, 2010	April 20, 2010

Dorothy Metcalf-Lindenburger became NASA's fourth educator astronaut and launched aboard Shuttle *Discovery* on STS-131 to the International Space Station (ISS). She explains:

> "I am the daughter of two teachers—a high school math teacher and a junior science teacher. So no big surprise that I love math and science. Their background in education, the environment of my native state, Colorado, and the excitement of space in the 1980s resulted in my passion to work at NASA."

Dorothy "Dottie" Marie Metcalf-Lindenburger was born on May 2, 1975, in Colorado Springs, Colorado. When she was a child, the family moved to Loveland, where she grew up and received her elementary education, and they finally settled for junior high and high school in Fort Collins, which she considers her hometown. "I think just growing up along the front range," she told a NASA interviewer, "provides you with a lot of opportunities. First of all, you take topography for granted until you move to a place like Houston and you don't have the mountains."

Geologist, astronomer, marathon runner, Dottie also likes scuba-diving, drawing, and singing. Attracted by the historic environment there, she frequently spent time with her sister Neva, just 3 years younger, digging fossils that were really abundant in her backyard. She adds:

> "I didn't know until later, when I became a geologist, just how lucky I was because when you're in geology, it's pretty hard to actually find fossils often, and you may spend a lot of time digging before you get one."

Dottie also recalls how she was impressed by the dinosaurs, mineral display, exotic animals, and planetarium that she saw when—before she started school—her parents brought her to visit the Denver Museum of Natural History. All this ignited her interest in geology and astronomy. Voyager's discoveries were just coming out as part of shows at the planetarium, with a lot of data coming back into museums. Dottie recalls that she asked lot of questions and became really interested in space:

> "As a result, I fell in love with our solar system. My parents continued to fuel this love, by taking me to see the film 'The Right Stuff' and talking about the shuttle launches in Florida. They helped me search for Halley's comet, and they watched lunar eclipses through the telescope Santa brought me for Christmas in 6th grade."

The launches of the Shuttle ignited her interest in manned space exploration: "I remember watching launches on the TV when I was in the elementary school. I just thought it was so amazing to watch these, and I never thought I was going to be on one." In 1989, she entered a Martin Marietta writing contest about space exploration. The first place was an all-expense-paid trip to Space Camp in Huntsville. She didn't win it, however, but came second, with a consolation prize of a nice T-shirt from NASA:

> "My parents knew that I really wanted to go, so they sent me to Space Camp the following year during the spring break in April. It was there that I realized if I keep working hard in math and science, it's a possibility that I could work at NASA. I kind of saw behind the scenes of what different people would do during a mission. Not just what astronauts do, but what the ground does as well, and I thought, 'I really would like to work at NASA'. So, that's why I wanted to go and study math and science in college."

It was the month the Hubble Space Telescope was launched. After graduating at the Fort Collins High School, Colorado, in 1993, Dottie joined the Whitman College, Walla Walla, near Washington: "It's a small liberal arts school in the middle of pretty much nowhere, in like a wheat field, and that was a really great place for me to learn, away from distractions." Her original plan was to study math, following in her mother's footsteps, but she ended up taking a geology course during her freshman year and enjoyed it so much that she pursued it as her major. She earned a bachelor's degree with honors in Geology and Earth Science in 1997. After graduating, she decided to pursue a career in education, or to do something to give back to society. She was accepted to the Peace Corps and was assigned to teach English in Kazakhstan. It was a time when there were some upheavals in that part of the world and so she decided to go teach in the US. In 1999, she went back and got her teaching certification in science and history at Central Washington University. For 5 years, she taught Earth Science and Astronomy at Vancouver School District at Hudson's Bay High School. In addition to teaching, she coached cross-country running and Science Olympiad, considering herself a science teacher even when she was outside of the classroom. "My husband and I built a telescope and took it on our summer vacation, and wherever we stopped, we showed people things like Jupiter or the moon," she said. "So many of the adults had never even looked through a telescope!"

In the spring of 2003, a question from a student would change her life forever:

> "When I was teaching astronomy at high school, one of my students asked, 'How do astronauts go to the bathroom in space?' It's kind of a common question we get when we're doing PRs. To find the answer that night I went to Google, which took me to the NASA Website and I found that they had posted that educators could become astronauts. So, I had the answer to my student's question, but I also got an answer to a dream that I had for a long time, and so I applied to be an educator astronaut."

Dottie was one of 8000 applicants in 2003. She recalls:

> "A little over one year after applying for the job, I received a call from Colonel Robert Cabana, head of the Astronaut Office. I was in my classroom and, once I realized he was calling to hire me, I burst into a shout of excitement. I called my mom and we cried on the phone together, it was just so exciting to have my dream fulfilled."

She entered the NASA Astronaut Corps as an educator mission specialist of Group 19 that in 2004—less than 1 year after the *Columbia* accident—selected three educators out of 8000 applicants, Dottie being the only woman. She moved to Houston. "Life certainly changed after we moved to Houston," she says. "I went from being a teacher to being a learner all over again." She flew in space in April 2010 on board the Shuttle *Discovery* STS-131:

> "As I look back at Space Camp in April 1990, it takes on more meaning. Not only did I decide then that I wanted to someday be a part of space exploration, but also it was the month the Hubble Space Telescope was launched. And, 20 years later, it would be during the same month that I took my first journey to space."

This mission set several records, not only because it marked the longest flight for Space Shuttle *Discovery* and had the most payloads since STS-107, but this mission—as we already mentioned—marked, among others, another first for the most women ever in orbit at the same time: when *Discovery* docked with the ISS carrying Dorothy Metcalf-Lindenburger together with Stephanie Wilson and Naoko Yamazaki, they found on the station Tracy Caldwell, who had arrived a couple of days before aboard Soyuz TMA-18 (Fig. 54.1).

Mission STS-131 was the ISS assembly flight 19A mission that delivered to the Space Station over 12 t of supplies, equipment, spare parts, experiments, the third and final MELFI (Minus Eighty Degree Laboratory Freezer for ISS), an extra sleeping compartment, and a fresh ammonia tank for the cooling system, whose installation was a cumbersome job requiring three spacewalks. On its return flight, the Italian Multi-Purpose Logistics Module (MPLM) Leonardo brought back to Earth three tons of results from scientific experiments and materials that were no longer needed on the station. After completing the space mission, Dottie served as "Cape Crusader" for the last three Shuttle missions.

Commemorative covers of launch and landing of mission STS-131, signed by Dorothy Metcalf-Lindenburger (from the collection of Jürgen P. Esders)

In June 2012, Dottie commanded the 12-day NEEMO 16 (NASA Extreme Environment Mission Operations) undersea exploration mission in the Aquarius underwater laboratory where techniques to explore an asteroid were tested. She left NASA in June 2014 and moved to Seattle; she is now Master Candidate in Applied Geosciences at the University of Washington. She has been a long-time lead singer with the all-astronaut rock band *Max Q*.

Fig. 54.1 For the first time in history, four women meet in space: Wilson, Dyson, Yamazaki, and Metcalf-Lindenburger in the ISS-Cupola. Credit: NASA

References

Antoun, C.; Antoun, M. "Dottie Metcalf-Lindenburger", *women.nasa.gov* (March 29, 2016).

Evans, B. "Education Can Take You Anywhere", *americaspace.com* (May 2, 2015).

"Interview: Dorothy M. Metcalf-Lindenburger", STS-131, *nasa.gov* (April 27, 2010).

"Metcalf-Lindenburger", *Encyclopedia Astronautica (astronautix.com)* (May 2004).

Official biography of Dorothy Metcalf-Lindenburger, *nasa.gov* (May 2014).

Sunseri, G. "Discovery Teacher-Astronaut Breaks the Mold", *abcnews.go.com* (April 5, 2010).

55

Naoko Yamazaki: Astronaut for 4088 Days

Credit: NASA

U. Cavallaro, *Women Spacefarers*, Springer Praxis Books, DOI 10.1007/978-3-319-34048-7_55

Mission	*Launch*	*Return*
STS-131	April 5, 2010	April 20, 2010

Naoko Sumino Yamazaki was born on December 27, 1970, in Matsudo City, Chiba Prefecture, Japan, and she spent 2 years of her childhood in Sapporo. She says:

> "Matsudo City is a small town, but because it is a suburb of Tokyo I had both of the best worlds, a small town in which I could enjoy nature, especially the Edo River, and also I could access Tokyo, a big city and its advantages. I also spent two years in Sapporo in Hokkaido in the northern area of Japan and I enjoyed skiing a lot in Sapporo and participating in the 'star-watching parties'."

As a young girl, Naoko was a fan of science-fiction movies. In her autobiographical book, she recalls: "I've always had a steady passion for space. I liked cartoons like 'Galaxy Express 999' or 'Space Battleship Yamato' on TV." She simply thought that anyone could go to space when they grew up. She remembers that she was fascinated by Matsudo's Planetarium, which she visited at every opportunity: "I could learn about the stars and the constellation for just 30 cents each visit. There I went very often with my older brother and it widened up my interest in space." She dreamed of becoming a school teacher. When she learned that Christa McAuliffe would fly aboard the Shuttle, she was fascinated: "Wow! A teacher goes into space …!" So she started to follow the event with much interest and realized "Oh, there are real space rockets in the world. It is not science fiction, but it is a real world!" It was then, she recalled later, that she started to think about the possibility of becoming an astronaut. She recalls that she was at grade 9 and, 1 night, she stood up to watch the launch of Shuttle STS-51L live (because of the time zone, in Japan, it was already 1:30 a.m.). In that flight, Ellison Shoji Onizuka, the first American-Japanese astronaut, was launched and this was the first time that, through a CNN channel, it was possible to watch at a Shuttle launch in Japan. Naoko writes in her book:

> "Suddenly I saw in the intense blue sky a bush tail of white smoke writhing like a big snake and stretching like a spiral …. I was in the middle of the room with a pen in hand and watching bated at the TV screen. What was being broadcast was not a cartoon or a science fiction movie, was the reality. The white smoke hanging in the sky, was sadly telling the truth. A sad truth of seven worthy lives that were torn apart in an instant. What may have thought Christa … who was a woman and a regular civilian?

> … Then I learned that she had left two sons: Scott of nine and Caroline of six. She, Christa was a woman, a wife, a mother, a teacher, and above all an astronaut."

The "Teacher in Space" became for Naoko a recurring thought:

> "Without my knowing it, her dreams and her hopes came little by little in my heart, in the heart of a girl of 15, who lived in a small town in the Far East. … The Challenger accident entered my heart and ended up to closely combine my two dreams: teaching and space."

When it was time to give a decisive swing to her life, Naoko remembered Christa McAuliffe's radiant smile while she was approaching the Shuttle and, following the example of "a person who already made it reality," she took the decision: "I'll become a teacher and go to space." She says:

> "More than once I've been hailed as 'superwoman', with a straight life path that has led me straight from childhood to the space, supported by parents who encouraged and fostered me. On the contrary nor I was born a genius, nor I was raised in a particularly favourable environment."

Naoko's was a modest family living on the salary of a father who was a military officer. She couldn't afford the after-school until the end of the eighth grade. She writes:

> "At home not even I had my own room where to study. I always studied on the dining room table, and only during the preparation for university entrance exam I managed to have a space for myself, obtained by placing a desk in a corner of the entrance, sheltered by a tent."

After graduating from the Ochanomizu University Senior High School in 1989, Naoko earned a bachelor's degree in Aerospace Engineering from the University of Tokyo in 1993 and a master's degree in 1996, specializing in space transportation systems and space robotics. She also spent 1 year in the US at the University of Maryland. "There was a huge water tank which was 30 ft in depth," she says, "so we put row boats in the water tank and we could simulate microgravity. So it was a very interesting experience to me."

Naoko learned about the call issued by NASDA (the Japanese Space Agency, now the Japan Aerospace Exploration Agency or JAXA) to recruit new astronauts and applied, although she was conscious of not having all the requirements, including the 3 years of employment after university. There weren't many opportunities to become astronauts in Japan and the calls of NASDA were rare: this was only the third call in the Agency's history. Besides the physical skills, the psycho-attitudinal, and the scientific training, they required a good level of general culture: the astronauts had to, she explains in her book:

> "have a good humanistic culture, possess a rich expressive ability to vividly convey their experiences, have a deep understanding of Japanese culture and of international society. I had the impression that the selection criteria were the same as a beauty contest, but in short, the candidates are people of science or engineering, who by participating as Japanese citizens in an international project also had to be able to promote Japan."

Naoko's first attempt to become an astronaut didn't succeed. She was invited instead to collaborate with NASDA to serve in the system integration department, helping to work on the development of Kibō, the Japanese space lab, and developed failure analysis and assembly/initial operation procedures for the module which had to be integrated into the International Space Station (ISS). From June 1998 to March 2000, she was also involved in the development of the ISS Centrifuge (life science experiment facility) and conducted conceptual framework and preliminary design in the Centrifuge Project Team. "This enhanced," she says, "my desire to become an astronaut and to go work on the International Space Station, especially on Kibō." In 1998, she applied again for the Japanese Astronaut Corps and this time she made the final cut. The following year, she was one of the three astronauts selected by NASDA. She met her future husband, Taichi Yamazaki, employed at Mitsubishi Space Software, a subcontractor working for Kibō, and they married in 2000.

After completing the basic astronaut training in Japan, Naoko attended specific training in the countries that are the major partners in the ISS program. Naoko recalls:

> "The basic training was held in many countries, in Japan, in Canada and United States, in Russia and in European countries because these countries are the major partners in International Space Station program. After I finished my basic training I went to Russia for seven months in the Star City to get qualified as a Flight Engineer of the Russian spacecraft Soyuz. After that, I came to Houston to join the NASA mission space training program."

She became a certified astronaut in September 2001. The following year, she gave birth to her first daughter, Yuki.

On October 1, 2003, NASDA merged with ISAS (Institute of Space & Astronautic Science) and NAL (National Aerospace Laboratory of Japan) and was renamed JAXA. Naoko went eventually back to Russia, to the Jury Gagarin Cosmonaut Training Center, Star City, to complete her training as a Soyuz-TMA Flight Engineer-1 and, in June 2004, she moved to NASA's Johnson Space Center (JSC) in Houston, where she qualified as a mission specialist and was assigned to the Robotic Branch of the Astronaut Office. She was in Japan for training when she learned that she had been assigned as mission specialist to the Shuttle mission STS-131, thus becoming the second Japanese woman astronaut, following Chiaki Mukai, and the first Japanese mum to fly into space.

This flight—that was initially scheduled for launch on the 30th anniversary of the launch of the first Shuttle mission STS-1—made history for many reasons, but especially for JAXA as, for the first time, two astronauts from the Land of the Rising Sun were in orbit at the same time: Naoko Yamazaki, arriving with the Shuttle, met on the station Soichi Noguchi, who had arrived 2 days before in a Russian Soyuz and was participating in the 22nd ISS expedition. In a pre-flight interview, Naoko said:

> "This is a very big step for Japan. Soichi and I are scheduled to do several tasks together like experimental rack transfer and installation on the space station so we are looking forward to working with each other and we are also looking forward to sharing some Japanese cultures among the crew members (Fig. 55.1)."

Fig. 55.1 Naoko Yamazaki's personal patch for STS-131. The logo features a seed encompassing life in space, in the hope that the technology and knowledge cultivated through International Space Station (ISS) missions will lead all life to a better future on both Earth and in space. The eight four-leafed clovers represent the eight Japanese astronauts flown in space. Credit: JAXA

The main objectives of STS-131 were to continue the assembly of the ISS and bring a payload of supplies and equipment. It delivered the Italian Leonardo Multi-Purpose Logistics Module (MPLM) module, full of 6 t of supplies, science experiments, equipment, and critical spare parts, including a new crew sleeping compartment, a tank full of ammonia for the cooling system, and three science racks to be transferred to the orbiting lab. This mission carried the most payloads since STS-107.

Naoko was the mission loadmaster, responsible for coordinating all the transfer activities from the MPLM to the station and operating the robotic arm (the SSRMS or Space Station Remote Manipulator System) to berth the MPLM to the station, to unload the payloads (that required more than 120 h), to put onto the Leonardo module 3 t of results of scientific experiments and unnecessary items to be brought back to Earth, such as equipment used in completed experiments, and to unberth the MPLM and put it back on the Shuttle cargo bay. She explained:

> "It is very challenging to orchestrate all the activities in order. Some hardware has constraints, so some of them need to be transferred in certain order and in a certain way. So I need to understand the hardware very well. It is a challenging part. It's like moving into a new house."

S131E007954

Fig. 55.2 For the first time in history, four women meet in space: Wilson, Dyson, Yamazaki, and Metcalf-Lindenburger in the ISS. Credit: NASA

Naoko also was responsible for operating the SSRMS to inspect the Space Shuttle's heat-resistant tiles for possible damage during ascent. This mission also made history—as already mentioned in previous chapters—for its new historical space record of the most women aboard the same spacecraft at the same time: four. In fact, when Naoko Yamazaki, Stephanie Wilson, and Dorothy Metcalf-Lindenburger entered the ISS, they found there Tracy Caldwell Dyson, who had arrived 2 days before with Expedition-23 (Fig. 55.2).

Naoko returned to Earth on April 20, 2010, after logging over 15 days in orbit. This was the last of the Shuttle crews to include any astronauts making their first spaceflight. STS-131's crew, together with the six Expedition-23 crewmembers, also tied the record for the most people on one spacecraft: 13.

To help the "Astronaut Mum"—as they called her in Japan—to fulfill her dream, her husband Taichi had decided to support her and to personally take care of their daughter Yuki. He temporarily put his dreams aside and left his job at Mitsubishi—where he had the prospect of becoming flight controller of the Japanese Kibō Laboratory, once connected to the ISS—thus voluntarily becoming a "full-time homemaker" and stay-at-home dad. The anomalous Yamazaki family, in which the traditional gender roles were reversed, came to the attention of newspapers and provoked heated discussions in the media. Japanese TV filmed Taichi while he was preparing rice, doing the laundry, and engaging in tasks rather unusual for a Japanese husband. In the super-chauvinist Japanese society, he was seen as an "extraterrestrial husband." In her book, Naoko admits that the last decade was for them a trying one, as the unusual redistribution of family responsibilities had not been easy, and she made no secret of having narrowly avoided divorce during her long training in the US.

Everything seemed to be on track until 2003, when Space Shuttle *Columbia* disintegrated over the sky of Texas during re-entry, pushing back Naoko's dream of spaceflight by years and it turned out that her plan was much harder for her family than she had originally expected. Another serious problem was her long stay in Russia without any support from the Japanese Space Agency. "When I tried to consult with my employer," Naoko says, "they told me these were problems that should be dealt with within the family" according to the Japanese mentality that sacrificing the family for work is unavoidable. In an interview, Taichi added: "If we were to come down with a cold and give it to Naoko, we would've been blamed for interfering with an important state project."

When Naoko moved to the US for her training at NASA, Taichi decided to follow his wife to make it possible for the family to stay together as much as possible and, after months of unsuccessfully trying to negotiate a way to keep his post while he was in the US, he quit his job in Japan, hoping to find another one in the space environment once they arrived in the US. In Houston, they were struck by the support given by NASA to the astronaut families. "They had an entire department dedicated to caring for astronauts' families," Naoko said. "It wasn't based on purely altruistic motivations: it was based on the understanding that a stable family life would improve astronaut productivity." At this point, however, new difficulties arose when, with his wife at work and his daughter in the childcare facility, Taichi realized that he could not acquire a new job because of his visa status as the spouse of a diplomatic visa holder: "I'd find myself at home," he said, "wondering what in the world I was doing." Naoko recounts in her book:

> "As he could not get a work permit, Taichi could not even get the 'Social Security Number'. Without this code in America one is not even considered as a person: not only you cannot open a bank account, but you cannot even buy a cell."

Taichi became seriously depressed, aggravated by the health problems of his elderly and ailing parents in Japan. It was then that he thought of divorce or suicide, seeing no other way out. To save the day, the assignment of Naoko to her space mission arrived in November 2008. But family relationships remained so sensitive that, in order to avoid hurting the feelings of her husband, Naoko specifically asked a reporter to avoid mentioning her as the "Astronaut Mum," which would have suggested that she was the only one working hard and juggling her work with child-rearing responsibilities.

Finally, after more than 10 years since her selection as an astronaut, Naoko and Taichi's dream came true in April 2010 and they could happily conclude their adventure. Naoko left JAXA in August 2011 to spend more time with her family and to start part-time aerospace engineering research at the University of Tokyo. Naoko and Taichi told their adventure in the book *Astronaut for Four Thousand and Eighty-Eight Days* (which unfortunately is only available in Japanese—a draft translation, still unpublished, was used for this chapter). The book concludes by calling for the creation of a society that does not require that, in a couple, one partner sacrifice his or her dreams for the other, and a woman's career is not pursued at the expense of her husband's.

References

Buerk, R. "Japan's First Female Astronaut Changes Gender Roles", *news.bbc.co.uk* (April 7, 2010).

Ikeuchi, T.; Saito, K. "Space Mom Wants Equal Opportunity for All", *japantimes.co.jp* (April 17, 2014).

Kobayashi, C. "Plight of Japanese Astronaut and Her Family", *The Mainichi Daily News*, Japan (April 19, 2011).

Kohtake, N. "EARTHLING Interview: Naoko Yamazaki", *thinktheearth.net* (June 1, 2011).

"Naoko Yamazaki, Getting Ready for First Spaceflight", *global.jaxa.jp* (March 9, 2010).

Nishiura, M. "Interview with Astronaut Naoko Yamazaki", *Jaxa Today*, No. 02 (August 2010), 3–7.

Official biography of Naoko Yamazaki, *jaxa.jp*.

Official biography of Naoko Yamazaki, *nasa.gov/externalflash* (April 2010).

"Preflight Interview: Naoko Yamazaki", STS-131, *nasa.gov* (March 8, 2010).

Yamazaki, N. *Astronaut for Four Thousand and Eighty-Eight Days*. Kadokawashoten, Tokyo (2013) (in Japanese – manuscript translated into English by Federica Cavallaro, pending publication).

56

Shannon Walker: The First Native Houstonian Astronaut to Fly into Space with Russians

Credit: NASA

U. Cavallaro, *Women Spacefarers*, Springer Praxis Books, DOI 10.1007/978-3-319-34048-7_56

Mission	*Launch*	*Return*
Soyuz TMA-19	June 15, 2010	November 25, 2010

Shannon Walker was the first native Houstonian to become an astronaut. She says:

> "Growing up in Houston and always having the astronauts and the Johnson Space Center in my backyard, I was always aware of the space program. So I just decided to pursue it, and was fortunate enough to work at the Johnson Space Center and then eventually become an astronaut."

The entire mission lasted for 163 days, 161 of them aboard the International Space Station (ISS), which she reached riding a Russian Soyuz.

Shannon was born in Houston on June 4, 1965: "I had just turned four, when we first walked on the moon, and I think that just set the seed right then." She went through the public-school system in Houston: Parker Elementary, Johnston Junior High, and Westbury Senior High, where she graduated in 1983. She received a Bachelor of Arts degree in Physics from Rice University in Houston in 1987. "Once I graduated I was extremely fortunate and was offered a job out here at the Johnson Space Center to work in the flight control center," she recalls. She began her professional career with the Rockwell Space Operations Company at the Johnson Space Center (JSC) in 1987 as a robotics flight controller for the Space Shuttle program. She worked on several Space Shuttle missions as a flight controller in the Mission Control Center, including STS-27, STS-32, STS-51, STS-56, STS-60, STS-61, and STS-66. At the beginning of the 1990s, she took a leave of absence to attend graduate school for 3 years and went back to Rice University to pursue a doctorate in Space Physics. In 1992 and 1993, respectively, she received a Master of Science degree and a Doctor of Philosophy degree in Space Physics, studying the solar wind's interaction with the Venusian atmosphere: "After I completed my doctorate I decided wanted to be part of the human spaceflight program, so I was able to come back to the Johnson Space Center and continue my work as a flight controller."

In 1995, Shannon Walker was hired by the NASA civil service and served on the ISS program at the JSC in the area of robotics integration, working with the ISS international partners in the design and construction of the robotics hardware for the space station: "Primarily I was working in the robotics area, working with the Canadians to build the Canadian robotic arm and our equipment that interfaces with it." After completing this task, she joined the ISS Mission Evaluation Room (MER) as a manager for coordinating on-orbit problem resolution for the ISS. In 1999, she moved to Moscow, Russia, to work with the Russian Space Agency and its contractors in the areas of avionics integration as well as integrated problem-solving for the ISS. "I took a little jog away from robotics and did avionics integration in Russia for a year, so it was making sure that our computer boxes were talking to their computer boxes, which is very important," she says. The following year, she returned to Houston and became the technical lead for the ISS MER as well as the deputy manager, and then acting manager, of the On-Orbit Engineering Office that was heavily involved in solving ISS problems from the ground. She explains:

> "I went back to the control center but not on the flight control side. I worked on the engineering side, and so I worked in the area and ultimately was the leader of the area in the control center which is responsible for what we call the technical health of the station. So we monitor all its systems and if something's not working properly, then it was our job to figure out not only what went wrong, but how we're going to fix it. Because sometimes what you need to fix a problem is not what you have on board, and so you have to be a little creative to get problems solved."

Shannon was selected as an astronaut candidate in May 2004 and, the following year, she married fellow astronaut Andrew (Andy) Thomas. In February 2006, she completed her Astronaut Candidate Training and her operational training on the ISS systems: "I know so much on the engineering side and not so much on the operations side," she says. "It's like, 'How do I turn this on?' But I can tell you all the problems that a component has had." After qualifying as mission specialist, she completed her training for EVAs (extravehicular activities). As a member of the support crew for the ISS Expedition-14, from September 2006 to April 2007, she served in the Houston Mission Control Center as CapCom (capsule communicator, the main contact point for interaction between the crew flying on the space station and the staff on ground). She was also CapCom during mission STS-118 in August 2007.

Just after that, Shannon began training for her long-duration mission on the ISS. After spending so much time at Mission Control, she started her new experience with great confidence in the support of the people on the ground:

> "I know how the control centers work. I know how to problem-solve. I know who's involved in working on the problems and can make a decision, so it is a big comfort factor. I have absolute confidence in the people in the ground, having known them for so many years."

Shannon was initially assigned as a member of the backup crew of Expedition-19, then as backup for Expedition-21/22, and finally as a member of the prime crew of Expedition-24/25. She lifted off from Baikonur on June 15, 2010, as the Soyuz TMA-19 flight engineer (co-pilot). The entire mission lasted for 163 days, 161 of them aboard the ISS.

Shannon was part of the crew that achieved the record for the longest continuous manned presence in space—a mark previously held by Mir, which was occupied for just under 10 years. She also flew, on behalf of the Ninety-Nines (the international organization of women pilots), with a watch owned by Amelia Earhart, who was their first president. "To me it represents how far women have come in the field of aviation and how far we can go," she says. She took part in almost 120 scientific experiments and in ongoing work to understand how the body reacts in space. She says:

> "When people live in space they lose bone density and they lose muscle mass. So a lot of the experiments that we are performing now are geared towards actually understanding the mechanisms of how that happens, so we can try and prevent it in the future. We don't want to lose too much bone density, and we certainly don't want to lose too much muscle mass, and one of the things that a lot of people don't realize or think about immediately is the heart is a muscle. So we actually lose mass in our heart, which is something we certainly want to understand."

Some of the experiments have to do with nutrition:

"We're doing very specific diets on orbit to try and understand what would be better because if we can prevent bone loss through something as simple as changing someone's diet that would be a very easy way to go and a very helpful way to go in space."

This research aims to prepare interplanetary trips but have important positive fall-outs here in our daily life. Shannon says:

"One of the most recent accomplishments was a salmonella study that they did on the Station and they were actually able to isolate the gene in the salmonella virus that causes it to be rather virulent. Right now the company that sponsored that research is working with the Food and Drug Administration to do clinical trials for a vaccine so the science is progressing, and it is contributing a lot to Earth."

Other research is concentrating on how to grow plants in space. Shannon explains:

"Some of the things we do are looking at how plants grow in space. There's a lot understanding the detailed mechanisms of how plants are reacting to the stimuli they see in space, be it the nutrients they receive or be it the light needs to be well understood. If we're going to depend on growing our own food in space we need to understand how they're going to grow, so we can best grow food in space."

She concludes:

"Our goal ultimately is living permanently off the planet. Now, whether it's close by on the moon or Mars, or if it's on a spacecraft in the solar system or somewhere else, I won't say, but what the space station does is help us figure out to do that. How do you have a life support system that can work for long periods of time and provide you with the water and the atmosphere that you need? How do you grow plants in space? Obviously we can't keep calling back to Earth for the next pizza delivery, so we've got to figure out how to be self-sufficient off the planet."

In October 2011, Shannon commanded the NEEMO 15 undersea exploration mission aboard the Aquarius underwater laboratory. The mission, originally scheduled to start on October 17, was delayed by stormy weather and high seas, and began on October 20, 2011. The mission then ended early on October 26 due to the approach of Hurricane Rina.

References

Official NASA biography of Shannon Walker *nasa.gov* (December 2010).
Contacts by e-mail with the Author in May 2016.
"Preflight Interview: Shannon Walker (Expedition24)", *nasa.gov* (June 1, 2010).
Williams, M. "Shannon Walker Preps for Six-Month Stint at International Space Station", *naturalsciences.rice.edu.*
Wilson, J. "Shannon Walker, Mission Specialist", *nasa.gov* (May 6, 2004).

57

Liu Yang: The First Female Taikonaut

Credit: Tksteven

U. Cavallaro, *Women Spacefarers*, Springer Praxis Books, DOI 10.1007/978-3-319-34048-7_57

Mission	*Launch*	*Return*
Shenzhou-9	June 16, 2012	June 29, 2012

Born on October 6, 1978, in Zhengzhou, the central Chinese province of Henan, a relatively poor but heavily populated agricultural region, Liu Yang became the most celebrated woman in contemporary China when, in June 2012, she was the first female taikonaut to fly in space, hailed as the "Chinese Valentina Tereshkova" or "Chinese Sally Ride."

An only daughter and member of the Chinese Communist Party, Liu, as per the China space program's requirements, is married: "Married women would be more physically and psychologically mature," Xinhua, the official Chinese Press Agency, states. Wu Bin, the director of the China Astronaut Centre at Jiuquan Satellite Launch Centre, was reported to have said: "We prefer married women because they are more likely to devote themselves to the hard training processes." According to the newspaper *China Daily*, female Chinese astronauts must be married and preferably be mothers because of concerns that higher levels of radiation in space would "harm their fertility." In February 2015, it was announced that Liu had given birth, but no further information was given about her child.

Liu is described as a sociable person, an avid reader, a lover of cooking, and an eloquent speaker, having a penchant for patriotic speeches; she also won first place in a military speech contest. As a child, she said in an interview, she first wanted to become "a lawyer like the ones in television series." Then the first time she sat on a bus with her mother she also thought that becoming a bus driver would be great, as she could ride the bus every day. In 1997, she enlisted in the People's Liberation Army Air Force (PLAAF). She says:

> "It was pure chance that I became a pilot. At that time, our country was recruiting the seventh batch of female pilots. The recruitment happens only every seven or eight years. So I didn't think that it would occur to me that I would one day be a pilot. Unknowingly, my class teacher signed me up for the pilot selection. She said I was qualified, with good test scores and good eyesight."

So, unwittingly, Liu embarked on a path to becoming a pilot: an extraordinary experience that most girls of her age could only dream of:

> "My first flight experience was totally amazing. I watched the ground get further and further away, and people below grow smaller and smaller, finally to the size of little

> ants. And I was increasingly closer to the sky. I didn't feel scared at all. It was a like miracle."

Liu is known for being tough. She was away from home for 4 years and didn't return there until she graduated in 2001, nor did she let her parents visit her during her 4 years at the college: "Baby eagles can never soar under their family's wing," she was reported to have said. When she made her first skydive, she did not immediately call her family to tell them that she was safe. Her parents broke down in tears when Liu finally called them, with her father saying to her: "It's good that you are safe." She graduated from an aviation college in Changchun, Jilin Province, in 2001 and became a cargo plane pilot for the Air Force aviation division:

> "I still remember the parachuting practice for new airborne troop recruits at an airport in Hubei. They were all teenagers—fit but sunburned due to years of hard training. They were seated in the cabin, and began singing when the door was opened. After the song, they raised their fists chanting 'Learn from pilots, Salute to pilots'. I felt so warm and honored!"

During 9 years in the Air Force, Liu became deputy commander of a military flight unit, with the rank of Major, and qualified on five types of aircraft, logging 1680 h of flight. She participated in several military exercises, emergency rescues, and disaster relief work. In 2008, she flew on a disaster relief mission in Guizhou. In 2009, she joined a drought relief task. She was praised for her cool and successful handling of a mid-air emergency when a flock of pigeons collided with the military transport plane she was piloting and disabled one engine; taking the precaution of sending out a mayday message, she managed, however, to retain control of the severely damaged aircraft and was still able to land safely. This kind of calmness under pressure would appeal to astronaut selectors and, in May 2010, she was recruited for the second group of taikonauts and was one of the two women who made the final cut of seven. "Female astronauts generally have better durability, psychological stability and ability to deal with loneliness," Wu Ping, spokeswoman for China's manned space program, said. Liu says:

> "I admired those involved in the launch missions of Shenzhou spacecraft. It never occurred to me that I would one day join their ranks. During the selection, we had photos taken with those astronauts, but I didn't imagine that I could be an astronaut myself."

The training for Liu was a very intense "total immersion" in the program. Once more, she lost sight of family and friends, who did not even know where she was. When she eventually reappeared, she said:

> "It was a disciplinary request. Our selection and training procedures were confidential. There were reports about me when I was a pilot. But I never mentioned my job to anyone while being an astronaut. Apart from my parents, nobody knew my whereabouts."

The training for astronauts in China is split into eight categories, including basic theory, comprehensive academic learning, physical training, psychological training, and training for adaptability to the astronautic environment, astronaut technique training, program assignments training, and large-scale cooperative exercises. It is very systematic and strict, and in many aspects resembles the Soviet space program during the Cold War. Liu said:

> "We had so much to learn. Sometimes it could be a bit dull. From the moment we joined the group, we have lived a routine life. We haven't had a break from the intense work schedule and heavy tasks. During the two years, I have never watched a film, gone shopping, or slept late. We even had classes on Saturdays. And we got up early on Sundays to review the lessons"
> "From day one I was told I am no different from the male astronauts .... Our training was the same as the male astronauts. If they do six-G in the centrifuge, we do too. We can only adapt to the space environment through hard training. The space environment won't change because you are female. They could lower requirements for us, but the space environment won't be that caring."

Even the official confirmation of her name as a member of the prime crew for the upcoming Shenzhou-9 mission arrived—as is the Chinese custom—just a few days before the launch. In contrast to the huge celebrity enjoyed by pioneering female astronauts in the Soviet Union and the US, the Chinese women have been virtually unknown for a long time. For 18 months, the names of the seven new taikonauts were kept as a closely guarded state secret until, at the end of 2011, Tony Quine, an Isle of Man-based British space memorabilia collector and space writer, discovered a "philatelic cover" listed for sale on a German space dealer's website. The stamped postcard was postmarked in China on May 10, 2010—the same day as the country's space program reportedly had appointed the candidates to its second group of taikonauts—and was signed by seven Chinese pilots, including two women. The cover had been accidentally released prematurely. Chinese online forum sources helped to verify the names and unravel the mystery of the names of the new taikonauts (Fig. 57.1).

Liu Yang blasted off on June 16, 2012 (on the 49th anniversary of the flight of Tereshkova) on board Shenzhou-9 ("Divine Vessel") on top of the Long March 2-F rocket, after saying to reporters in the style of the patriotic rhetoric—according to the official Press Agency—a heartfelt "Thank you for the confidence put in me by the Motherland and by the people: I feel honored to fly into space on behalf of hundreds of millions of female Chinese citizens and earn their trust and support." The mission lifted off from Jiuquan, the Chinese spaceport located on the edge of the Gobi Desert in Inner Mongolia. The team also included Commander Jing Haipeng, the first Chinese taikonaut to fly twice (he had already flown in 2008 with the Shenzhou-7 mission, the first Chinese mission with three men on board), and the military pilot Liu Wang, a rooky taikonaut who had waited for his first flight for 14 years, having been selected in the first group of Chinese astronauts in 1998. Much attention was paid to the "mental compatibility" of the crew. Wu Bin, the director of the China Astronaut Centre at Jiuquan Satellite Launch Centre, said:

> "This is our first attempt to send a female astronaut into space; it will pose problems for the team's mental compatibility. They will be asked to work as a team during training and we will observe and evaluate how well they cope with each other."

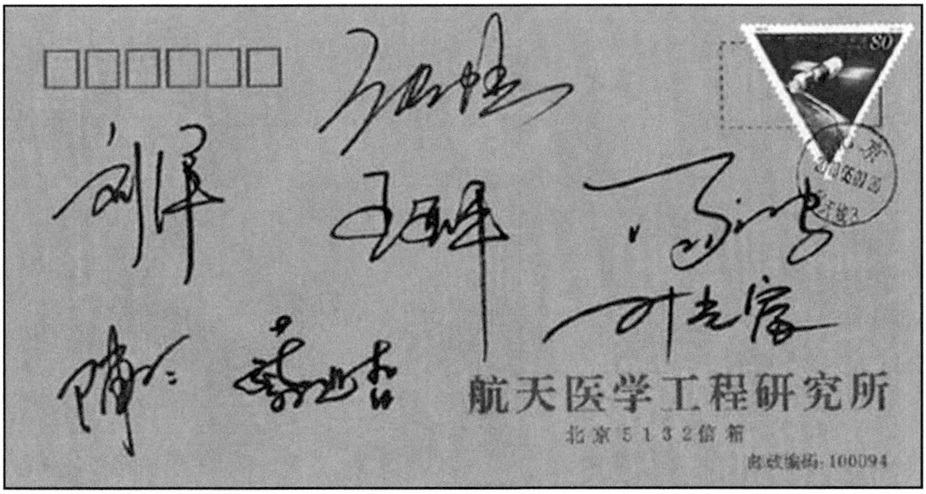

Fig. 57.1 The names of the seven new taikonauts selected in May 2010 were kept secret for 18 months. They were accidentally released in December 2011 when an autographed postcard was listed for sale on a German space dealer's website. The cover was signed by seven Chinese pilots, including two women. Research by the British space writer Tony Quine, with contributions from international experts, helped to solve the mystery and reveal the names. Credit: Spaceflori.com/collectspace.com

Shenzhou-9 was the first manned mission to perform a manual docking procedure—emphatically presented as "another outstanding contribution by the Chinese people to humanity's efforts to explore and use space"—with the orbiting space lab module Tiangong-1 ("Heavenly Palace") that was put into orbit in September 2011: the mission marked the first time China had transferred astronauts between two orbiting craft—a milestone in an effort to acquire the technological and logistical skills to run, in around 2020, a full space station that can house people for long periods. (Automated rendezvous and docking between unmanned Shenzhou-8 spacecraft and Tiangong-1 had already been performed in November 2011.) Jing Haipeng and Liu Yang had rehearsed the maneuver 1500 times in simulation. On the great day, which coincided with the Dragon Boat Festival, docking exercises with Tiangong were broadcast live on state television and met with an outpouring of national pride.

During an interview at the International Astronautical Congress (IAC) held in Naples in October 2012, Liu said that she was astonished by the beauty of Earth as seen from space, and that she was pleasantly surprised by the toy panda that had been left aboard Tiangong-1 by the ground crew. Applause was heard from China's Mission Control room at the Beijing Aerospace Control Center as she boarded the module. Despite the Tiangong Space Lab being only 15 cubic meters inside, Liu had a separate toilet and a soundproofed bedroom to protect her privacy; the compartment, specially designed for females, was equipped with cosmetics specially made for space and a water supply for the bathroom was twice that given to male colleagues.

During the mission, 10 biomedical experiments were carried out: the first Chinese medical experiments in space. It was a week of scientific achievement for China because, that very week, the submersible Jiaolong was exploring inner space, diving 6095 m into the Marianne trench in the Pacific. The space mission lasted for 12 days and 15 h—more than twice as long as the previous longest Chinese mission. "Every minute in space I felt like a fish swimming freely in water," Liu said. "Everything floats and flies because of the weightlessly. Compared with the Earth, it seems that everything in space has got a life!"

China has been barred from the International Space Station (ISS) since 2011, when US Congress passed a law prohibiting official American contact with the Chinese space program due to concerns about national security and banned NASA from engaging in bilateral agreements and coordination with China. China is not an ISS partner and no Chinese nationals have been aboard. While the European Space Agency (ESA) is open to China's inclusion, the US is against it due to concerns over the transfer of technology that could be used for military purposes. Liu says:

> "International cooperation is very necessary. Chinese have the saying, 'When all the people collect the wood, you will make a great fire.' International cooperation can help us to join our efforts together to have a better exploration of the universe and accelerate our exploration steps."

There is a general consensus amongst space analysts that Liu Yang, China's first woman in space, will remain grounded in the future and will no longer be eligible to fly. Although she remains an active astronaut in an official sense, she has gained a place in history and is probably judged too valuable to be sent on any more risky space launches.

References

Coppinger, R. "China's First Woman in Space: Q&A with Astronaut Liu Yang", *space.com* (October 4, 2012).

Dan, Z. "US Media Focuses on Chinese Female Astronaut", *english.cntv.cn* (June 18, 2012).

Duncan, M. "Liu Yang Draws Cheers as First Chinese Woman Set for Space Voyage", *www.reuters.com* (June 15, 2012).

Gibson, K.B. *Women in Space: 23 Stories of First Flights, Scientific Missions and Gravity-Breaking Adventures*, pp. 197–200. Chicago Review Press, Inc., Chicago (2014).

Harvey, B. *China in Space: The Great Leap Forward*, pp. 1–11. Springer Praxis, New York (2013).

Huei, P.S. "Countdown Starts for Chinese Woman Taikonauts", *www.thejakartapost.com* (June 13, 2012).

Martina, M. "China Hails Space Mission's Success as Crew Returns to Earth", *www.reuters.com* (June 29, 2012).

McDonald, M. "No Fanfare for China's Female Astronauts", *nytimes.com* (June 11, 2011).

Moskowitz, C. "Chinese Astronauts Enter Orbiting Space Module", *www.nbcnews.com* (June 18, 2012).

Pearlman, R. "Names of China's Secret Astronauts Revealed by Autographed Envelope", *collectspace.com* (December 7, 2011).
Qiuyuan, L. "Exclusive Interview: Astronauts Selection Process", *english.cntv.cn* (June 16, 2012).
Rui, Z. "Liu Yang: From Pilot to Astronaut", *english.cntv.cn* (June 15, 2012).
Xiaodan, D. "China Launches Spaceship with First Female Astronaut", *english.cntv.cn* (June 16, 2012).

58

Wang Yaping: A Live Lesson from the "Heavenly Palace"

Credit: Chinanews.com

U. Cavallaro, *Women Spacefarers*, Springer Praxis Books, DOI 10.1007/978-3-319-34048-7_58

Mission	*Launch*	*Return*
Shenzhou-10	June 11, 2013	June 26, 2013

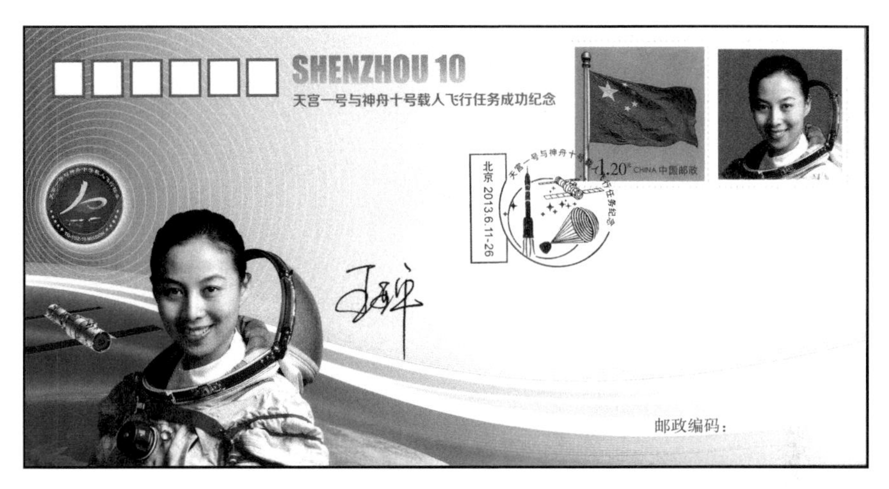

Commemorative cover of mission Shenzou-10 (from the collection of Umberto Cavallaro)

When the first Chinese astronaut made his debut in the nation's maiden manned space mission Shenzhou-5, 10 years before, Wang Yaping was watching it on TV. She says:

> "I was so proud and also very excited. But as I watched it, I thought: We have male pilots and female pilots. And now a male astronaut. When will there be a female astronaut? And today, it's me becoming one of the first few."

Captain Wang Yaping was born probably on January 27, 1980, in the prefecture of Yantai, in east China's Shandong Province, the hometown of China's most famous educationist, Confucius. Her parents are farmers and she is reported to have two sisters. "The experience of doing farm work since an early age has made her strong," reported the Xinhua official news agency broadcast, "and the habit of long-distance running tempered her will." A pilot in the People's Liberation Army Air Force (PLAAF), she is married to another PLAAF pilot, Zhao Peng, and is likely to have a child, as Chinese officials had previously said that only women who have already given birth would be considered for the taikonaut program. According to the official Shenzhou-10 report, her birth date is stated as "January 1980" but this does not tally with the Shenzhou-9 report, in which her birth date is reported as "April 1978."

Wang was recruited in August 1997 as a transport pilot in the PLAAF and is one of 37 members of the so-called "7th Generation" of female pilots, the same as Liu Yang. Fervent and enthusiastic, she has always been ready to take on new challenges as a pilot. She says:

> "I remember the first time I flew a plane on my own. I turned around, and found my trainer was not with me. I was really thrilled and had a good shout in the cockpit. It's like, I could finally do this on my own."

Wang became member of the Communist Party of China in May 2000 and graduated from the Air Force University in 2001. She was one of six female pilots who took part in relief flights that operated uninterruptedly day and night for 1 week following the Wenchuan Earthquake—one of the major earthquakes in Sichuan Province in 2008. Later that year, she was reported to have been involved in the Air Force flights for dispelling clouds for the opening ceremony of the Beijing Olympic Games. She has over 1600 flying hours in her log book.

It has been a long and winding road to the launch pad. After undergoing a strict selection process in April 2009, Wang was selected as one of the only 15 female candidates to become a Chinese astronaut. In the dense atmosphere of secrecy that for a long time shrouded the selection of the seven taikonauts of Group-2 in 2010, Wang was the first to be identified, having been the first female astronaut to be named by the China National Space Administration (CNSA), but was the second Chinese woman in space. She says:

> "It takes a great deal to become an astronaut. You have to be outstanding overall, have great specialty knowledge, go through lots of rigorous training to adapt to the space environment, and take very strict tests that allow almost no errors or mistakes."

After serving as a member of the backup team of the Shenzhou-9 mission in 2012, Wang flew into space on board the Shenzhou-10, the fifth Chinese manned space mission, "carrying the space dream of the Chinese nation," as emphatically stated by President Xi Jinping. Unusually, Wang was the first member of the team to be announced in an official news release in April 2013, 2 months before the lift-off from Jiuquan, on June 11, 2013 (while, as usual, the other crewmembers were confirmed 2 days before the launch)—even though it may be argued that the assignment had already been decided long in advance, so much so that Liu Yang, almost 1 year beforehand, had left her a surprise on the Tiangong-1 laboratory, as Liu told a reporter. While the rest of the crew of Shenzhou-10 was kept mostly in the shadows, also their choice was not really a surprise, since both Nie Haisheng and Zhang Xiaoguang, flanked by Wang Yaping, were already in the backup crew of the previous Shenzhou mission.

Nie was the veteran of the crew. He had been one of three candidates to train for the Shenzhou-5 flight, China's first piloted space mission. Yang Liwei was picked for that flight, with Zhai Zhigang ranked second ahead of Nie. Nie flew eventually into orbit, along with Commander Fei Junlong, as flight engineer of the Shenzhou-6 flight on October 12,

2005. He is only the second spacefarer in history, after Apollo-Soyuz crewman Tom Stafford, to have reached general-officer rank before a space mission. Zhang was a rookie from the original 1998 group of astronauts recruited for the Shenzhou program. He waited 15 years before this assignment. What was uncommon was the way of communicating the crew details, since China traditionally announced the entire crews in one statement, shortly before the launch. This fact left room for many speculations. For sure, releasing Wang Yaping's name early resulted in drumming up interest in the mission.

After referring to "partial technical modifications" to the Shenzhou-10 spacecraft and the Long-March-2F rocket to improve their reliability, few further details were provided by the Chinese authorities on such modifications, as used to happen in the former USSR during the Cold War. Vague references were made to experiments to be performed on Tiangong-1—the 8.5-t Chinese space lab which had already been circling Earth since September 2011—including medical, scientific and technological experiments, "to gain experience for longer space missions."

The rather modest first Chinese space station was just less than half the mass of the world's first space station, the Soviet Union's Salyut-1 launched in 1971. During the early stages of their mission, the crew replaced part of the laboratory's interior cladding, removing the previous flexible, trampoline-like floor "to stabilize astronauts' body in the microgravity environment," stated the official release, without providing further details. Also, some seal rings inside the Tiangong-1 were replaced. No details were given.

After 15 days in space and two docking and re-docking procedures with the prototype space station module, the spacecraft landed on target in northern China's remote Inner Mongolia Autonomous Region on June 26, 2013. It was the longest ever manned mission of the Chinese space program.

Wang was one of the two women in space on the 50th anniversary of Vostok-6, the first historic spaceshot by a woman, Valentina Tereshkova, the other being the American astronaut Karen Nyberg, who in those days was traveling aboard the International Space Station (ISS). During the pre-launch press conference, Wang told reporters that she was a post-1980s woman with hobbies varying from photography, music, and basketball to traveling. Looking forward to her upcoming trip into space, she was happy to be sharing her experiences and thoughts with the public. She added:

> "As a first-time 'spacewoman' aside from scientific experiments and other work-related duties, I will fully enjoy life in the zero-gravity environment and am going in with much curiosity, eager to explore and feel the magic and splendour of space with young friends."

Amid a full slate of science and technology development experiments, during her mission in space, Wang beamed down China's first educational live video lesson on physics in zero-gravity to children in 330 elementary and middle schools in Beijing, thus becoming China's first teacher in space. "We are all students in facing the vast universe. We are looking forward to joining our young friends to learn and explore the mystical and beautiful universe," Wang had said in the pre-flight press conference. The American astronaut Barbara Morgan, who first taught a class from the ISS in 2007, sent her a letter before her lesson:

> "On Behalf of teachers and students around the world, I send you greetings of honor and love as you orbit our Earth and prepare to teach your lessons from space. We are proud of you. We wish you and your crewmates safety and success. You will be very busy up there, but please remember to take time to look out the window. China and all of this world are beautiful (Fig. 58.1)."

More than 60 million students and teachers watched the televised broadcast on giant screens placed around in schools, according to Xinhua. As emphatically highlighted by the rhetoric official press release, "the Heavenly Palace" (that's the name of the Tiangong-1 space lab) "became the China's highest teaching post." During her 41-min lecture, Wang used different experiments to demonstrate the peculiarities of the microgravity environment and the concepts of weight and mass in space, and showed the students how water behaves in zero-gravity, creating a bubble of liquid to demonstrate the properties of surface tension while in microgravity. "Okay everybody, this is where magic happens," Wang said as she held up a bubble of water trapped within a metal ring. "In a weightless environment, we are very skillful marshal artists," she said after Nie floated around the lab in various positions. Then she engaged the students by asking questions. Students discussed how they weigh themselves on Earth before the taikonauts demonstrated how the space flyers weigh objects in microgravity. After showing how normal scales do not work in space, she used a special scale to measure the mass of crew commander Nie Haisheng using Newton's second law of motion and measuring the mass of an object through force and acceleration. Answering the questions asked by the "young pioneers," the youth organization run by China's ruling Communist Party, dressed in a white shirt and red scarf, Wang said:

> "From the window we can see the beautiful Earth and the sun, the moon and the stars, but we haven't seen UFOs. As we are now in outer space without the atmosphere, we can see the stars shining brightly, but they do not twinkle. The sky we see isn't blue, but black. And every day, we can see the sun rise 16 times because we circle the Earth every 90 minutes."

At the end of the lecture came the recommendation of Commander Nie: "I wish all of you will study hard, learn more and contribute to the Chinese dream." Chinese President Xi Jinping—who had earlier been at the Jiuquan Satellite Launch Centre to witness Shenzhou-10's lift-off—connected by video link, congratulated the crew on the progress of their mission and upon the successful docking exercise. "The space dream is part of the dream to make China stronger," he told them. "With the development of space programs, the Chinese people will take bigger strides to explore further into space." "This mission," explained Wang in a press conference upon her return, "allowed me to make two of my dreams come true: exploring space and teaching students."

After Wang's flight, analysts expect that no Chinese woman taikonaut will fly into space for a long time. The structure of the Chinese Astronaut Corps and the space program scheduled for the forthcoming years suggests that the country has no need to fly other women. It's a fact that the Chinese manned missions are quite sporadic and many male taikonauts still have not yet flown.

The Shenzhou-10 mission was the second and last mission to dock with the space lab Tiangong-1. The next mission will be directed to the Tiangong-2, which was originally

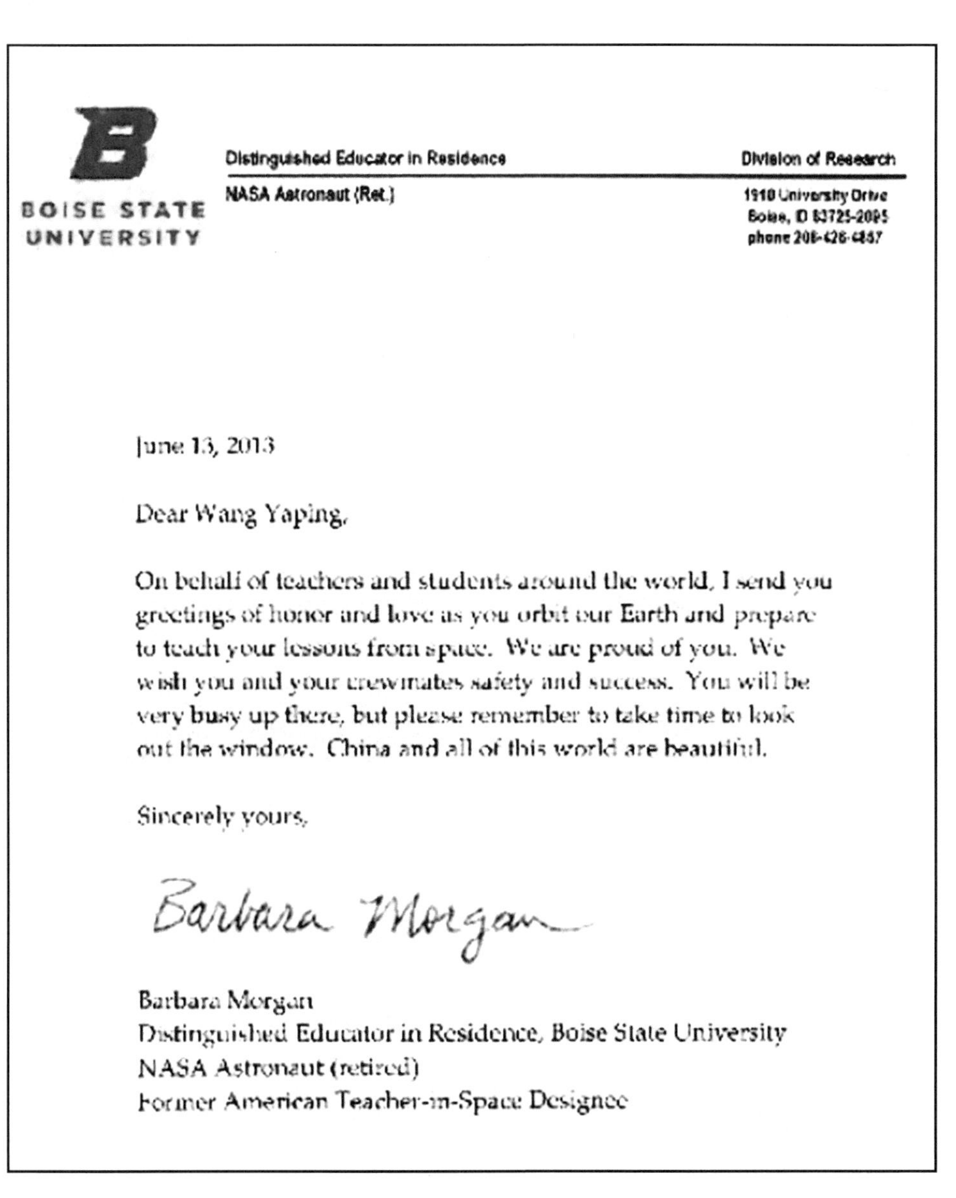

BOISE STATE UNIVERSITY

Distinguished Educator in Residence
NASA Astronaut (Ret.)

Division of Research
1910 University Drive
Boise, ID 83725-2095
phone 208-426-4457

June 13, 2013

Dear Wang Yaping,

On behalf of teachers and students around the world, I send you greetings of honor and love as you orbit our Earth and prepare to teach your lessons from space. We are proud of you. We wish you and your crewmates safety and success. You will be very busy up there, but please remember to take time to look out the window. China and all of this world are beautiful.

Sincerely yours,

Barbara Morgan

Barbara Morgan
Distinguished Educator in Residence, Boise State University
NASA Astronaut (retired)
Former American Teacher-in-Space Designee

Fig. 58.1 The American astronaut, Barbara Morgan (see Chap. 49, pages 313–319), who had been the backup of the first "Teacher in Space" in 1986 and first taught a class from the ISS in 2007, sent this letter to Wang Yaping before she gave her lesson, wishing her safety and success. Credit: Barbara Morgan

scheduled to be launched by 2015. The new Chinese space module will offer expanded capabilities—including a second docking port—and the ability to sustain taikonaut crews for up to 20 days at a time. A subsequent Tiangong-3 may then include axial ports for additional modules, preparing the road for a fully fledged Chinese modular space station in 2020 or shortly thereafter.

References

Evans, B. "China Prepares for Future Exploration as Shenzhou-10 Mission Ends", *www.americaspace.com* (June 26, 2013).
Harwood, W. "Chinese Astronauts Complete Space Mission, Return to Earth", *cbsnews.com* (June 25, 2013).
Jackson. "Wang Yaping 10:00 Space Teaching Students the Opportunity to Conduct World Dialogue", *51jiwo.com* (June 20, 2013).
Jones, M. "China's Next Women Astronauts", *spacedaily.com* (March 26 2013).
Jones, M. "Final Countdown for Shenzhou-10", *spacedaily.com* (June 11, 2013).
Jones, M. "Shenzhou's Shadow Crew", *spacedaily.com* (April 3, 2013).
Kramer, M. "Chinese Astronauts Beam 1st Science Lesson from Space" (video), *Space.com* (June 20, 2013).
Pearlman, R. "Names of China's Secret Astronauts Revealed by Autographed Envelope", *collectspace.com* (December 7, 2011).
Quine, T. "Identity of One of the Chinese Female Taikonaut Candidates Revealed", *www.nasaspaceflight.com* (November 14, 2010).
"Shenzhou-10: Chinese Astronaut Gives Lecture from Space", *www.bbc.com* (June 20, 2013).
"Wang Yaping: China's First Teacher in Space", *chinadaily.com.cn* (June 20, 2013).
"Wang Yaping: China's Second Female Astronaut to Enter Space", *english.cntv.cn* (June 10, 2013).
Yi, Y. "First Space Lecture Opens New Horizons for China", *xinhuanet.com* (June 20, 2013).

59

Elena Serova: The First Russian Woman Cosmonaut on the International Space Station

Credit: NASA

U. Cavallaro, *Women Spacefarers*, Springer Praxis Books, DOI 10.1007/978-3-319-34048-7_59

Mission	*Launch*	*Return*
Soyuz TMA-14M	September 26, 2014	March 11, 2015

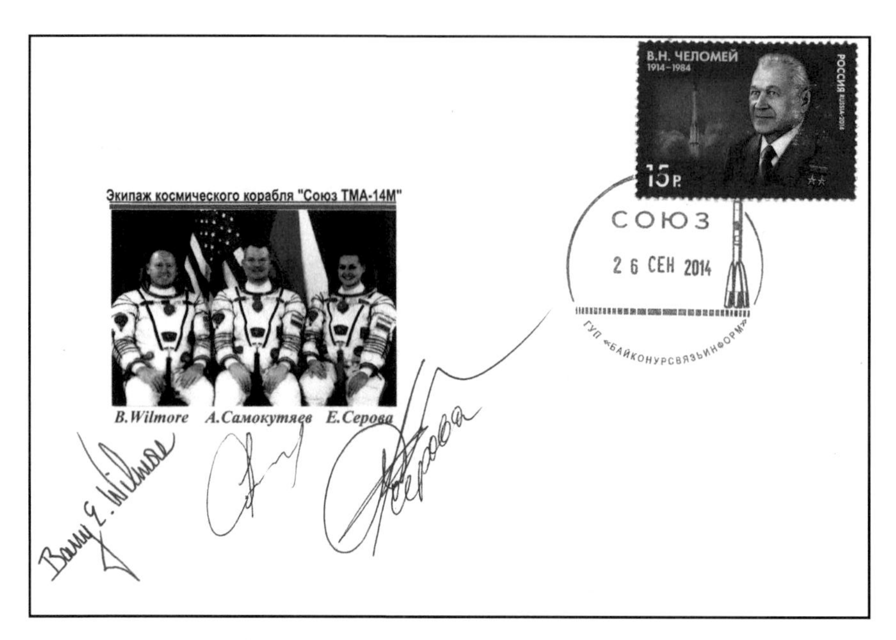

Commemorative cover of mission Soyuz TMA-14M, signed by Elena Serova (from the collection of Umberto Cavallaro)

Elena Serova and her crewmates, NASA astronaut Barry "Butch" Wilmore and Russia's Alexander Samokutyaev, blasted off from the Baikonur Cosmodrome in Kazakhstan on Friday September 26, 2014, at 02:26 a.m. local time (20:25 GMT, on Thursday September 25). About 9 min after lift-off, the soft toy with an attached tiny Russian tricolor flag carried by Elena started floating in weightlessness. And 6 h later, at about 08:15 a.m. (02:15 GMT), the Soyuz safely and punctually docked to the International Space Station (ISS) despite a malfunctioning in one of the solar arrays of the space capsule that failed to deploy.

Elena is currently the only female cosmonaut in service in the Russian Cosmonauts Corps, and—in more than 50 years of history of space exploration—she is only the fourth Russian woman to fly into space, following Valentina Tereshkova (the *first* woman in space), Svetlana Savitskaya (the world's *first* woman spacewalker and the *first* woman ever to fly in space twice), and Elena Kondakova (the *first* woman to live in space for a long-duration mission). Elena is the first Russian woman to visit the ISS in its 13 years of life. But, apart from this, she is not aiming at setting new records:

> "I don't think I'm doing anything extraordinary. I will do things that have already been done by colleagues cosmonauts and astronauts who have preceded me. Space is what I do for work, and that's what I think about it: It's my work."

Perhaps this is a sign that the Russian space program is finally integrating women: another candidate cosmonaut, Anna Kikina, has been included in the program and is following the training process to qualify as a cosmonaut.

In an interview, Elena said that she had long dreamed about proving that Russian women are able to return to spaceflights: "If I was to choose, I had taken 'Phoenix' as my personal call-sign for this mission," she said, recalling in some way the name chosen 45 years before by the Apollo-7 astronauts for their return to flight. She says, however, that the term "woman cosmonaut" is not one she prefers: "There's no such profession. You can be a space test engineer or a space researcher, but the gender makes no difference," she says. And she stresses: "My flight is my job. I feel a huge responsibility towards the people who taught and trained us and I want to tell them: we won't let you down."

Integration is her dream, but there is still a long way to go. Despite her long preparation, her vision, and her commitment, at the news conference before lift-off in Baikonur, a Russian reporter overwhelmed her with irrelevant questions focusing on gender and parenting. At first she humored the reporters, even offering to give a demonstration of washing her hair in space, but her patience seemed to run out when she was repeatedly asked about how she would look after her hair on board the space station. "Can I ask a question, too? Aren't you interested in the hair styles of my colleagues?" she replied at the televised news conference, flanked by the male astronauts who would accompany her in space.

Elena was personally chosen by the then-head of Russia's space agency, Vladimir Popovkin. "We are doing this flight for Russia's image," he was reported to say. "She will manage it, but the next woman won't fly out soon." The editor of Russian magazine *Space News*, Igor Marinin, who referred to this said that the Soviet Union had wanted to win the space race with Mrs. Tereshkova, but then rejected the use of women for many years because they were not seen as physically strong enough. "In space, it's men's work," he concluded. "The leadership then were military, they decided not to take women as cosmonauts any more." And added: "Six months with five men in a confined space is complicated. Elena is a charming, attractive woman." Needless to say, this sentence has caused surprise and controversy in those who did not expect such a comment from a specialist of space news.

Elena Olegovna Serova was born on April 22, 1976, in Vozdvizhenka, a village near the town of Ussurijsk in the region of Primorsky Krai, Eastern Russia, where her father was stationed in the military. When she was young, her father moved the family to Germany for a while. They then returned to Russia and settled in Moscow, where Elena graduated in 2001 as a test engineer in the aerospace department of the prestigious Moscow Aviation Institute and, in 2003, graduated in Economics from the National Academy of Machinery and Informatics of Moscow. She worked as a test engineer at the Rocket Space Corporation RSC Energia, the Russian spacecraft manufacturer, and served in the Mission Control Center in Moscow. She says:

> "When I was a little girl, my father was in the military. He used to take me to the airfield where I could feel the atmosphere of the military, of the air force. I could feel everything that was related to flight and missions. I also was impressed by the starry sky. Once I felt myself drowning in the starry night, and I felt that the stars were so close. It was so near, and I felt like I was flying when I was looking into the starry sky. That impression, that feeling, overwhelmed me; many times, in my dreams, I would fly. Also, when I was in school I liked astronomy a lot, so after I graduated from high school I decided to join the Moscow Aviation Institute. I graduated there, and I also met my future husband, Mark Serov, there. Since we shared the same interests, we decided to go and apply for work for the rocket and science corporation Energia. I never thought that I would join the cosmonaut corps, but even unconsciously all my life I was making steps to actually one day fly to space. So when I had a chance, and when we heard that we could join the new cosmonaut corps, I applied."

Elena was selected in October 2006 and, in December of the same year—with a resolution of the Ministry of Defence—she was included in the 15th Group of the Cosmonaut Corps as a cosmonaut candidate and began basic training in February 2007. At the end of the 2-year training, on June 9, 2009, the Interdepartmental Committee certified her as a test cosmonaut of the RSC Energia. With the recent unification of the three independent cosmonaut organizations previously existing in Russia, she is now part of the Cosmonaut Corps of the Russian Federal Space Agency, known as Roscosmos, or RKK. Elena was one of the five cosmonauts chosen to carry the Russian flag during the opening ceremony of the Winter Olympics in Sochi in 2014.

Elena was assigned to Expedition-41/42 on the ISS and was launched—as mentioned—on September 26, 2014, aboard the Soyuz TMA-14M to begin her long-term mission on the ISS, where, for 170 days, she was busy with experiments and observations. In a prelaunch conference, she explained:

> "We have to perform several medical and biological experiments. Also are planned several Earth observations ranging from the water surface to the forests and several sites of ecological interest. Observations will be important not only for our country but also for the international partners. In preparation for longer space trips that humanity will embark in the future to explore deep space and venture outside the solar system, we also need to better understand the changes that occur in our body when we travel through space. So we continue to do more experiments to understand how change our blood, our skin, the bone mass and density."

Elena shared Expedition-42 with the Italian woman astronaut Samantha Cristoforetti, who arrived on the ISS on November 23, 2014 (Fig. 59.1).

On March 11, 2015, after 167 days in space, Elena successfully returned to Earth to a remote area near the town of Zhezkazgan, Kazakhstan. She is married to Mark Serov, who, in May 2003, entered the Cosmonaut Corps as a candidate for the 13th Group (RKKE-13) but retired for health reasons without flying into space and now works as a manager for the RSC Energia. They have a daughter, Elena Markovna, born in 2004.

Fig. 59.1 Elena Serova and Samantha Cristoforetti (2015). Credit: NASA

References

Agence France-Presse in Moscow "Female Cosmonaut Bats Back Questions about Hair and Parenting", *theguardian.com* (September 25, 2014).
Burks, R. "Watch a Female Cosmonaut Respond to Sexist Questions about Her Hair during a Press Conference", *techtimes.com* (September 26, 2014).
Garcia, M. "Preflight Interview: Elena Serova", *nasa.gov* (August 1, 2014).
Official biography of Elena Serova, *energia.ru* (March 2014).
Official NASA biography of Elena Serova, *nasa.gov* (January 2013).
Solovyov, D. "First Russian Woman Lifts Off to International Space Station", *reuters.com* (September 26, 2014).
Stallard, K. "Female Cosmonaut Angry over Hair Questions", *news.sky.com* (September 25, 2014).

60

Samantha Cristoforetti: A New Record for Endurance in Space for a Woman

Credit: ESA

U. Cavallaro, *Women Spacefarers*, Springer Praxis Books, DOI 10.1007/978-3-319-34048-7_60

Mission	*Launch*	*Return*
Soyuz TMA-15M	November 23, 2014	June 11, 2015

Cover sent by Commander Anton Shkaplerov from on board the ISS and signed by the crew of Expedition 42, including Samantha Cristoforetti (from the collection of Umberto Cavallaro)

Samantha Cristoforetti is one of six European Space Agency (ESA) astronauts, selected in May 2009—from 8413 candidates—in the "Shenanigans" class, together with the Italian astronaut Luca Parmitano, and is currently the only woman in the ESA Astronaut Corps. Being the first Italian woman astronaut built upon her a symbolic, almost mythical, image that Samantha would rather avoid. Everyone knows by now that she does not like to be considered a representative of the gender, her success being built on merit, as well as on a good dose of luck, as Samantha herself concedes: "The opportunity to become an astronaut are few and infrequent, and you must be there at the right time." She does not like this topic, but confronts it herself with determination, and also with a bit of clever irony. She says:

> "At Star City, Russia astronauts usually enter the hall of the Soyuz simulator with blue suit and sneakers. And when it was my turn to go to pick up the sneakers they had assigned to me, I opened the box and I found a pair of pink shoes. Pink is actually a nice colour like so many others, and the gesture was done with the best of intentions. However I found it as an unusual gesture, because very infrequently, in

> my professional life and in my training, it happens me to highlight a difference or a peculiarity with respect of my colleagues. But I realize that, to the outsider, the fact that I am a woman can raise questions and curiosity."

And to answer this kind of curiosity, she created the "Avamposto42" blog, hosted by *www.esa.int*. On the first page, she immediately tackles the core of the issue:

> "Who looks at our group picture, usually immediately notices one thing: I am the only woman. To be honest, as I am inside the picture and not outside, I do not pay much attention. First of all, let me clearly tell that there's really nothing outstanding in the deal! Many astronauts have been in space, and there is no activity in this profession that was not already done before by somebody else. Nobody expects me to prove anything. I am an engineer, a military pilot, an astronaut: there is nothing in my training that makes me an expert on gender issues. Nor I have an eye trained to grasp these issues, or the mind trained to think about it. In short, it's unlikely that I have something smart or original to add on this subject!"

Samantha understands, however, that, within certain limits, some curiosity about those who are out of the picture is also legitimate. She admits:

> "Basically me too when I was a young girl and I looked at the pictures and read the stories of those who flew into space, I was especially interested in the female astronauts. Maybe because they were a minority, or they were more like me. If I had met any of them, I believe that I had even some specific questions, that I would not ask their male colleagues."

Born in Milan, Italy, on 26 April 1977, Samantha Cristoforetti spent her childhood in Malè, in Val di Sole, Trentino Region, Italy. She recalls that she wanted to become an astronaut since she was a young girl. After completing her high-school education at the Liceo Scientifico in Trento, she spent 1 year in the US on a student-exchange program. She then worked on an experimental project in aerodynamics at the École Nationale Supérieure de l'Aéronautique et de l'Espace in Toulouse, France (Erasmus program, 2000) and performed research for her doctorate on solid rocket propellants at the Mendeleev

Russian University of Chemistry and Technology in Moscow, Russia (2000–2001). In 2001, she graduated from the Technical University of Munich, Germany, with a degree in Mechanical Engineering. Besides Italian (her mother tongue), she speaks English, French, German, and Russian, and she also is studying Chinese—as a "hobby," she clarifies, Captain Cristoforetti has logged more than to avoid misunderstandings.

Samantha was one of the first Italian women to become a lieutenant and fighter pilot in the Italian Air Force. Her career began in 2001 at the Italian Accademia Aeronautica in Pozzuoli, Naples, Italy, where she attended the 4-year "Boreas V" course and, in 2005, graduated with honors in Mechanical Engineering and Aeronautical Science at the University Federico II of Naples and was awarded the Honour Sword for best academic achievement, being the first woman to receive such an award. In 2005–2006, she completed the ENJJPT (Euro-NATO Joint Jet Pilot Training) at Sheppard Air Force Base in Wichita Falls, Texas, and became a fighter pilot.

In her short but intense military career, in 2007, Samantha was assigned to the 212th Squadron, 61st Flight Training Wing of Galashiels, where she completed the training "Introduction to Fighter Fundamentals"; from 2007 to 2008, she piloted the MB-399 and served in the Section of Planning and Operations for the 51st Bomber Wing based in Istrana, Treviso, Italy. In 2008, she joined the 101st Squadron, 32nd Bomber Squadron based at Amendola, Foggia, Italy, where she attended the operational conversion training for the AM-X ground attack fighter.

Captain Cristoforetti has logged more than 500 h of flying in six types of military aircraft: SF-260, USAF T-37, T-38, Aermacchi MB-339, MB-339CD, and AM-X. She was selected as an ESA astronaut in May 2009. She joined ESA in September 2009 and completed the basic astronaut training in November 2010. In July 2012, she was assigned to an Italian Space Agency (ASI) mission aboard the International Space Station (ISS) and was launched on a Soyuz spacecraft from Baikonur Cosmodrome in Kazakhstan on November 23, 2014. This is the second long-duration ASI mission and the eighth long-duration mission for an ESA astronaut.

In 2011, Samantha was awarded Reserve Astronaut status and started a challenging training that involved travels to the US, Canada, Japan, and the European Astronaut Centre (EAC) in Cologne, Germany, to familiarize with the ISS systems developed by different partner countries: a long and arduous undertaking that, in her interview with *AD*ASTRA*, she likened to a "marathon." She states:

> "Arriving to the launch requires years-long preparation. It's a sort of endurance race and you have to dose and measure out your energies. It's not a sprint race, nor is it a high hurdle. You have no peaks of difficulties to overcome. Really it's a matter of keeping up the pace."

In July 2012, Samantha was assigned to "Futura," the ASI mission on board the ISS. "Training," as she recalled in the above interview, "is a duty which requires that you live out of a suitcase." Sixteen different countries contribute to the ISS. The astronaut must know all the complex technologies and on-board systems operating in the station and those of the cargo spacecraft that will deliver supplies. Astronauts therefore move alternatively from the US to Russia, the main contributors to the ISS, but also they train in the Japanese and Canadian Astronaut Training Centre to get acquainted with the different technologies they have to live with during their long-term mission and, in the EAC of Cologne, Germany, to practice with the European Laboratory Columbus.

Samantha was launched aboard the Soyuz TMA-15M with astronauts Terry Virts (US) and Anton Shkaplerov (Russia) on November 23, 2014, at 21:01 CET (3:01 on November 24 local time in Baikonur, Kazakhstan). Just 5 h and 48 min after lift-off, the crew arrived at the ISS, where they were welcomed aboard by Commander Barry Wilmore (NASA) and cosmonauts Elena Serova (the fourth Russian woman in space) and Alexander Samokutyaev (Roscosmos). This was the third long-duration mission of an Italian astronaut on the ISS. It was acquired by Italy within the frame of the flight opportunities provided according to the bilateral ASI/NASA agreement as a counterpart for the development of the pressurized MPLM (Multi-Purpose Logistics Module) that Italy delivered to the US, along with the Leonardo living module, now permanently attached to the ISS, known as PMM (Permanent Multipurpose Module), built by Thales Alenia Space in Turin, Italy. During launch and return, she played the role of flight engineer and co-pilot on the Soyuz spacecraft. In the aforementioned interview, she said:

> "Most of the training I do in Russia is spent on the Soyuz. We will use it for the launch and the return. Few hours altogether, but they will be the hours potentially more at risk. And I must gain a thorough knowledge of the spacecraft, as I'll have to help the commander to manage the critical phases."

Samantha's mission was named "Futura" to highlight the science and technology research she ran in weightlessness to help shape our future. During her 6 months on the ISS, she was involved in Earth observation experiments and performed various tests in physics, human physiology in weightlessness, and telemedicine, with important impacts on Earth. Experiments sponsored by the ASI also included technology demonstrators such as three-dimensional printers, "which could be revolutionary in the future and allow to print on board the necessary spare parts." She says:

> "European astronauts have often chosen a 'leitmotif', a 'key-subject', to focus the communication on their mission. Looking at my personal story I have chosen to focus on Wellbeing and Nutrition, which is something that concerns everybody. Food is essential to stay healthy over time, to keep us in a state of serenity and well-being in our daily lives and to allow us to perform at our best. What I discovered is that just a few basic knowledge on the interaction between food and our health allows us to make informed choices that make a huge difference in our ability to stay in shape. The mission 'Futura' wants to explore the topic of interaction between food and body that, if wrong, can generate imbalances that in a long term may generate problems, such as cardiovascular disease or diabetes. Those remain for years as subclinical aspects, because they act in more dilated times. In space we can investigate such issues from a privileged point of view, since astronauts offer an accelerated aging model."

In order to create a link between space, as one of the most advanced sectors, and the concrete issues of a healthy lifestyle on Earth, Samantha set up the "Avamposto-42" (*Outpost-42*) website as a meeting point allowing fans and the curious to closely follow the "Futura" mission and its name refers to the number of the long-duration Expedition-42 that Samantha participated in. "The International Space Station," outlined Samantha, referring to the name of her blog, "is a laboratory where we do science, but I also consider it as an outpost of humanity in space." But the name of the site also refers to *The Hitchhiker's Guide to the Galaxy*, the comedy science-fiction series created by Douglas Adams. She explains:

> "I've always loved science fiction and I found it funny that the number of my mission, '42', coincides with the number which in the book is, ironically, the answer—provided by the supercomputer Deep Thought after a computation which took 7½ years—to 'the Ultimate Question of Life, the Universe and Everything'."

When I sent Samantha my first e-mail, a few days after the lift-off, I got back an automatic out-of-office reply: "Off the planet, returning May 2015!"

During the first in-flight call of the "Futura" mission, connected with the ASI, she said: "Day after day I'm becoming a true Space being."

Among the many visits received during Samantha's long expedition on the ISS, one of the most desired arrivals was that of the SpaceX Dragon CRS-6, which she captured by extending the robotic arm on April 17, 2015. For her, and for the team on the station, this was a historical day on which the long-awaited "ISSpresso machine" finally arrived on board—the space expresso maker built by the engineering firm Argotec (Turin, Italy), who teamed up with Turin-based Lavazza and the ASI (Italian Space Agency) and they worked together for more than a year to bring into space authentic Italian expresso coffee. Through Twitter, Samantha shared a picture of herself drinking the coffee from a special cup designed for use in zero-gravity, whilst wearing her favorite *Star Trek* outfit (Figs. 60.1 and 60.2).

Samantha's return to Earth was delayed by the loss of the Russian unmanned Progress-59 resupply ship, which failed to arrive at the ISS and burned up in the atmosphere in April. The investigation into the failure pushed back the flight schedule for the return crew. When the delay was announced, recalled Anton Shkaplerov in a post-flight tour conference held in Milan, Italy, few months later, a dazzling smile appeared on Samantha's face. As a result of the delay, on June 6, she surpassed the record of 194+ days of endurance for a female astronaut in space for a single mission, set by NASA Sunita Williams in 2007. Samantha, with Terry Virts and mission commander Anton Shkaplerov, returned on June 11, 2015, after spending 199 days, 16 h, and 42 min in space. The Soyuz landed on schedule about 90 miles south-east of Dzhezkazgan, Kazakhstan. The landing was "hard and quick," said Shkaplerov, the commander of the Soyuz spacecraft. "For some reason, we were spinning," he added in comments broadcast on Russian television. No other details were provided. "I'm fine but ... but heavy," confirmed Samantha in her first @ESA TV interview.

Fig. 60.1 The ISSpresso machine installed on the International Space Station (ISS). Italian companies and the Italian Space Agency (ASI) teamed up and worked together for more than a year to bring into space authentic Italian expresso coffee. Credit: Argotec/Lavazza/ASI

Fig. 60.2 Samantha in the "Cupola" of the International Space Station (ISS), drinking coffee from a special cup designed for use in zero-gravity whilst wearing her favorite *Star Trek* outfit. Credit: NASA/ESA

References

Cavallaro, U. "The Marathon that leads to space", *AD*ASTRA, ASITAF Quarterly Journal*, **18**(September 2013), pp. 7–8.

Gibson, K.B. *Women in Space: 23 Stories of First Flights, Scientific Missions and Gravity-Breaking Adventures*, pp. 201–203. Chicago Review Press, Inc., Chicago (2014).

Interviews by the Author in September 2013 and October 2014.

Official biography of Samantha Cristoforetti, *aeronautica.difesa.it*, *asi.it*, and *esa.int*.

Outpost42.esa.int/blog/, (June 2014).

Pilello, A. "ISSpresso: An espresso machine for space", *AD*ASTRA, ASITAF Quarterly Journal*, 22 (September 2014), P. 6

Appendix 1: List of space missions participated in by women spacefarers

		Mission	Launch
1	Valentina Tereskova	Vostok 6	June 16, 1963
2	Svetlana Savitskaya	Soyuz T-7	August 19, 1982
		Soyuz T-12	July 17, 1984
3	Sally Ride	STS-7	June 18, 1983
		STS-41-G	October 5, 1984
4	Judith Resnik	STS-41-D	August 30, 1984
		STS-51-L	January 28, 1986
5	Christa McAuliffe	STS-51-L	January 28, 1986
6	Kathryn D. Sullivan	STS-41-G	October 5, 1984
		STS-31	April 24, 1990
		STS-45	March 24, 1992
7	Anna Lee Fisher	STS-51-A	November 8, 1984
8	Margaret Rhea Seddon	STS-51-D	April 12, 1985
		STS-40	June 5, 1991
		STS-58	October 18, 1993
9	Shannon Lucid	STS-51-G	June 17, 1985
		STS-34	October 18, 1989
		STS-43	August 2, 1991
		STS-58	October 18, 1993
		STS-76	March 22, 1996
10	Bonnie J. Dunbar	STS-61-A	October 30, 1985
		STS-32	January 9, 1990
		STS-50	June 25, 1992
		STS-71	June 27, 1995
		STS-89	January 22, 1998
11	Mary L. Cleave	STS-61-B	November 26, 1985
		STS-30	May 4, 1989
12	Ellen S. Baker	STS-34	October 18, 1989
		STS-50	June 25, 1992
		STS-71	June 27, 1995

U. Cavallaro, *Women Spacefarers*, Springer Praxis Books, DOI 10.1007/978-3-319-34048-7

		Mission	Launch
13	Kathryn C. Thornton	STS-33	November 22, 1989
		STS-49	May 7, 1992
		STS-61	December 2, 1993
		STS-73	October 20, 1995
14	Marsha Ivins	STS-32	January 9, 1990
		STS-46	July 31, 1992
		STS-62	March 4, 1994
		STS-81	January 12, 1997
		STS-98	February 7, 2001
15	Linda M. Godwin	STS-37	April 5, 1991
		STS-59	April 9, 1994
		STS-76	March 22, 1996
		STS-108	December 5, 2001
16	Helen Sharman	Soyuz TM-12	May 18, 1991
17	Tamara E. Jernigan	STS-40	June 5, 1991
		STS-52	October 22, 1992
		STS-67	March 2, 1995
		STS-80	November 19, 1996
		STS-96	May 27, 1999
18	Millie Hughes-Fulford	STS-40	June 5, 1991
19	Roberta Bondar	STS-42	January 22, 1992
20	Jan Davis	STS-47	September 12, 1992
		STS-60	February 3, 1994
		STS-85	August 7, 1997
21	Mae Jemison	STS-47	September 12, 1992
22	Susan J. Helms	STS-54	January 13, 1993
		STS-64	September 9, 1994
		STS-78	June 20, 1996
		STS-101	May 19, 2000
		STS-102	March 8, 2001
23	Ellen Ochoa	STS-56	April 8, 1993
		STS-66	November 3, 1994
		STS-96	May 27, 1999
		STS-110	April 8, 2002
24	Janice E. Voss	STS-57	June 21, 1993
		STS-63	February 3, 1995
		STS-83	April 4, 1997
		STS-94	July 1, 1997
		STS-99	February 11, 2000
25	Nancy Currie	STS-57	June 21, 1993
		STS-70	July 13, 1995
		STS-88	December 4, 1998
		STS-109	March 1, 2002
26	Chiaki Mukai	STS-65	July 8, 1994
		STS-95	October 29, 1998
27	Yelena Kondakova	Soyuz TM-20	October 3, 1994
		STS-84	May 15, 1997

		Mission	Launch
28	Eileen Collins	STS-63	February 3, 1995
		STS-84	May 15, 1997
		STS-93	July 23, 1999
		STS-114	July 26, 2005
29	Wendy Lawrence	STS-67	March 2, 1995
		STS-86	September 25, 1997
		STS-91	June 2, 1998
		STS-114	July 26, 2005
30	Mary E. Weber	STS-70	July 13, 1995
		STS-101	May 19, 2000
31	Catherine Coleman	STS-73	October 20, 1995
		STS-93	July 23, 1999
		Soyuz TMA-20	December 15, 2010
32	Claudie Deshais Haigneré	Soyuz TM-24	August 17, 1996
		Soyuz TM-33	October 21, 2001
33	Susan Still Kilrain	STS-83	April 4, 1997
		STS-94	July 1, 1997
34	Kalpana Chawla	STS-87	November 19, 1997
		STS-107	January 16, 2003
35	Kathryn P. Hire	STS-90	April 17, 1998
		STS-130	February 8, 2010
36	Janet L. Kavandi	STS-91	June 2, 1998
		STS-99	February 11, 2000
		STS-104	July 12, 2001
37	Julie Payette	STS-96	May 27, 1999
		STS-127	July 15, 2009
38	Pamela Melroy	STS-92	October 11, 2000
		STS-112	October 7, 2002
		STS-120	October 23, 2007
39	Peggy Whitson	STS-111	June 5, 2002
		Soyuz TMA-11	October 10, 2007
40	Sandra Magnus	STS-112	October 7, 2002
		STS-126	November 14, 2008
		STS-135	July 8, 2011
41	Laurel B. Clark	STS-107	January 16, 2003
42	Stephanie Wilson	STS-121	July 4, 2006
		STS-120	October 23, 2007
		STS-131	April 5, 2010
43	Lisa Nowak	STS-121	July 4, 2006
44	Heidemarie M. Stefanyshyn-Piper	STS-115	September 9, 2006
		STS-126	November 14, 2008
45	Anousheh Ansari	Soyuz TMA-9	September 18, 2006
46	Sunita Williams	STS-116	December 9, 2006
		Soyuz TMA-05M	July 15, 2012
47	Joan Higginbotham	STS-116	December 9, 2006
48	Tracy Caldwell Dyson	STS-118	August 8, 2007
		Soyuz TMA-18	April 2, 2010

		Mission	Launch
49	Barbara Morgan	STS-118	August 8, 2007
50	Yi So-yeon	Soyuz TMA-12	April 8, 2008
51	Karen L. Nyberg	STS-124	May 31, 2008
		Soyuz TMA-19	April 2, 2010
52	K. Megan McArthur	STS-125	May 11, 2009
53	Nicole P. Stott	STS-128	August 28, 2009
		STS-133	February 24, 2011
54	Dorothy Metcalf-Lindenburger	STS-131	April 5, 2010
55	Naoko Yamazaki	STS-131	April 5, 2010
56	Shannon Walker	Soyuz TMA-19	June 15, 2010
57	Liu Yang	Shenzhou 9	June 16, 2012
58	Wang Yaping	Shenzhou 10	June 11, 2013
59	Yelena Serova	Soyuz TMA-14M	September 25, 2014
60	Samantha Cristoforetti	Soyuz TMA-15M	November 23, 2014

List updated July 1st, 2016

Appendix 2: Extravehicular activities

	Number of EVAs	Total time spent in outer space (hh:mm)
Svetlana Savitskaya	1	03:34
Kathryn D. Sullivan	1	03:27
Kathryn C. Thornton	3	21:11
Linda M. Godwin	2	10:14
Tamara E. Jernigan	2	08:41
Susan J. Helms	1	08:56
Peggy Whitson	6	39:44
Heidemarie M. Stefanyshyn-Piper	5	33:42
Sunita Williams	7	50:40
Nicole P. Stott	1	06:35
Tracy Caldwell Dyson	2	22:49

List updated July 2016

U. Cavallaro, *Women Spacefarers*, Springer Praxis Books, DOI 10.1007/978-3-319-34048-7

Appendix 3: Military astronauts

	Corps	Year of selection
Svetlana Savitskaya	Russian Air Force	1980
Kathryn C. Thornton	US Army	1984
Millie Hughes-Fulford	US Army	1983
Susan J. Helms	US Air Force	1990
Nancy Currie	US Army	1990
Eileen Collins	US Air Force	1990
Wendy Lawrence	US Navy	1992
Catherine Coleman	US Air Force	1992
Susan Still Kilrain	US Navy	1995
Kathryn P. Hire	US Navy	1995
Pamela Melroy	US Air Force	1995
Laurel B. Clark	US Navy	1996
Lisa Nowak	US Navy	1996
Heidemarie M. Stefanyshyn-Piper	US Navy	1996
Sunita Williams	US Navy	1998
Liu Yang	People's Liberation Army Air Force	2010
Wang Yaping	People's Liberation Army Air Force	2010
Samantha Cristoforetti	Italian Air Force	2009

List updated July 1st, 2016

U. Cavallaro, *Women Spacefarers*, Springer Praxis Books, DOI 10.1007/978-3-319-34048-7

Appendix 4: Women spacefarers married to astronauts (country and year of selection in brackets)

- Valentina Tereshkova (USSR—1962) and Andriyan Nikolayev (USSR—1960): married in 1963 during a solemn ceremony headed by Nikita Khrushchev; divorced in 1982.
- Sally K. Ride (USA—1978) and Steven Hawley (USA—1978): married in 1982; divorced in 1987.
- Anna L. Fisher (USA—1978) and William F. Fisher (USA—1980): already married before joining NASA; now divorced.
- M. Rhea Seddon (USA—1978) and Robert L. Gibson (USA—1978): married in 1981; the first to wed among NASA astronauts.
- Bonnie J. Dunbar (USA—1980) and Ronald M. Sega (USA—1990): married in 1988; now divorced. The news that her husband Ron Sega had been selected for the Astronaut Corps was released while Bonnie was in space (STS-32).
- Linda M. Godwin (USA—1985) and Steven R. Nagel (USA—1978): married in 1995, after Steven had left the Astronaut Office. He passed away aged 67 in August 2014.
- Tamara E. Jernigan (USA—1985) and Peter Jeff K. Wisoff (USA—1990): married in 1999.
- Claudie Deshays Haigneré (France—1985) and Jean-Pierre Haigneré (France—1985): married in 1999.
- Nancy Jan Davis (USA—1987) and Mark C. Leand (USA—1984): married in 1992; they flew together that year in mission *Endeavour* STS-47; now divorced.
- Heike Walpot (Germany—1987—unflown) and Hans Schlegel (Germany—1998).
- Elena Kondakova (Russia—1989) and Valeri Ryumin (USSR—1973): married in 1985.
- Nancy Sherlock-Currie (USA—1990) and David W. Currie (USA—1990): married in 1995. David passed away aged 55 in 2011.
- Marianne Merchez (Belgium—1992—unflown) and Maurizio Cheli (Italy—1992).
- Karen L. Nyberg (USA—2000) and Douglas G. Hurley (USA—2000): married in 2009.
- K. Megan McArthur (USA—2000) and Robert L. Behnken (USA—2000): married in 2008.
- Shannon Walker (USA—2004) and Andrew S.W. Thomas (USA—1992): married in 2005.
- Elena Serova (Russia—2006) and Mark Serov (Russia—2003): married in 1985. Mark retired from Cosmonaut Corps without flying in space.

U. Cavallaro, *Women Spacefarers*, Springer Praxis Books, DOI 10.1007/978-3-319-34048-7

Glossary

Airlock A chamber that allows people and objects to pass between a pressure vessel and its surroundings while maintaining the pressure difference and minimizing loss of air from the vessel. It is used, for example, in space to permit astronauts to exit and enter for extravehicular activities, or underwater (e.g. NEEMO expeditions) to allow passage between an air environment in a pressure vessel and the water environment outside.

Astronaut A person who travels into space.

CapCom Or *Capsule Communicator* in the Mission Control Center (MCC) is the individual responsible for communicating directly with the crew of a manned spaceflight during a mission. During much of the US manned space program, NASA felt it important for all communication with the astronauts in space to pass through a single individual in the MCC. NASA believes that an astronaut is most able to understand the situation in the spacecraft and pass information in the clearest way. Until 2011 (the end of the Shuttle program), the role was filled by another astronaut, often one of the backup or support crewmembers. After 2011, for long-duration expeditions on the ISS, also instructors or trainers may act as CapComs.

Cape Crusaders The team of American Astronaut Support Personnel (ASP) (five to eight) who—during the Shuttle program—served as the crew's point of contact between NASA Johnson Space Center (JSC) in Houston, Texas, and NASA Kennedy Space Center (KSC), in Cape Kennedy, Florida. They were "the eyes and ears to the Shuttle vehicle." Their duties included setting the Shuttle orbiter's cockpit switches to the appropriate settings to make sure that they were properly configured for lift-off, loading the vehicle with the crew equipment, participating in communication checks between the orbiter and the ground, strapping in the flight crew for launch, assisting with landing operations, helping the crew exit after landing, and carrying some of their equipment out of the Orbiter. According to astronaut Jerry Ross, "the only assignment that beats being a Cape Crusader is to do the flight yourself."

Similarly, at NASA, the term **Russian Crusader** was used to refer to the team of American astronauts in charge of supporting activities to be performed in Russia for the joint Russian–American operations like the Shuttle–Mir program or participating in the testing

U. Cavallaro, *Women Spacefarers*, Springer Praxis Books, DOI 10.1007/978-3-319-34048-7

and integration of Russian hardware and software products developed for the International Space Station (ISS) to ensure mutual compatibility of the US and Russian equipment.

Cosmonaut A Soviet or Russian spacefarer.

Docking and Rendezvous A *space rendezvous* is an orbital maneuver during which two separate free-flying spacecraft, one of which is often a space station, arrive at the same orbit and approach at a very close distance (e.g. within visual contact). Rendezvous requires a precise match of the orbital speed and position vectors of the two spacecraft. The ISS orbits Earth at 27,000 kilometers per hour.

Rendezvous may or may not be followed by *docking or berthing*—procedures that bring the spacecraft into physical contact and create a link between the two.

IUS Inertial Upper Stage: a large Boeing-built booster rocket used to lift satellites into their final orbit or to accelerate space probes out of Earth orbit.

Mission Specialists Career astronauts who train for a specific mission.

NEEMO (*NASA Extreme Environment Mission Operations*) is the program run by NASA in the Aquarius underwater laboratory—the world's only undersea research station, located 3.5 miles (5.6 kilometers) off Key Largo, Florida, in the Florida Keys National Marine Sanctuary, 62 feet (19 meters) below the surface.

NASA sends groups of astronauts, engineers, and scientists (called aquanauts) to live in Aquarius for space exploration simulation missions in preparation for future space exploration, since the underwater lab is a convincing analog for space exploration: much like space, the undersea world is a hostile, alien place for humans to live.

Payload In our context, *payload* is the carrying capacity of a spacecraft or launch vehicle and can be cargo or equipment sent or carried into space to reach the mission goals. It can be a scientific satellite launched atop a rocket that will put it in Earth orbit or a space probe carried to the target inside or outside the Solar System, or a set of equipment carried during a manned flight to perform an experiment in microgravity or structures to be mounted or left in outer space for different purposes.

Payload Commander The astronaut responsible for the overall success of the experiments during a mission.

Payload Specialist Scientists who know the topic and are experts on a particular mission's experiments or "payload." Although they undergo rigorous training, payload specialists aren't career astronauts who expect to make space travel their life's work. Instead, their goal is to carry out a group of experiments for other scientists who remain on Earth.

RMS (*Remote Manipulator System*): robotic arm. The first robotic arm was built for the Shuttle (SRMS or Shuttle Remote Manipulator System, also known as Canadarm). A second robotic arm was installed on the ISS (SSRMS or Space Station Remote Manipulator System, also known as Canadarm2). The Japanese Experiment Module Remote Manipulator System (JEM-RMS) is primarily used on the ISS to service the JEM Exposed Facility. An additional robotic arm, the European Robotic Arm (ERA), is scheduled to launch alongside the Russian-built Multipurpose Laboratory Module during 2017.

SAIL (*Shuttle Avionics Integration Laboratory*) was a facility at Lyndon B. Johnson Space Center in Houston, Texas, that supported the entire Space Shuttle program and performed integrated verification tests and the testing and integration of orbiter hardware and flight software in a simulated flight environment.

Taikonaut A Chinese spacefarer.

Bibliography

Books

Bondar, R. *Shaping our School, Shaping Our Future: Environmental Education in Ontario Schools*, Report of the Working Group of the Environmental Education, Giugno (2007), 22 pp.

Briggs, C.S. *Women Space Pioneers*. Lerner Publications, Minneapolis (2005), 112 pp.

Cavallaro, U. *Propaganda e Pragmatismo, in gara per la conquista della Luna* [*Propaganda and Pragmatism, in the Race to the Moon*]. Impremix, Turin, Italy (2011), 166 pp.

Chien, P. *Columbia: Final Voyage—The Last Flight of NASA's First Space Shuttle*, Springer/Praxis, New York (2006), 439 pp.

Cunningham, W. *The All-American Boys*. iBooks, Inc., New York (2004), 488 pp.

Dickson, P. *A Dictionary of the Space Age*. Johns Hopkins University Press (2009).

Evans, R. *Tragedy and Triumph in Orbit: The Eighties and Early Nineties*. Springer-Praxis, New York (2012), 614 pp.

Gibson, K.B. *Women in Space: 23 Stories of First Flights, Scientific Missions and Gravity-Breaking Adventures*. Chicago Review Press, Inc., Chicago (2014), 234 pp.

Gueldenpfenning, S. *Women in Space Who Changed the World*. The Rosen Publishing Group, New York (2012), 118 pp.

Harland, D.M. *The Story of the Space Shuttle*. Springer Praxis Books, New York (2004), 470 pp.

Harrison, P. *The Edge of Time: The Authoritative Biography of Kalpana Chawla*. Harrison Publishing, Los Gatos, CA, USA (2011), 236 pp.

Harvey, B. *China in Space: The Great Leap Forward*. Springer Praxis, New York (2013), 416 pp.

Hasday, J.L. *Ellen Ochoa*, Chelsea House Pub., New York (2007), 106 pp.

Haven, K. *Women at the Edge of Discovery: 40 True Science Adventures*. Libraries Unlimited, Westport, London (2003), 273 pp.

Kevles, T.H. *Almost Heaven: The Story of Women in Space*. The MIT Press, Cambridge, MA, and London, UK (2006), 280 pp.

Malerba, F. *The Summit*. Tormena, Genova (1993), 257 pp.

U. Cavallaro, *Women Spacefarers*, Springer Praxis Books, DOI 10.1007/978-3-319-34048-7

Morgan, C. *Shuttle-MIR: The United States and Russia Share History's Highest Stage.* NASA SP-4225, Houston (2001), 223 pp.

Muir, E.G. *Canadian Women in the Sky: 100 Years of Flight.* Dundurn Group, Toronto, Canada (2015), 176 pp.

Mullane, M. *Riding Rockets: The Outrageous Tales of a Space Shuttle Astronaut.* Scribner, New York (2006), 368 pp.

Norberg J. *Wings of Their Dreams: Purdue in Flight.* Purdue University Press, West Lafayette, Indiana (2003), 451 pp.

O'Shaughnessy, T. *Sally Ride: a Photobiography of American Pioneering Woman in Space*, Roaring Brook Press, New York (2015), 154 pp.

Seddon, R. *Go for Orbit.* Your Space Press, Murfreesboro, TN (2015), 464 pp.

Shayler, D.; Hall, R. *Soyuz: A Universal Spacecraft.* Springer, London (2003), 460 pp.

Shayler, D.J.; Moule, I. *Women in Space—Following Valentina.* Springer/Praxis Publishing, Chichester, UK (2005), 410 pp.

Weitekamp, M.A. *Right Stuff, Wrong Sex: America's First Women in Space Program*, p. 78, Johns Hopkins University Press, Baltimore, MD (2006).

Woodmansee, L.S. *Women Astronauts.* Apogee Books, Burlington, Ontario, Canada (2002), 168 pp.

Yamazaki, N. *Astronaut for Four Thousand and Eighty-Eight Days.* Kadokawashoten, Tokyo (2013) (in Japanese—manuscript translated into English by Federica Cavallaro, pending publication).

Index

U. Cavallaro, *Women Spacefarers*, Springer Praxis Books, DOI 10.1007/978-3-319-34048-7